T0310223

Guided Waves in Structures for SHM

Guided Waves in Structures for SHM

The Time-Domain Spectral Element Method

Wieslaw Ostachowicz
Pawel Kudela
Marek Krawczuk
Arkadiusz Zak

Polish Academy of Sciences,
Institute of Fluid Flow Machinery

A John Wiley & Sons, Ltd., Publication

This edition first published 2012

Registered Office
John Wiley & Sons Ltd, The Atrium, Southern Gate, Chichester, West Sussex, PO19 8SQ, United Kingdom

For details of our global editorial offices, for customer services and for information about how to apply for permission to reuse the copyright material in this book please see our website at www.wiley.com.

Library of Congress Cataloging-in-Publication Data:

Guided waves in structures for SHM : the time-domain spectral element method / [edited by] Wieslaw Ostachowicz . . . [et al.].

p. cm.
Includes bibliographical references and index.
ISBN 978-0-470-97983-9 (hardback)
1. Elastic analysis (Engineering) 2. Elastic wave propagation–Mathematical models. 3. Composite materials–Analysis. 4. Finite element method. I. Ostachowicz, W. M. (Wieslaw M.)
TA653.G85 2012
531′.1133–dc23

2011043928

A catalogue record for this book is available from the British Library.

Print ISBN: 978-0-470-97983-9; oBook 9781119965855; ePDF 9781119965862; ePub 9781119966746; Mobi 9781119966753

Set in 10/12.5pt Palatino by Aptara Inc., New Delhi, India
Printed in Singapore by Ho Printing Singapore Pte Ltd

Contents

Preface ix

1 Introduction to the Theory of Elastic Waves **1**
1.1 Elastic Waves 1
1.1.1 Longitudinal Waves (Compressional/Pressure/Primary/P Waves) 2
1.1.2 Shear Waves (Transverse/Secondary/S Waves) 2
1.1.3 Rayleigh Waves 3
1.1.4 Love Waves 4
1.1.5 Lamb Waves 4
1.2 Basic Definitions 5
1.3 Bulk Waves in Three-Dimensional Media 10
1.3.1 Isotropic Media 10
1.3.2 Christoffel Equations for Anisotropic Media 12
1.3.3 Potential Method 14
1.4 Plane Waves 15
1.4.1 Surface Waves 16
1.4.2 Derivation of Lamb Wave Equations 17
1.4.3 Numerical Solution of Rayleigh–Lamb Frequency Equations 26
1.4.4 Distribution of Displacements and Stresses for Various Frequencies of Lamb Waves 29
1.4.5 Shear Horizontal Waves 32
1.5 Wave Propagation in One-Dimensional Bodies of Circular Cross-Section 35
1.5.1 Equations of Motion 35
1.5.2 Longitudinal Waves 36

1.5.3 Solution of Pochhammer Frequency Equation 39
1.5.4 Torsional Waves 42
1.5.5 Flexural Waves 43
References 45

2 Spectral Finite Element Method 47
2.1 Shape Functions in the Spectral Finite Element Method 53
2.1.1 Lobatto Polynomials 54
2.1.2 Chebyshev Polynomials 56
2.1.3 Laguerre Polynomials 60
2.2 Approximating Displacement, Strain and Stress Fields 62
2.3 Equations of Motion of a Body Discretised Using Spectral Finite Elements 67
2.4 Computing Characteristic Matrices of Spectral Finite Elements 72
2.4.1 Lobatto Quadrature 75
2.4.2 Gauss Quadrature 76
2.4.3 Gauss–Laguerre Quadrature 78
2.5 Solving Equations of Motion of a Body Discretised Using Spectral Finite Elements 81
2.5.1 Forcing with an Harmonic Signal 82
2.5.2 Forcing with a Periodic Signal 83
2.5.3 Forcing with a Nonperiodic Signal 84
References 92

3 Three-Dimensional Laser Vibrometry 93
3.1 Review of Elastic Wave Generation Methods 94
3.1.1 Force Impulse Methods 94
3.1.2 Ultrasonic Methods 94
3.1.3 Methods Based on the Electromagnetic Effect 97
3.1.4 Methods Based on the Piezoelectric Effect 98
3.1.5 Methods Based on the Magnetostrictive Effect 102
3.1.6 Photothermal Methods 103
3.2 Review of Elastic Wave Registration Methods 104
3.2.1 Optical Methods 106
3.3 Laser Vibrometry 109
3.4 Analysis of Methods of Elastic Wave Generation and Registration 114
3.5 Exemplary Results of Research on Elastic Wave Propagation Using 3D Laser Scanning Vibrometry 116
References 121

4 **One-Dimensional Structural Elements** 125
4.1 Theories of Rods 125
4.2 Displacement Fields of Structural Rod Elements 127
4.3 Theories of Beams 133
4.4 Displacement Fields of Structural Beam Elements 135
4.5 Dispersion Curves 141
4.6 Certain Numerical Considerations 143
4.6.1 Natural Frequencies 144
4.6.2 Wave Propagation 147
4.7 Examples of Numerical Calculations 155
4.7.1 Propagation of Longitudinal Elastic Waves in a Cracked Rod 156
4.7.2 Propagation of Flexural Elastic Waves in a Rod 158
4.7.3 Propagation of Coupled Longitudinal and Flexural Elastic Waves in a Rod 162
References 164

5 **Two-Dimensional Structural Elements** 167
5.1 Theories of Membranes, Plates and Shells 167
5.2 Displacement Fields of Structural Membrane Elements 169
5.3 Displacement Fields of Structural Plate Elements 175
5.4 Displacement Fields of Structural Shell Elements 181
5.5 Certain Numerical Considerations 184
5.6 Examples of Numerical Calculations 189
5.6.1 Propagation of Elastic Waves in an Angle Bar 189
5.6.2 Propagation of Elastic Waves in a Half-Pipe Aluminium Shell 192
5.6.3 Propagation of Elastic Waves in an Aluminium Plate 195
References 198

6 **Three-Dimensional Structural Elements** 201
6.1 Solid Spectral Elements 202
6.2 Displacement Fields of Solid Structural Elements 202
6.2.1 Six-Mode Theory 202
6.2.2 Nine-Mode Theory 203
6.3 Certain Numerical Considerations 204
6.4 Modelling Electromechanical Coupling 208
6.4.1 Assumptions 213
6.4.2 Linear Constitutive Equations 213
6.4.3 Basic Equations of Motion 214

6.4.4 Static Condensation 215
6.4.5 Inducing Waves 216
6.4.6 Recording Waves 216
6.4.7 Electrical Boundary Conditions 216
6.5 Examples of Numerical Calculations 220
6.5.1 Propagation of Elastic Waves in a Half-Pipe Aluminium Shell 220
6.5.2 Propagation of Elastic Waves in an Isotropic Plate – Experimental Verification 222
6.6 Modelling the Bonding Layer 227
References 230

7 Detection, Localisation and Identification of Damage by Elastic Wave Propagation 233
7.1 Elastic Waves in Structural Health Monitoring 235
7.2 Methods of Damage Detection, Localisation and Identification 247
7.2.1 Energy Addition Method 253
7.2.2 Phased Array Method 255
7.2.3 Methods Employing Continuous Registration of Elastic Waves within the Analysed Area 263
7.2.4 Damage Identification Algorithms 266
7.3 Examples of Damage Localisation Methods 269
7.3.1 Localisation Algorithms Employing Sensor Networks 269
7.3.2 Algorithms Based on Full Field Measurements of Elastic Wave Propagation 275
References 288

Appendix: *EWavePro* Software 295
A.1 Introduction 295
A.2 Theoretical Background and Scope of Applicability (Computation Module) 296
A.3 Functional Structure and Software Environment (Pre- and Post-Processors) 298
A.4 Elastic Wave Propagation in a Wing Skin of an Unmanned Plane (UAV) 312
A.5 Elastic Wave Propagation in a Composite Panel 320
References 333

Index 335

Preface

This book is aimed at professionals whose scientific interests are directly associated with propagation of elastic waves in structural elements. This book may be useful not only for students of technical universities but also for researchers and engineers who solve practical problems involving propagation of elastic waves in structural elements made of isotropic materials or laminated composites.

Waves propagating in elastic media have been known for many centuries and have been the subject of scientific research of many scholars. Elastic waves result from stresses acting within the media and are associated with volume (compression and tension) and shape (shear) deformations. Better recognition and understanding of the complex phenomena behind the propagation of elastic waves in structural elements have promoted various novel and practical applications in many fields of technology. One such field is diagnostics of structural elements, where the use of elastic waves increases rapidly each year. Local methods employing elastic waves have been employed successfully for many years, but attempts to apply elastic waves in a global sense for diagnosing structural elements are still at an early stage of development. The measure of success in these attempts comes from various achievements made in parallel in several different fields. The first of them is the development of numerical simulation methods and tools aimed at modelling and analysing the phenomena associated with propagation of elastic waves in structural elements. The second, independent, one is the development of appropriate experimental methods and techniques allowing verification and validation results of numerical simulations. Recently these goals have become achievable in practice thanks to employing the most advanced measuring techniques based on three-dimensional (3D) laser scanning vibrometry.

This book is intended to report on the challenges associated with numerical simulation methods, analyses and experimental investigations related to the propagation of elastic waves in structural elements made of isotropic materials or composite laminates. For the first time the full spectrum of theoretical and practical issues associated with the propagation of elastic waves are presented and discussed in one study.

The first part of the book, devoted to various modelling and analysis issues associated with propagation of elastic waves, is focused on the Spectral Finite Element Method, which in the authors' opinion is the most suitable modelling technique out of a variety of numerical methods used nowadays to solve wave propagation-related problems. This part of the book gives a broad overview of the existing state of the art and knowledge concerning modelling of elastic wave propagation in structural elements, while emphasising the problems associated with developing efficient numerical methods and tools and verifying them. Original solutions developed by the authors, suitable for constructing appropriate numerical models for simulating propagation of elastic waves in 1D, 2D and 3D structural elements made of isotropic and laminated composites are presented and discussed. Based on the developed spectral finite elements, a range of numerical tests has been carried out in order to verify the accuracy of the models, beginning from wave propagation in simple rods, beams, membranes and plates, and ending with shells or 3D structures.

The second part of the book, devoted to experimental measurements, presents the application of 3D laser scanning vibrometry for measuring, investigating and visualising the propagation of elastic waves in real-life structural elements. This part of the book naturally complements the theoretical and numerical investigations of the earlier part. Numerous scenarios and results of experimental measurements carried out on 1D, 2D and 3D structures are presented and discussed.

The last part of the book is concerned with various practical applications associated with wave propagation phenomena in structural elements. Problems of damage detection and location are discussed and investigated here. These problems are a part of a wide multidisciplinary research subject known as Structural Health Monitoring. Several damage detection methods developed or/and implemented by the authors and their practical applications in the context of Structural Health Monitoring are described in great detail, based on the results of either numerical or experimental investigations. The results of experimental studies included in this book make use of excitation and registration of elastic waves within structural elements using piezoelectric transducers. Additionally, and in parallel, independent registration of propagating elastic waves employs advanced 3D laser scanning vibrometry. These two techniques have been applied and investigated in order to qualitatively and quantitatively characterise the wave propagation phenomena.

The authors would like to underline the unique character of this book resulting from its complex and multidisciplinary character. Various acclaimed books dedicated to wave propagation phenomena in elastic media are usually theoretical in nature, while the question of appropriate verification of the developed numerical methods is addressed in a very limited manner. The authors of the studies mentioned often use analytical models of the wave propagation phenomena and/or apply different numerical methods based on either the finite element method or spectral methods in the frequency domain. The intention of the authors of this book is to present for the first time in one place new models of spectral finite elements defined in the time domain developed to facilitate analysis of propagation of elastic waves in structural elements. Originality of the material presented in this book comes from the attempt to connect together the results of both numerical and experimental investigations, as well as to indicate their practical implications. Until now, the original numerical models discussed in the book as well as the results of experimental studies using 3D laser scanning vibrometry have had no equivalents in published books dedicated to this field. Therefore the level of scientific research of this book, in the opinion of the authors, closely follows the latest trends in this area. It is worth noting that this book is accompanied by a demonstration version of software employing methods of analysing elastic wave propagation in structural elements using spectral finite elements. It should be emphasised that this software has been developed by the authors of this book.

Guided Waves in Structures for SHM: The Time-Domain Spectral Element Method is accompanied by a website (www.wiley.com/go/ostachowicz). The website contains and describes the ***EWavePro*** (Elastic Wave Propagation) software, which can be used for analysing phenomena of propagation of longitudinal, shear and flexural waves in two and three–dimensional thin–walled structures composed of isotropic materials or composite laminates. The abbreviation ***EWavePro*** is used here to distinguish the software developed by the authors. The software is developed in order to facilitate better understanding of elastic wave propagation phenomena and to be used as a tool in designing structural health monitoring systems based on changes in the elastic wave propagation patterns.

The authors want to thank their colleagues from the Department of Mechanics of Intelligent Structures: Dr P. Malinowski, Dr M. Radzienski and Dr T. Wandowski for assistance with writing Chapters 3 and 7, as well as Dr L. Murawski for involvement in writing the Appendix. Their efforts contributed to the development of the mentioned parts of this book cannot be overstated.

1

Introduction to the Theory of Elastic Waves

1.1 Elastic waves

Elastic waves are mechanical waves propagating in an elastic medium as an effect of forces associated with volume deformation (compression and extension) and shape deformation (shear) of medium elements. External bodies causing these deformations are called wave sources. Elastic wave propagation involves exciting the movement of medium particles increasingly distant from the wave source. The main factor differentiating elastic waves from any other ordered motion of medium particles is that for small disturbances (linear approximation) elastic wave propagation does not result in matter transport.

Depending on restrictions imposed on the elastic medium, wave propagation may vary in character. Bulk waves propagate in infinite media. Within the class of bulk waves one can distinguish longitudinal waves (compressional waves) and shear waves. A three-dimensional medium bounded by one surface allows for propagation of surface waves (Rayleigh waves and Love waves). Propagation of bulk waves and surface waves is used for describing seismic wave phenomena. Bounding the elastic medium with two equidistant surfaces causes compressional waves and shear waves to interact, which results in the generation of Lamb waves. One can say that a free boundary restricting an elastic body guides and drives waves; therefore the term

Guided Waves in Structures for SHM: The Time-Domain Spectral Element Method, First Edition.
Wieslaw Ostachowicz, Pawel Kudela, Marek Krawczuk and Arkadiusz Zak.

guided waves is also used. Lamb waves and guided waves are used in broadly considered diagnostics and nondestructive testing. There are also waves that propagate on media boundary (interface waves) with names derived from their discoverers: in the interface between two solids Stoneley waves propagate, while in the one between a solid and a liquid Scholte waves propagate.

1.1.1 Longitudinal Waves (Compressional/Pressure/Primary/P Waves)

Longitudinal waves are characterised by particle motion alternately of compression and stretching character. The direction of medium point motion is parallel to the direction of wave propagation (i.e. longitudinal).

1.1.2 Shear Waves (Transverse/Secondary/S Waves)

Shear waves are characterised by transverse particle movements in alternating direction. The direction of medium particle motion is perpendicular to

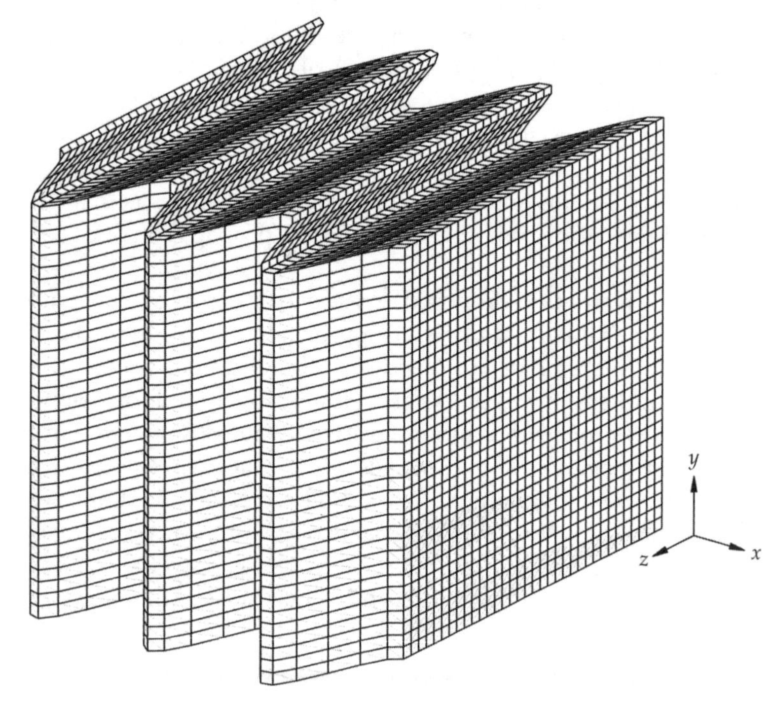

Figure 1.1 Distribution of displacements for the horizontal shear wave

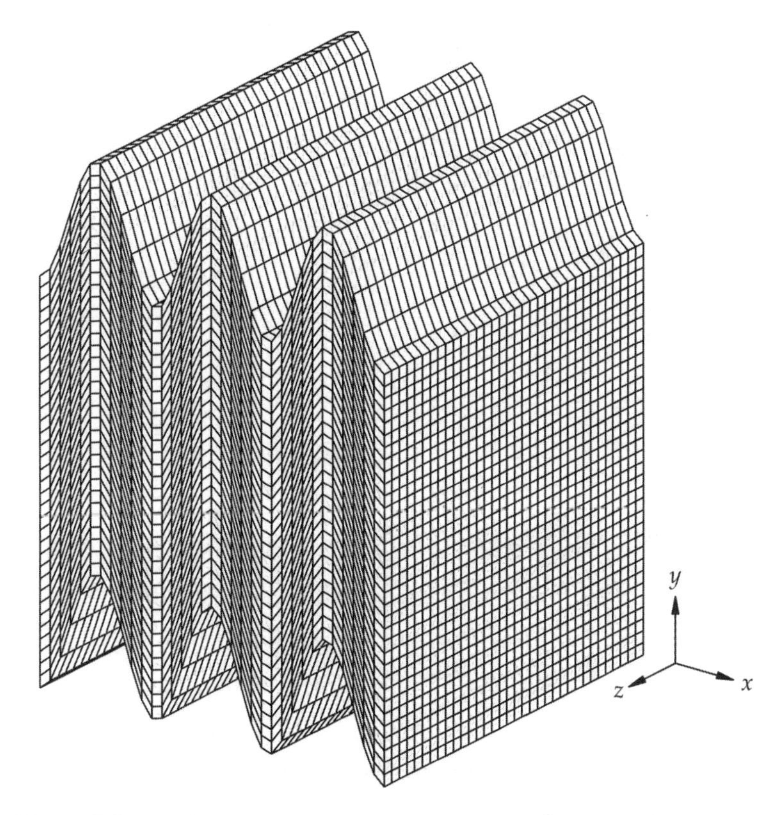

Figure 1.2 Distribution of displacements for the vertical shear wave

the propagation direction (transverse). The transverse particle movement can occur horizontally (horizontal shear wave, SH; see Figure 1.1) or vertically (vertical shear wave, SV; see Figure 1.2).

1.1.3 Rayleigh Waves

Rayleigh waves (Figure 1.3) are characterised by particle motion composed of elliptical movements in the xy vertical plane and of motion parallel to the direction of propagation (along the x axis). Wave amplitude decreases with depth y, starting from the wave crest. Rayleigh waves propagate along surfaces of elastic bodies of thickness many times exceeding the wave height. Sea waves are a natural example of Rayleigh waves.

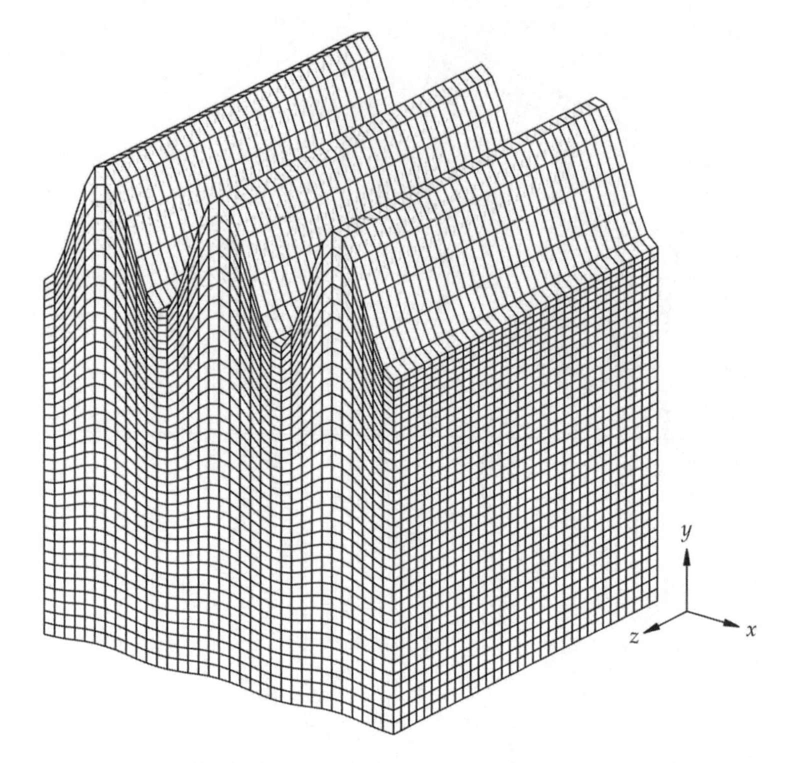

Figure 1.3 Distribution of displacements for the Rayleigh wave

1.1.4 Love Waves

Love waves (Figure 1.4) are characterised by particle oscillations involving alternating transverse movements. The direction of medium particle oscillations is horizontal (in the *xz* plane) and perpendicular to the direction of propagation. As in the case of Rayleigh waves, wave amplitude decreases with depth.

1.1.5 Lamb Waves

These waves were named after their discoverer, Horace Lamb, who developed the theory of their propagation in 1917 [1]. Curiously, Lamb was not able to physically generate the waves he discovered. This was achieved by Worlton [2], who also noticed their potential usefulness for damage detection. Lamb waves propagate in infinite media bounded by two surfaces and arise as a result of superposition of multiple reflections of longitudinal P waves

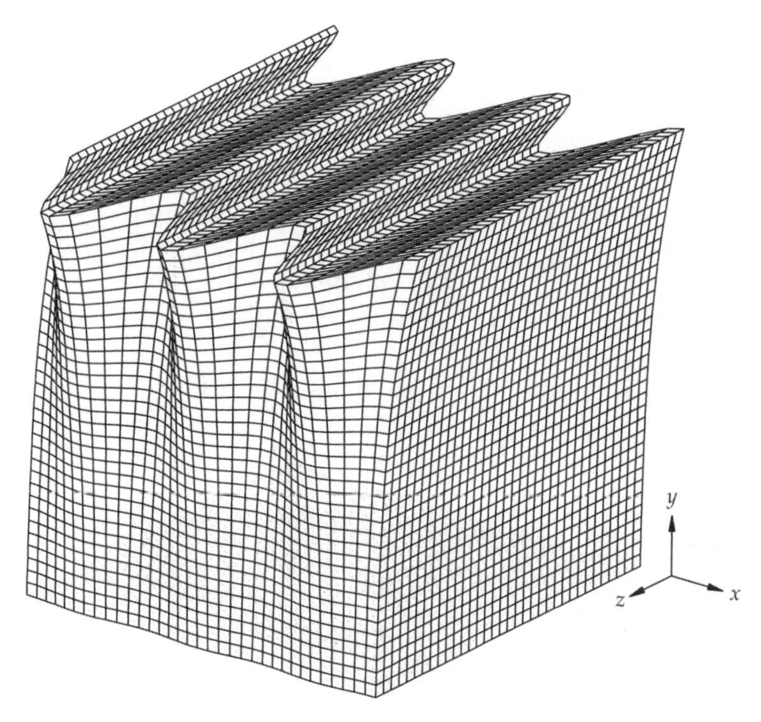

Figure 1.4 Distribution of displacements for the Love wave

and shear SV waves from the bounding surfaces. In the case of these waves medium particle oscillations are very complex in character. Depending on the distribution of displacements on the top and bottom bounding surface, two forms of Lamb waves appear: symmetric, denoted as $S_0, S_1, S_2, \ldots$, and antisymmetric, denoted as $A_0, A_1, A_2, \ldots$. It should be noted that numbers of these forms are infinite. Displacement fields of medium points for the fundamental symmetric mode S_0 and fundamental antisymmetric mode A_0 of Lamb waves are illustrated in Figures 1.5 and 1.6, respectively.

1.2 Basic Definitions

A specific case of waves as harmonic initial perturbations is considered here:

$$u(x, 0) = U_0 \cos(kx) \tag{1.1}$$

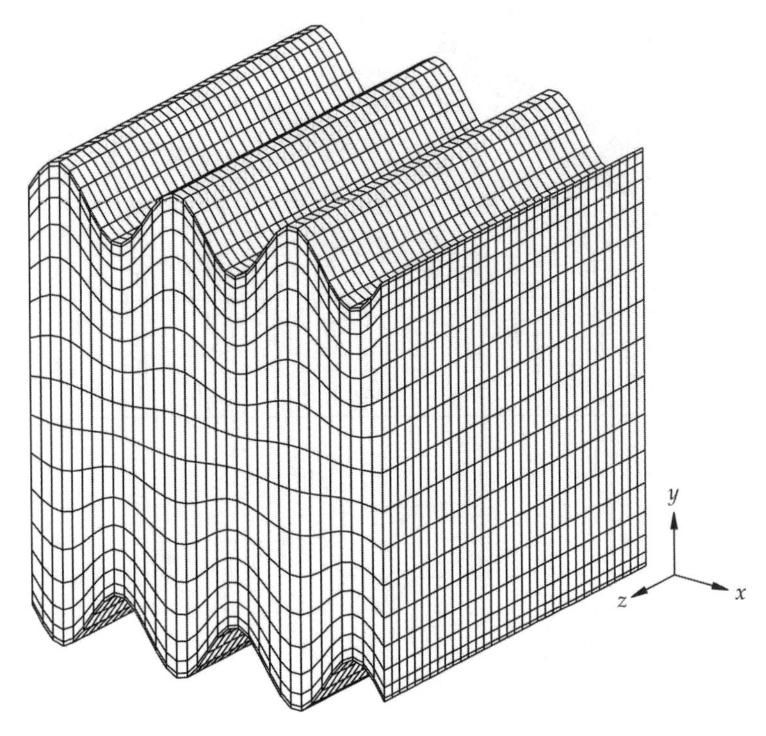

Figure 1.5 Distribution of displacements for the fundamental symmetric mode of Lamb waves

The notions of wavenumber k and wavelength λ are common for waves of every type. Wavenumber k refers to the spatial frequency of perturbations. Wavelength λ refers to the spatial period of perturbations (Figure 1.7) and is expressed by the following formula:

$$\lambda = \frac{2\pi}{k} \tag{1.2}$$

Solution of Equation (1.1) can be expressed in a general form as:

$$u(x, t) = \frac{U_0}{2}\left[\cos(kx - \omega t) + \cos(kx + \omega t)\right] \tag{1.3}$$

where U_0 is wave amplitude and ω is angular velocity. The first term in square brackets is associated with wave propagation to the right (or

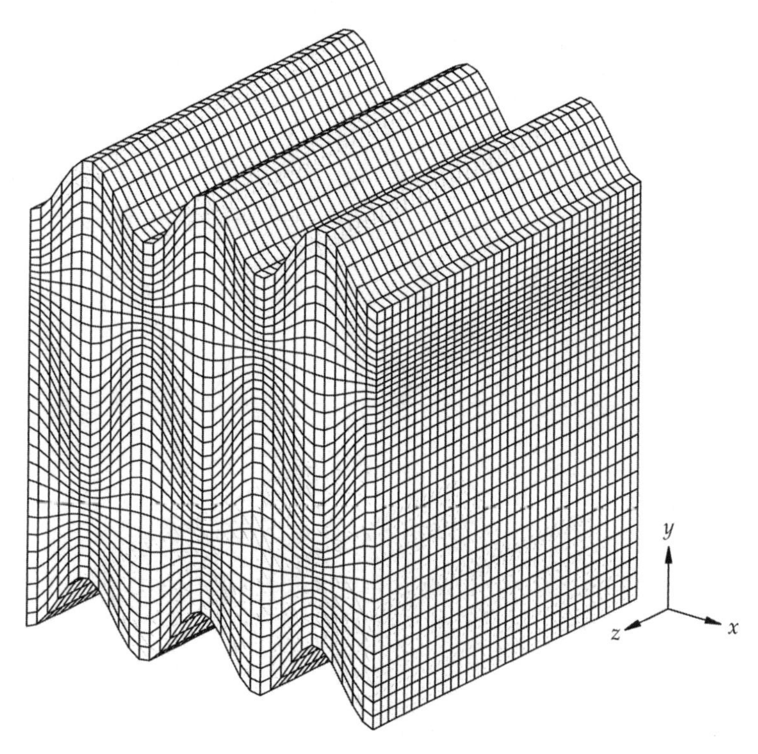

Figure 1.6 Distribution of displacements for the fundamental antisymmetric mode of Lamb waves

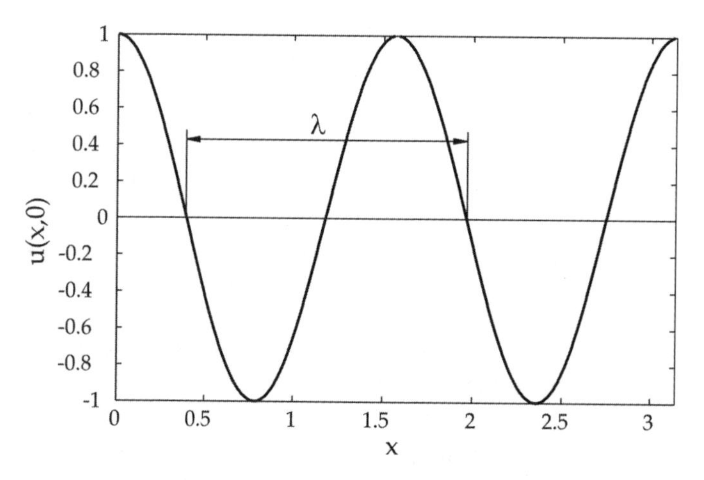

Figure 1.7 Harmonic wave of length λ

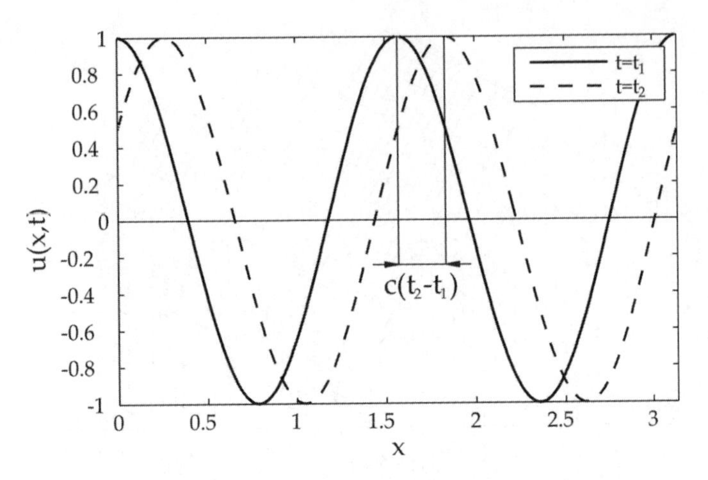

Figure 1.8 Harmonic wave propagating with velocity c

forwards), while the second term is associated with wave propagation to the left (or backwards). Considering a wave propagating to the right, this can be written as:

$$u^R(x, t) = \frac{U_0}{2} \cos(kx - \omega t) \tag{1.4}$$

The phase of this wave is $\phi = kx - \omega t$. For the constant phase $kx - \omega t = \text{const}$ it is $x = (\omega/k)t + \text{const}$. Thus, a point of constant phase moves with velocity:

$$c = \frac{\omega}{k} \tag{1.5}$$

The harmonic wave propagating to the right with velocity c is presented in Figure 1.8.

The phase velocity of a wave describes the relationship between spatial frequency k and temporal frequency ω of the propagating waves. The dependency $\omega = \omega(k)$ is called the dispersion relationship. If this relation is linear, that is $\omega = ck$, the wave is nondispersive. In a nondispersive medium, the phase velocity is constant for all velocities.

Besides phase velocity, the term of group velocity is also associated with wave propagation. Group velocity refers to propagation of a group of waves called a wave packet. In order to understand the term of wave group velocity two waves propagating to the right, having the same amplitudes, but different

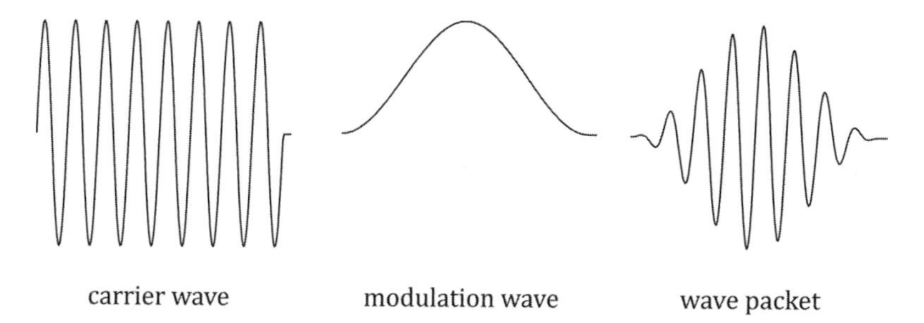

Figure 1.9 Wave packet as the superposition of a carrier wave and a modulating wave

frequencies and wavenumbers are considered:

$$u(x, t) = U_0 \left[\sin(k_1 x - \omega_1 t) + \sin(k_2 x - \omega_2 t)\right] \tag{1.6}$$

Application of universally known trigonometric identities for the sum of sinus functions leads to:

$$u(x, t) = 2U_0 \cos\left(\frac{k_1 - k_2}{2}x - \frac{\omega_1 - \omega_2}{2}t\right) \sin\left(\frac{k_1 + k_2}{2}x - \frac{\omega_1 + \omega_2}{2}t\right) \tag{1.7}$$

In formula (1.7) one can distinguish a term associated with modulation and one associated with a carrier wave:

$$u(x, t) = 2U_0 \underbrace{\cos(\Delta k x - \Delta\omega t)}_{\text{modulation}} \underbrace{\sin(k_0 x - \omega_0 t)}_{\text{carrier wave}} \tag{1.8}$$

The wave packet is a superposition of a carrier wave and a modulating wave in the form of a window, as presented in Figure 1.9.

The propagation velocity of a modulating wave defines the propagation velocity of a wave packet. For a constant phase $\Delta k x - \Delta\omega t = \text{const}$, this is $x = (\Delta\omega/\Delta k)t + \text{const}$. Thus, group velocity at the limit transition $\Delta\omega \to 0$, $\Delta k \to 0$ is defined as:

$$c_g = \frac{\mathrm{d}\omega}{\mathrm{d}k} \tag{1.9}$$

One should note that for the nondispersive media the group velocity is equal to the phase velocity. In the dispersive media these velocities differ, which manifests directly as wave packet deformation during propagation. First of all, the wave packet amplitude decreases and the packet stretches.

1.3 Bulk Waves in Three-Dimensional Media

1.3.1 Isotropic Media

In infinite elastic medium waves propagate freely in every direction and are called bulk waves. The basis for discussing bulk waves is the three-dimensional theory of elasticity. The full set of equations is as follows:

$$\sigma_{ij,j} + \rho f_i = \rho \ddot{u}_i, \quad i, j = 1, 2, 3 \tag{1.10}$$

$$\varepsilon_{ij} = \frac{1}{2}(u_{i,j} + u_{j,i}), \quad i, j = 1, 2, 3 \tag{1.11}$$

$$\sigma_{ij} = \lambda \delta_{ij} \varepsilon_{kk} + 2\mu \varepsilon_{ij}, \quad i, j = 1, 2, 3 \tag{1.12}$$

where $\varepsilon_{kk} = \varepsilon_{11} + \varepsilon_{22} + \varepsilon_{33}$ (Einstein summation convention) and δ_{ij} is the Kronecker delta. Equation (1.10) covers three motion equations, Equation (1.11) describes linear relationships between deformations and displacements (six independent equations) and Equation (1.12) covers six independent constitutive equations for the isotropic case. In Equation (1.12) Lame constants have been used; these are defined as:

$$\begin{aligned} \lambda &= \frac{\nu E}{(1+\nu)(1-2\nu)} \\ \mu &= G = \frac{E}{2(1+\nu)} \end{aligned} \tag{1.13}$$

Equations (1.10) to (1.12) may be expanded using Cartesian notation. Thus, the motion equation can be written as:

$$\begin{aligned} \frac{\partial \sigma_{xx}}{\partial x} + \frac{\partial \sigma_{xy}}{\partial y} + \frac{\partial \sigma_{xz}}{\partial z} + \rho f_x &= \rho \frac{\partial^2 u_x}{\partial t^2} \\ \frac{\partial \sigma_{yx}}{\partial x} + \frac{\partial \sigma_{yy}}{\partial y} + \frac{\partial \sigma_{yz}}{\partial z} + \rho f_y &= \rho \frac{\partial^2 u_y}{\partial t^2} \\ \frac{\partial \sigma_{zx}}{\partial x} + \frac{\partial \sigma_{zy}}{\partial y} + \frac{\partial \sigma_{zz}}{\partial z} + \rho f_z &= \rho \frac{\partial^2 u_z}{\partial t^2} \end{aligned} \tag{1.14}$$

where ρ is the mass density. The relationships between stress components are governed by symmetry, that is $\sigma_{yx} = \sigma_{xy}, \sigma_{zy} = \sigma_{yz}, \sigma_{xz} = \sigma_{zx}$. The deformation–displacement equations take the following form:

$$\begin{aligned} \varepsilon_{xx} &= \frac{\partial u_x}{\partial x}, & \varepsilon_{xy} &= \frac{1}{2}\left(\frac{\partial u_x}{\partial y} + \frac{\partial u_y}{\partial x}\right) \\ \varepsilon_{yy} &= \frac{\partial u_y}{\partial y}, & \varepsilon_{yz} &= \frac{1}{2}\left(\frac{\partial u_y}{\partial z} + \frac{\partial u_z}{\partial y}\right) \\ \varepsilon_{zz} &= \frac{\partial u_z}{\partial z}, & \varepsilon_{zx} &= \frac{1}{2}\left(\frac{\partial u_z}{\partial x} + \frac{\partial u_x}{\partial z}\right) \end{aligned} \tag{1.15}$$

They are also subject to symmetry, that is $\varepsilon_{yx} = \varepsilon_{xy}, \varepsilon_{zy} = \varepsilon_{yz}, \varepsilon_{xz} = \varepsilon_{zx}$. The constitutive Equation (1.10) in Cartesian notation is as follows:

$$\begin{aligned} \sigma_{xx} &= (\lambda + 2\mu)\varepsilon_{xx} + \lambda\varepsilon_{yy} + \lambda\varepsilon_{zz}, & \sigma_{xy} &= 2\mu\varepsilon_{xy} \\ \sigma_{yy} &= \lambda\varepsilon_{xx} + (\lambda + 2\mu)\varepsilon_{yy} + \lambda\varepsilon_{zz}, & \sigma_{yz} &= 2\mu\varepsilon_{yz} \\ \sigma_{zz} &= \lambda\varepsilon_{xx} + \lambda\varepsilon_{yy} + (\lambda + 2\mu)\varepsilon_{zz}, & \sigma_{zx} &= 2\mu\varepsilon_{zx} \end{aligned} \tag{1.16}$$

Equations (1.14) and (1.15) remain valid for any continuous medium; the specific type of the discussed medium is introduced by Equations (1.16) – in this case it is isotropic. Elimination of stresses and deformations from Equations (1.14) to (1.16) leads to:

$$(\lambda + \mu)u_{j,ji} + \mu u_{i,jj} + \rho f_i = \rho \ddot{u}_i \tag{1.17}$$

Motion Equations (1.17) containing only particle displacements are displacement-type partial differential equations. These equations are also known as Navier equations and in Cartesian notation take the following form [3]:

$$\begin{aligned} (\lambda + \mu)\frac{\partial}{\partial x}\left(\frac{\partial u_x}{\partial x} + \frac{\partial u_y}{\partial y} + \frac{\partial u_z}{\partial z}\right) + \mu\left(\frac{\partial^2 u_x}{\partial x^2} + \frac{\partial^2 u_x}{\partial y^2} + \frac{\partial^2 u_x}{\partial z^2}\right) + \rho f_x = \rho\frac{\partial^2 u_x}{\partial t^2} \\ (\lambda + \mu)\frac{\partial}{\partial y}\left(\frac{\partial u_x}{\partial x} + \frac{\partial u_y}{\partial y} + \frac{\partial u_z}{\partial z}\right) + \mu\left(\frac{\partial^2 u_y}{\partial x^2} + \frac{\partial^2 u_y}{\partial y^2} + \frac{\partial^2 u_y}{\partial z^2}\right) + \rho f_y = \rho\frac{\partial^2 u_y}{\partial t^2} \\ (\lambda + \mu)\frac{\partial}{\partial z}\left(\frac{\partial u_x}{\partial x} + \frac{\partial u_y}{\partial y} + \frac{\partial u_z}{\partial z}\right) + \mu\left(\frac{\partial^2 u_z}{\partial x^2} + \frac{\partial^2 u_z}{\partial y^2} + \frac{\partial^2 u_z}{\partial z^2}\right) + \rho f_z = \rho\frac{\partial^2 u_z}{\partial t^2} \end{aligned} \tag{1.18}$$

If the area where the solution is sought is infinite, these equations are sufficient for describing elastic wave propagation. If the area is finite, on the other hand, boundary conditions are necessary for the problem to be well-posed. These boundary conditions take the form of imposed stresses and/or displacements at area boundaries.

1.3.2 Christoffel Equations for Anisotropic Media

Wave propagation in infinite anisotropic elastic solids is governed by the full set of equations of the three-dimensional theory of elasticity. Compared to the isotropic case, the difference lies in a more general constitutive equation. The full set of equations of the theory of elasticity for homogeneous anisotropic media is as follows:

$$\sigma_{ik,k} + \rho f_i = \rho \ddot{u}_i, \quad i, k = 1, 2, 3 \tag{1.19}$$

$$\varepsilon_{lm} = \frac{1}{2}(u_{l,m} + u_{m,l}), \quad l, m = 1, 2, 3 \tag{1.20}$$

$$\sigma_{ik} = C_{iklm}\varepsilon_{lm}, \quad i, k, l, m = 1, 2, 3 \tag{1.21}$$

By combining Equations (1.19), (1.20) and (1.21) and ignoring external forces the motion equations are obtained:

$$\frac{1}{2}C_{iklm}\left(u_{l,km} + u_{m,kl}\right) = \rho \ddot{u}_i \tag{1.22}$$

The tensor of elasticity constants C_{iklm} is symmetric with regard to l and m, and therefore:

$$C_{iklm} = C_{ikml} = C_{kilm} \tag{1.23}$$

A flat harmonic plane wave propagating forwards is assumed:

$$u_{\mathrm{i}} = A_{\mathrm{i}} \mathrm{e}^{\mathrm{i}(k_j x_j - \omega t)} \tag{1.24}$$

where $\mathrm{i} = \sqrt{-1}$ is the imaginary unit, k_j is the wavenumber, A_{i} is a vector of wave amplitudes and ω is angular frequency. Substitution of Equation (1.24) into Equation (1.22) leads to:

$$C_{iklm} k_k k_l u_m = \rho \omega^2 u_i \tag{1.25}$$

It can be seen that $u_i = u_m\delta_{im}$; therefore:

$$(\rho\omega^2\delta_{im} - C_{iklm}k_k k_l)u_m = 0 \tag{1.26}$$

This is the Christoffel equation for an anisotropic medium. The Christoffel tensor can be defined as:

$$\lambda_{im} = \Gamma_{im} = C_{iklm}n_k n_l \tag{1.27}$$

where n_k are direction cosines normal to the wavefront. Furthermore, taking into account the relationships:

$$k_k = kn_k, \quad k_l = kn_l \tag{1.28}$$

leads to:

$$\left(\Gamma_{im}k^2 - \rho\omega^2\delta_{im}\right)u_m = 0 \tag{1.29}$$

By recalling the definition of phase velocity:

$$c = \frac{\omega}{k} \tag{1.30}$$

Equation (1.29) is brought to the following form:

$$\left(\Gamma_{im} - \rho c^2\delta_{im}\right)u_m = 0 \tag{1.31}$$

This is a uniform system of three equations. The system has a nontrivial solution if the determinant of the coefficient matrix is equal to zero. This is a classic eigenvalue problem. The solution is composed of three velocities (eigenvalues with regard to c^2) and the corresponding eigenvectors. Depending on the arrangement of eigenvectors in space, one can be dealing with: a P wave together with SH and SV waves, a quasi-P wave together with SH and SV waves, a P wave together with quasi-SH and quasi-SV waves or a quasi-P wave together with quasi-SH and quasi-SV waves [4]. One should note that phase velocities depend on the direction of propagation, which results from the definition of the Christoffel tensor (Equation (1.27)). In an isotropic medium there are always pure waves: a longitudinal one and two shear ones, the phase velocities of which do not depend on the direction of propagation.

1.3.3 Potential Method

Bulk waves connected with wave propagation in an isotropic infinite media are considered in this section. When no external forces f are present, Equation (1.18) can be expressed in vector form as:

$$(\lambda + \mu)\nabla(\nabla \cdot \boldsymbol{u}) + \mu\nabla^2\boldsymbol{u} = \rho\ddot{\boldsymbol{u}} \tag{1.32}$$

The motion Equation (1.32) can be simplified further by applying Helmholtz decomposition and the potential method [4–6]. Such an operation is only possible for isotropic media. It is assumed that the displacement vector $\boldsymbol{u}$ can be expressed through two potential functions: the scalar potential Φ and the vector potential $\boldsymbol{H} = H_x\boldsymbol{i} + H_y\boldsymbol{j} + H_z\boldsymbol{k}$, that is:

$$\boldsymbol{u} = \nabla\Phi + \nabla \times \boldsymbol{H} \tag{1.33}$$

Equation (1.33) is known as the Helmholtz solution complemented by the condition:

$$\nabla \cdot \boldsymbol{H} = 0 \tag{1.34}$$

By applying Equation (1.33), components of Equation (1.32) can be expressed as:

$$\nabla \cdot \boldsymbol{u} = \nabla(\nabla\Phi + \nabla \times \boldsymbol{H}) = (\nabla \cdot \nabla)\Phi + \underbrace{\nabla \cdot (\nabla \times \boldsymbol{H})}_{=0} = \nabla^2\Phi \tag{1.35}$$

$$\nabla^2\boldsymbol{u} = \nabla^2(\nabla\Phi + \nabla \times \boldsymbol{H}) = \nabla^2\nabla\Phi + \nabla^2\nabla \times \boldsymbol{H} \tag{1.36}$$

$$\ddot{\boldsymbol{u}} = \nabla\ddot{\Phi} + \nabla \times \ddot{\boldsymbol{H}} \tag{1.37}$$

By substituting Equations (1.35), (1.36) and (1.37) into Equation (1.32) the following formula is obtained:

$$(\lambda + \mu)\nabla(\nabla^2\Phi) + \mu\left(\nabla^2\nabla\Phi + \nabla^2\nabla \times \boldsymbol{H}\right) = \rho\left(\nabla\ddot{\Phi} + \nabla \times \ddot{\boldsymbol{H}}\right) \tag{1.38}$$

Noting that $\nabla\nabla^2 = \nabla^2\nabla$ (commutativity of differentiation), Equation (1.38) after transformations yields:

$$\nabla\left((\lambda + 2\mu)\nabla^2\Phi - \rho\ddot{\Phi}\right) + \nabla \times \left(\mu\nabla^2\boldsymbol{H} - \rho\ddot{\boldsymbol{H}}\right) = \boldsymbol{0} \tag{1.39}$$

Equation (1.39) is satisfied for any point in space at any time, if the terms in parentheses vanish, that is:

$$(\lambda + 2\mu)\nabla^2\Phi - \rho\ddot{\Phi} = 0 \tag{1.40}$$

$$\mu\nabla^2\boldsymbol{H} - \rho\ddot{\boldsymbol{H}} = \boldsymbol{0} \tag{1.41}$$

After dividing by ρ and ordering, Equations (1.40) and (1.41) become wave equations for the scalar potential Φ and the vector potential $\boldsymbol{H}$, that is:

$$c_L\nabla^2\Phi = \ddot{\Phi} \tag{1.42}$$

$$c_S\nabla^2\boldsymbol{H} = \ddot{\boldsymbol{H}} \tag{1.43}$$

where c_L is the longitudinal wave velocity, defined as:

$$\boxed{c_L = \sqrt{\frac{\lambda + 2\mu}{\rho}}} \tag{1.44}$$

and c_S is the shear wave velocity, defined as:

$$\boxed{c_S = \sqrt{\frac{\mu}{\rho}}} \tag{1.45}$$

As a result, the motion Equation (1.32) was decomposed into two simplified wave Equations (1.42) and (1.43). Assuming that the rotational part $\nabla \times \boldsymbol{H}$ of Equation (1.33) is equal to zero, the longitudinal wave equation is obtained:

$$c_L\nabla^2\boldsymbol{u} = \ddot{\boldsymbol{u}} \tag{1.46}$$

Assuming that displacements in Equation (1.33) contain the rotational part only, the shear wave equation is obtained as:

$$c_S\nabla^2\boldsymbol{u} = \ddot{\boldsymbol{u}} \tag{1.47}$$

1.4 Plane Waves

A specific case of three-dimensional waves are plane waves. These waves are invariant in one direction along the wave crest. Such a situation happens when the wave crest is parallel to the z axis (cf. Figures 1.1 to 1.6). Moreover, the normal vector of the wave crest is perpendicular to the z axis. Invariance in the direction of the z axis means that all wave functions are independent of z, and therefore their derivatives with respect to z are equal to zero, that is:

$$\frac{\partial}{\partial z} \equiv 0 \quad \text{and} \quad \nabla = \boldsymbol{i}\frac{\partial}{\partial x} + \boldsymbol{j}\frac{\partial}{\partial y} \tag{1.48}$$

After substituting Equation (1.48) into Equation (1.33) and expanding, the expression for displacement is obtained:

$$u = \underbrace{\left(\frac{\partial \Phi}{\partial x} + \frac{\partial H_z}{\partial y}\right)}_{u_x} \boldsymbol{i} + \underbrace{\left(\frac{\partial \Phi}{\partial y} - \frac{\partial H_z}{\partial x}\right)}_{u_y} \boldsymbol{j} + \underbrace{\left(\frac{\partial H_y}{\partial x} - \frac{\partial H_x}{\partial y}\right)}_{u_z} \boldsymbol{k} \tag{1.49}$$

Although movement is invariant with respect to the z axis, Equation (1.49) indicates that displacement components appear in all three directions (x, y and z). It is noteworthy that the displacement component u_z depends only on potentials H_x and H_y that are associated with the horizontally polarised shear wave (SH wave). Displacement components u_x and u_y depend on potentials Φ and H_z associated with the longitudinal wave (P wave) and vertically polarised shear wave (SV wave), respectively. Thanks to these relations one can seek solutions of wave equations separately for the SH wave and the P + SV wave combination.

1.4.1 Surface Waves

An example of waves propagating near a solid surface are Rayleigh waves, the amplitude of which decreases rapidly with depth. The effective penetration depth for Rayleigh waves is smaller than their wavelength. Particle movement takes place in the vertical plane and is independent of the z direction; therefore one can seek solutions in terms of P + SV type plane waves. The following assumptions are made:

$$\frac{\partial}{\partial z} = 0, \quad u_x \neq 0, \quad u_y \neq 0, \quad u_z = 0 \tag{1.50}$$

The wave Equations (1.42) and (1.43) can then be expressed in terms of potentials Φ and H_z:

$$c_L \nabla^2 \Phi = \ddot{\Phi}, \quad c_S \nabla^2 H_z = \ddot{H}_z \tag{1.51}$$

Assuming potentials of the following forms:

$$\Phi(x, y, t) = f(y)e^{i(kx-\omega t)}, \quad H_z(x, y, t) = h_z(y)e^{i(kx-\omega t)} \tag{1.52}$$

where k denotes the wavenumber, and then imposing boundary conditions (stresses vanishing on the bounding surface):

$$\sigma_{yy}\Big|_{y=0} = 0, \quad \sigma_{xy}\Big|_{y=0} = 0 \tag{1.53}$$

After transformations the characteristic equation is obtained:

$$\left(\beta^2 + k^2\right)^2 - 4\alpha\beta k^2 = 0 \tag{1.54}$$

where α and β depend on the frequency ω and the wavenumber k:

$$\alpha^2 = k^2 - \frac{\omega^2}{c_L^2}, \quad \beta^2 = k^2 - \frac{\omega^2}{c_S^2} \tag{1.55}$$

Even though Equation (1.54) has three double roots, only one of them is real. The real root corresponds to the surface Rayleigh wave velocity: $c_R = \omega/k_R$. By transforming Equation (1.54) one can show that the Rayleigh wave velocity depends on the Poisson coefficient ν and the shear wave velocity c_S [5]. A universally accepted Rayleigh wave velocity approximation formula is:

$$c_R(\nu) = c_S\left(\frac{0.87 + 1.12\nu}{1+\nu}\right) \tag{1.56}$$

Particle movement for Rayleigh waves can be described as follows:

$$\begin{aligned} u_x(y) &= Ai\left(ke^{-\alpha y} - \frac{\beta^2 + k^2}{2k}e^{-\beta y}\right) \\ u_y(y) &= A\left(-\alpha e^{-\alpha y} + i\frac{\beta^2 + k^2}{2\beta}e^{-\beta y}\right) \end{aligned} \tag{1.57}$$

where A is any constant.

1.4.2 Derivation of Lamb Wave Equations

The P + SV wave combination that leads to Lamb wave equations is considered in this section. Lamb wave propagation in a free plate of shape as

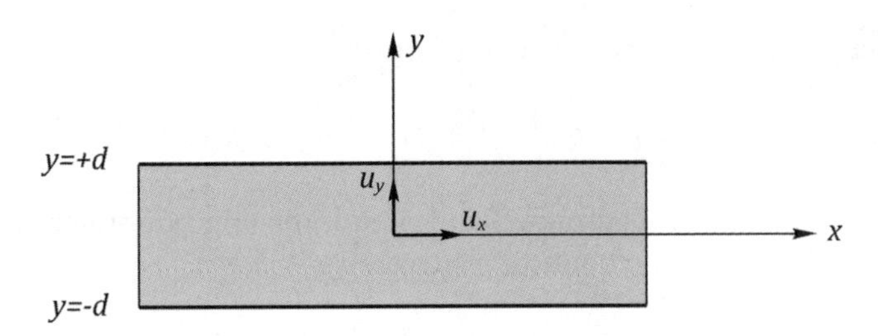

Figure 1.10 Shape of the free plate in the Lamb wave problem

presented in Figure 1.10 is investigated. The problem is described by the motion Equations (1.42) and (1.43) with boundary conditions imposed on the plate surfaces. It is assumed that no stress is present on surfaces of coordinates $y = d$ and $y = -d$. Exciting the plate at any point causes the propagating excited impulse to encounter the top and bottom surfaces of the plate edge. By means of multiple reflections from the top and bottom plate surfaces as well as constructive and destructive interference of P and SV waves, Lamb waves are generated. Lamb waves are composed of waves standing in the thickness direction y (Lamb wave modes) and propagating in the x direction.

In the case of the solution applicable to the P + SV wave combination, motion is contained in the vertical (x, y) plane and the following conditions are observed:

$$u_x \neq 0, \quad u_y \neq 0, \quad \frac{\partial}{\partial z} = 0, \quad \Phi \text{ and } H_z \text{ only} \tag{1.58}$$

In order to simplify the notation, the two potential functions Φ and H_z are denoted as ϕ and ψ, respectively. Substituting conditions (1.58) into Equations (1.42) and (1.43) the following expressions are obtained:

$$\begin{aligned} &\frac{\partial^2 \phi}{\partial x^2} + \frac{\partial^2 \phi}{\partial y^2} = \frac{1}{c_L^2}\frac{\partial^2 \phi}{\partial t^2} \quad \text{longitudinal waves} \\ &\frac{\partial^2 \psi}{\partial x^2} + \frac{\partial^2 \psi}{\partial y^2} = \frac{1}{c_S^2}\frac{\partial^2 \psi}{\partial t^2} \quad \text{shear waves} \end{aligned} \tag{1.59}$$

Referring to the displacement field expressed through potentials (1.49):

$$u_x = \frac{\partial \phi}{\partial x} + \frac{\partial \psi}{\partial y}, \quad u_y = \frac{\partial \phi}{\partial y} - \frac{\partial \psi}{\partial x}, \quad u_z = 0 \tag{1.60}$$

and substituting Equations (1.60) into the deformation–displacement relationship (1.15), the strain relations are obtained as:

$$\begin{aligned}
\varepsilon_{xx} &= \frac{\partial^2 \phi}{\partial x^2} + \frac{\partial^2 \psi}{\partial x \partial y} \\
\varepsilon_{yy} &= \frac{\partial^2 \phi}{\partial y^2} - \frac{\partial^2 \psi}{\partial x \partial y} \\
\varepsilon_{yx} &= \frac{1}{2}\left(2\frac{\partial^2 \phi}{\partial x \partial y} - \frac{\partial^2 \psi}{\partial x^2} + \frac{\partial^2 \psi}{\partial y^2}\right) \\
\varepsilon_{zz} &= 0 \\
\varepsilon_{yz} &= 0 \\
\varepsilon_{zx} &= 0
\end{aligned} \tag{1.61}$$

Stress, in turn, can be expressed according to Equations (1.16) as:

$$\begin{aligned}
\sigma_{xx} &= \lambda\left(\frac{\partial^2 \phi}{\partial x^2} + \frac{\partial^2 \phi}{\partial y^2}\right) + 2\mu\left(\frac{\partial^2 \phi}{\partial x^2} + \frac{\partial^2 \psi}{\partial x \partial y}\right) \\
\sigma_{yy} &= \lambda\left(\frac{\partial^2 \phi}{\partial x^2} + \frac{\partial^2 \phi}{\partial y^2}\right) + 2\mu\left(\frac{\partial^2 \phi}{\partial y^2} - \frac{\partial^2 \psi}{\partial x \partial y}\right) \\
\sigma_{zz} &= 0 \\
\sigma_{yx} &= \mu\left(2\frac{\partial^2 \phi}{\partial x \partial y} - \frac{\partial^2 \psi}{\partial x^2} + \frac{\partial^2 \psi}{\partial y^2}\right) \\
\sigma_{yz} &= 0 \\
\sigma_{zx} &= 0
\end{aligned} \tag{1.62}$$

The solution of Equations (1.59) is assumed in the following form:

$$\begin{aligned}
\phi &= \Phi(y)\mathrm{e}^{\mathrm{i}(kx-\omega t)} \\
\psi &= \Psi(y)\mathrm{e}^{\mathrm{i}(kx-\omega t)}
\end{aligned} \tag{1.63}$$

It should be noticed that these solutions represent waves propagating in the x direction and waves standing in the y direction. The complex term of the exponential function includes a time variable depending on x, which is associated with wave propagation. On the other hand, the unknown functions Φ and Ψ are 'static' functions that only depend on y. In other words, these functions describe the stress distribution in the crosswise direction (across the plate thickness). Substituting the relationships (1.63) into Equations (1.59) leads to a system of differential equations with regard to functions Φ and Ψ:

$$\begin{aligned}
\frac{\partial^2 \Phi}{\partial y^2} + \left(\frac{\omega^2}{c_L^2} - k^2\right)\Phi = 0 \\
\frac{\partial^2 \Psi}{\partial y^2} + \left(\frac{\omega^2}{c_S^2} - k^2\right)\Psi = 0
\end{aligned} \tag{1.64}$$

In the same fashion displacements and stress can be directly obtained from Equations (1.60) and (1.62). By ignoring the $e^{i(kx-\omega t)}$ term in all expressions, the displacements and stresses can be expressed as:

$$\begin{aligned}
u_x &= ik\Phi + \frac{d\Psi}{dy}, \quad u_y = \frac{d\Phi}{dy} - ik\Psi \\
\sigma_{xx} &= \lambda\left(-k^2\Phi + \frac{d^2\Phi}{dy^2}\right) + 2\mu\left(-k^2\Phi + ik\frac{d\Psi}{dy}\right) \\
\sigma_{yy} &= \lambda\left(-k^2\Phi + \frac{d^2\Phi}{dy^2}\right) + 2\mu\left(\frac{d^2\Phi}{dy^2} - ik\frac{d\Psi}{dy}\right) \\
\sigma_{yx} &= \mu\left(2ik\frac{d\Phi}{dy} + k^2\Psi + \frac{d^2\Psi}{dy^2}\right)
\end{aligned} \tag{1.65}$$

After the following symbols are introduced:

$$p^2 = \frac{\omega^2}{c_L^2} - k^2, \quad q^2 = \frac{\omega^2}{c_S^2} - k^2 \tag{1.66}$$

Equations (1.64) are brought to the following form:

$$\begin{aligned}
\frac{\partial^2 \Phi}{\partial y^2} + p^2\Phi = 0 \\
\frac{\partial^2 \Psi}{\partial y^2} + q^2\Psi = 0
\end{aligned} \tag{1.67}$$

Equations (1.67) are fulfilled by a general solution:

$$\begin{aligned}
\Phi &= A_1 \sin(py) + A_2 \cos(py) \\
\Psi &= B_1 \sin(qy) + B_2 \cos(qy)
\end{aligned} \tag{1.68}$$

Derivatives of potentials with regard to y are as follows:

$$\frac{d\Phi}{dy} = A_1 p\cos(py) - A_2\sin(py), \quad \frac{d^2\Phi}{dy^2} = -A_1 p^2\sin(py) - A_2 p^2\cos(py)$$
$$\frac{d\Psi}{dy} = B_1 q\cos(qy) - B_2\sin(qy), \quad \frac{d^2\Psi}{dy^2} = -B_1 q^2\sin(qy) - B_2 q^2\cos(qy) \tag{1.69}$$

As field variables contain sine and cosine functions with argument y, which are odd and even with regard to $y = 0$, respectively, the solution can be sorted into two sets of modes: symmetric modes and antisymmetric ones. Specifically, the distribution of displacements in the direction of the x axis will be symmetric with respect to the middle plane of the plate when u_x contains cosines and antisymmetric when u_x contains sines. This is reversed for displacements in the direction of the y axis. Thus, equation systems for individual wave propagation modes are as follows.

1.4.2.1 Symmetric Modes

$$\begin{aligned}
\Phi^S &= A_2\cos(py) \\
\Psi^S &= B_1\sin(py) \\
u_x^S &= A_2 ik\cos(py) + B_1 q\cos(qy) \\
u_y^S &= -A_2 p\sin(py) + B_1 ik\sin(qy) \\
\sigma_{xx}^S &= -A_2(\lambda p^2 + (\lambda + 2\mu)k^2)\cos(py) - B_1 2\mu ikq\cos(qy) \\
\sigma_{yy}^S &= -A_2(\lambda k^2 + (\lambda + 2\mu)p^2)\cos(py) - B_1 2\mu ikq\cos(qy) \\
\sigma_{yx}^S &= \mu\left[-A_2 2ikp\sin(py) + B_1(k^2 - q^2)\sin(qy)\right]
\end{aligned} \tag{1.70}$$

1.4.2.2 Antisymmetric Modes

$$\begin{aligned}
\Phi^A &= A_1\sin(py) \\
\Psi^A &= B_2\cos(py) \\
u_x^A &= A_1 ik\sin(py) - B_2 q\sin(qy) \\
u_y^A &= A_1 p\cos(py) - B_2 ik\cos(qy) \\
\sigma_{xx}^A &= -A_1(\lambda p^2 + (\lambda + 2\mu)k^2)\sin(py) - B_2 2\mu ikq\sin(qy) \\
\sigma_{yy}^A &= -A_1(\lambda k^2 + (\lambda + 2\mu)p^2)\sin(py) + B_2 2\mu ikq\sin(qy) \\
\sigma_{yx}^A &= \mu\left[A_1 2ikp\cos(py) + B_2(k^2 - q^2)\cos(qy)\right]
\end{aligned} \tag{1.71}$$

Using the relationship:

$$\begin{aligned}\lambda k^2+(\lambda+2\mu)p^2 &= \lambda k^2+(\lambda+2\mu)\left(\frac{\omega^2}{c_L^2}-k^2\right)\\ &= \lambda k^2-k^2(\lambda+2\mu)+(\lambda+2\mu)\left(\frac{\omega^2}{c_L^2}\right)\\ &= -2k^2\mu+\omega^2\rho=\mu\left(-2k^2+\frac{\omega^2}{c_S^2}\right)\\ &= \mu\left[\left(\frac{\omega^2}{c_S^2}-k^2\right)-k^2\right]=\mu(q^2-k^2)\end{aligned} \tag{1.72}$$

gives:

$$\begin{aligned}\sigma_{yy}^S &= -A_2\mu(q^2-k^2)\cos(py)-B_1 2\mu ikq\cos(qy)\\ \sigma_{yy}^A &= -A_1\mu(q^2-k^2)\sin(py)+B_2 2\mu ikq\sin(qy)\end{aligned} \tag{1.73}$$

One should note that the waves can be separated into symmetric and anti-symmetric modes only in specific cases, when structure symmetry is present. Such separation is impossible in the case of analysis of anisotropic plates, unless the wave propagates along the symmetry plane of the plate.

Constants A_1, A_2, B_1, B_2, as well as dispersion equations, still remain unknown. They can be obtained by imposing free boundary conditions.

1.4.2.3 Symmetric Solution

The symmetric solution of Lamb wave equations is obtained when displacements and stresses are assumed to be symmetrical with respect to the middle plane (see Figure 1.11):

$$\begin{aligned}u_x(x,-d) &= u_x(x,d), \qquad \sigma_{yx}(x,-d)=-\sigma_{yx}(x,d)\\ u_y(x,-d) &= -u_y(x,d), \qquad \sigma_{yy}(x,-d)=\sigma_{yy}(x,d)\end{aligned} \tag{1.74}$$

One should note that positive shear stresses have the same directions on the top and bottom surfaces, and thus the opposite signs in Equations (1.74). Symmetric boundary conditions are as follows:

$$\begin{aligned}\sigma_{yx}(x,-d) &= -\sigma_{yx}(x,d)=0\\ \sigma_{yy}(x,-d) &= \sigma_{yy}(x,d)=0\end{aligned} \tag{1.75}$$

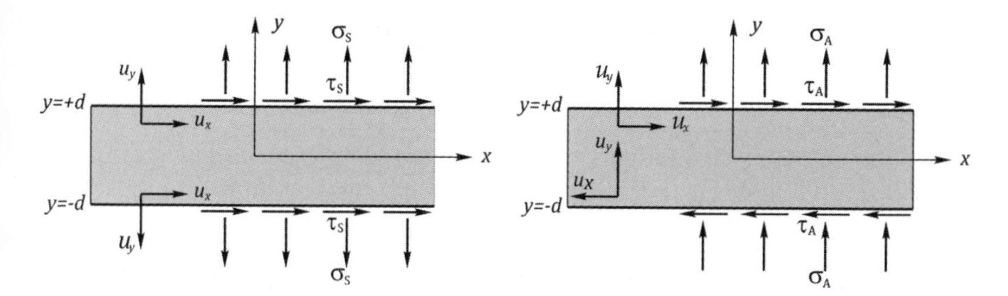

Figure 1.11 Symmetric and antisymmetric analysis

After substituting boundary conditions into the stress relationships described by formulas (1.70) and (1.73) a set of linear equations is obtained:

$$\begin{bmatrix} -2ikp\sin(pd) & (k^2-q^2)\sin(qd) \\ (k^2-q^2)\cos(pd) & -2ikq\cos(qd) \end{bmatrix} \begin{bmatrix} A_2 \\ B_1 \end{bmatrix} = \begin{bmatrix} 0 \\ 0 \end{bmatrix} \tag{1.76}$$

A uniform system of linear equations can be solved when its determinant is equal to zero:

$$D_S = (k^2-q^2)^2\cos(pd)\sin(qd) + 4k^2pq\sin(pd)\cos(qd) = 0 \tag{1.77}$$

After transformations:

$$\boxed{\frac{\tan(qd)}{\tan(pd)} = -\frac{4k^2pq}{(k^2-q^2)^2}} \quad \text{symmetric modes} \tag{1.78}$$

One should note that p and q depend on the wavenumber k as well as on the frequency $f = \omega/2\pi$. Equation (1.78) is known as the Rayleigh–Lamb frequency equation or the dispersion equation. Using this equation, one can compute the symmetric mode (S) velocity, with which waves propagate in a plate of thickness d for a chosen frequency f. A numerical solution of Equation (1.78) is a set of symmetric eigenvalues $k_0^S,\ k_1^S,\ k_2^S,\ \ldots$. Substituting the eigenvalues into the uniform system of Equations (1.76) allows thecoefficients (A_2, B_1) to be determined in the form:

$$A_2 = 2ikq\cos(qd), \quad B_1 = (k^2-q^2)\cos(pd) \tag{1.79}$$

Substitution of the coefficients (A_2, B_1) into Equations (1.70) yields symmetric modes of the Lamb waves:

$$\boxed{\begin{aligned} u_x^S &= -2k^2 q\cos(qd)\cos(py) + q(k^2 - q^2)\cos(pd)\cos(qy) \\ u_y^S &= -2ikpq\cos(qd)\sin(py) - ik(k^2 - q^2)\cos(pd)\sin(qy) \end{aligned}} \tag{1.80}$$

Stress distribution is obtained by means of substituting the coefficients (A_2, B_1) into Equations (1.70) and (1.73):

$$\begin{aligned} \sigma_{xx}^S &= -(2ikq)\big[(\lambda p^2 + (\lambda + 2\mu)k^2)\cos(qd)\cos(py) \\ &\quad + \mu(q^2 - k^2)\cos(pd)\cos(qy)\big] \\ \sigma_{yy}^S &= -2\mu(ikq)(k^2 - q^2)\left[\cos(qd)\cos(py) - \cos(pd)\cos(qy)\right] \\ \sigma_{yx}^S &= \mu\left[4k^2 pq\cos(qd)\sin(py) + (k^2 - q^2)^2\cos(pd)\sin(qy)\right] \end{aligned} \tag{1.81}$$

1.4.2.4 Antisymmetric Solution

An antisymmetric solution of the Lamb wave equations is obtained when displacements and stresses are assumed to be antisymmetrical with respect to the middle plane (see Figure 1.11):

$$\begin{aligned} u_x(x, -d) &= -u_x(x, d), \qquad & \sigma_{yx}(x, -d) &= \sigma_{yx}(x, d) \\ u_y(x, -d) &= u_y(x, d), \qquad & \sigma_{yy}(x, -d) &= -\sigma_{yy}(x, d) \end{aligned} \tag{1.82}$$

One should note that positive shear stresses have opposite directions on the top and bottom surfaces, and thus they are antisymmetric. Antisymmetric boundary conditions are as follows:

$$\begin{aligned} \sigma_{yx}(x, -d) &= \sigma_{yx}(x, d) = 0 \\ \sigma_{yy}(x, -d) &= -\sigma_{yy}(x, d) = 0 \end{aligned} \tag{1.83}$$

After substituting boundary conditions into the stress relationships expressed by formulas (1.71) and (1.73) a set of linear equations is obtained:

$$\begin{bmatrix} 2ikp\cos(pd) & (k^2 - q^2)\cos(qd) \\ (k^2 - q^2)\sin(pd) & 2ikq\sin(qd) \end{bmatrix} \begin{bmatrix} A_1 \\ B_2 \end{bmatrix} = \begin{bmatrix} 0 \\ 0 \end{bmatrix} \tag{1.84}$$

A uniform system of linear equations can be solved when its determinant is equal to zero:

$$D_A = (k^2 - q^2)^2 \sin(pd)\cos(qd) + 4k^2 pq\cos(pd)\sin(qd) = 0 \tag{1.85}$$

After transformations:

$$\boxed{\frac{\tan(qd)}{\tan(pd)} = -\frac{(k^2 - q^2)^2}{4k^2 pq}} \quad \text{antisymmetric modes} \tag{1.86}$$

One should note that p and q depend on the wavenumber k as well as on the frequency $f = \omega/2\pi$. Equation (1.86) is known as the Rayleigh–Lamb frequency equation or the dispersion equation. Using this equation, one can compute the antisymmetric mode (A) velocity, with which waves propagate in a plate of thickness d for a chosen frequency f. A numerical solution of Equation (1.86) is a set of antisymmetric eigenvalues $k_0^A,\ k_1^A,\ k_2^A, \ldots$. Substituting the eigenvalues into the uniform system of Equations (1.86) allows the coefficients $(A_1,\ B_2)$ to be determined in the form:

$$A_1 = 2ikq\sin(qd), \quad B_2 = -(k^2 - q^2)\sin(pd) \tag{1.87}$$

Substitution of the coefficients $(A_1,\ B_2)$ into Equations (1.71) yields antisymmetric modes of the Lamb waves:

$$\boxed{\begin{aligned} u_x^A &= -2k^2 q\sin(qd)\sin(py) + q(k^2 - q^2)\sin(pd)\sin(qy) \\ u_y^A &= -i[2kpq\sin(qd)\cos(py) + k(k^2 - q^2)\sin(pd)\cos(qy)] \end{aligned}} \tag{1.88}$$

Stress distribution is obtained by means of substituting the coefficients $(A_1,\ B_2)$ into Equations (1.71) and (1.73):

$$\begin{aligned} \sigma_{xx}^A &= -(2ikq)(\lambda p^2 + (\lambda + 2\mu)k^2)\sin(qd)\sin(py) \\ &\quad + \mu(q^2 - k^2)\sin(pd)\sin(qy) \\ \sigma_{yy}^A &= 2\mu(ikq)(k^2 - q^2)\,[\sin(qd)\sin(py) - \sin(pd)\sin(qy)] \\ \sigma_{yx}^A &= -\mu\left[4k^2 pq\sin(pd)\cos(py) + (k^2 - q^2)^2\sin(qd)\cos(py)\right] \end{aligned} \tag{1.89}$$

1.4.3 Numerical Solution of Rayleigh–Lamb Frequency Equations

Solving the Rayleigh–Lamb frequency Equations (1.78) and (1.86) is not easy, because the parameters p and q are also dependent on the wavenumber. These equations can be analysed as relationships $\omega(k)$ or $c(\omega)$ describing dispersion curves, where ω is the angular frequency and c is the phase velocity. The phase velocity is defined by the following formula:

$$c = \frac{\omega}{k} \tag{1.90}$$

For the given frequency there is an infinite number of solutions in the form of wavenumbers fulfilling Equations (1.78) and (1.86). These solutions can be real, imaginary or complex. However, in the case of the nonloaded plate problem it is sufficient to consider real wavenumber k values only. This can be achieved by means of the following system of equations:

$$\begin{aligned} \frac{\tan(qd)}{q} + \frac{4k^2 p \tan(pd)}{(q^2 - k^2)^2} &= 0 \quad \text{symmetric modes} \\ q\tan(qd) + \frac{(q^2 - k^2)^2 \tan(pd)}{4k^2 p} &= 0 \quad \text{antisymmetric modes} \end{aligned} \tag{1.91}$$

Further transformations aimed at introducing phase velocity and the product of frequency and thickness as analysis parameters lead to the following relationships:

$$\begin{aligned} \boldsymbol{LHS}^S &= \frac{\tan(\hat{q}\omega d)}{\hat{q}} + \frac{4\hat{p}\tan(\hat{p}\omega d)}{c^2(\hat{q}^2 - 1/c^2)^2} \\ \boldsymbol{LHS}^A &= \hat{q}\tan(\hat{q}\omega d) + \frac{(\hat{q}^2 - 1/c^2)^2 \tan(\hat{p}\omega d)}{4\hat{p}c^2} \end{aligned} \tag{1.92}$$

where:

$$\omega d = 2\pi f d, \qquad \hat{p} = \sqrt{\frac{1}{c_L^2} - \frac{1}{c^2}}, \qquad \hat{q} = \sqrt{\frac{1}{c_S^2} - \frac{1}{c^2}} \tag{1.93}$$

An algorithm that can be used for solving Equation (1.92) is presented below [4]:

1. Choose an initial value of the product of frequency and thickness $(\omega d)_0$.
2. Estimate the value of the phase velocity c_0.
3. Investigate the signs of the left-hand sides $\boldsymbol{LHS}^S$ and $\boldsymbol{LHS}^A$ (assuming they are nonzero).
4. Repeat steps 3 and 4 until signs in one of the equations $\boldsymbol{LHS}^S$ or $\boldsymbol{LHS}^A$ change.
5. Use the bisection method to locate the value of phase velocity precisely within the $c_n < c < c_{n+1}$ range, where $n+1$ is the step at which sign changes occur.
6. Continue iterating until the left-hand side of the desired equation is close to zero.
7. After locating the root continue searching for the value of the frequency–thickness product ωd, in order to locate other conceivable roots by repeating steps 2 to 6.
8. Choose a different value of the ωd product and repeat steps 2 to 7.

The procedure presented above is run for such number of ωd values as to achieve the desired accuracy.

One should note that despite the investigated functions being continuous, sign changes accompanying a zero crossing may pass unnoticed if increments of ωd are too large. This is due to the fact that the left-hand sides of Equations (1.91) have plots with peaks passing through zero in narrow ranges of the frequency–thickness product. That is why one needs to enhance the algorithm with an additional rule that would account for missed roots. This can be achieved by extrapolating the phase velocity curve. In case a root is lost, extrapolation allows the dispersion curve to be complemented with the missing root in the analysed frequency range. Sample results of root extraction from Equations (1.92) in the form of dispersion curves for an aluminium plate are presented in Figure 1.12. One should note that the procedure described above can be programmed in such a fashion that roots are classified into families of symmetric modes $S_0, S_1, S_2, \ldots$ and antisymmetric modes $A_0, A_1, A_2, \ldots$.

From the shear horizontal mode (SHM) point of view, another important property of Lamb waves are the group velocity dispersion curves. Group velocity is defined as:

$$c_g = \frac{\mathrm{d}\omega}{\mathrm{d}k} \tag{1.94}$$

However, in order to reduce computation time and complexity of the code that computes dispersion curves, group velocity can be derived from phase

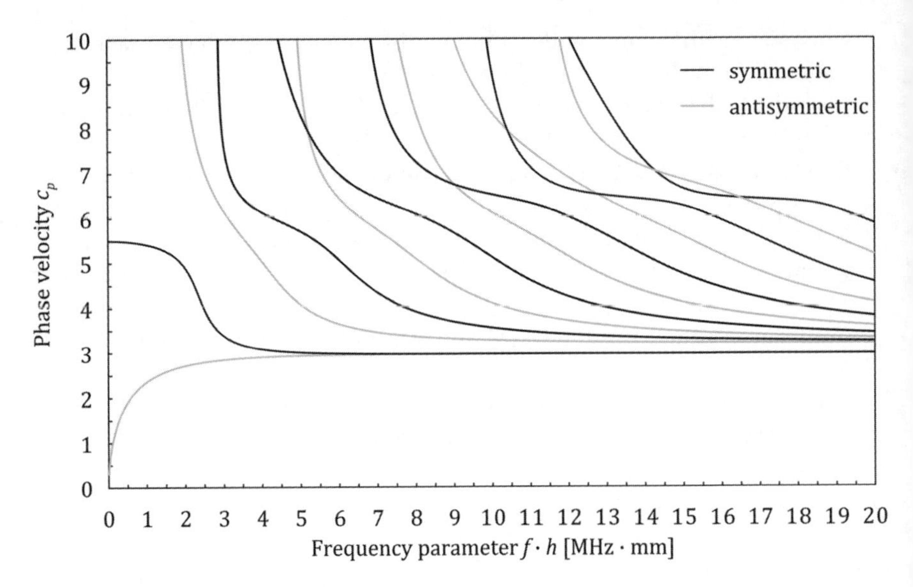

Figure 1.12 Phase velocity dispersion curves for symmetric and antisymmetric modes of Lamb waves ($c_L = 6.3$ km/s, $c_S = 3.2$ km/s)

velocity. After substituting $k = \omega/c$ into Equation (1.94) the group velocity is defined as [4]:

$$c_g = \frac{\mathrm{d}\omega}{\mathrm{d}\left(\frac{\omega}{c}\right)} = \frac{\mathrm{d}\omega}{\frac{\mathrm{d}\omega}{c} - \omega\frac{\mathrm{d}c}{c^2}} = \frac{c^2}{c - \omega\frac{\mathrm{d}c}{\mathrm{d}\omega}} \tag{1.95}$$

After taking $\omega = 2\pi f$ into account, the third equality can be written as:

$$c_g = \frac{c^2}{c - (fd)\frac{\mathrm{d}c}{\mathrm{d}(fd)}} \tag{1.96}$$

where fd denotes the product of frequency and thickness. One should note that when the derivative of phase velocity with regard to fd is equal to zero, then $c_g = c$. One should also note that when the derivative of phase velocity with regard to fd approaches infinity (i.e. at the cut-off frequency), the group velocity approaches zero. Numerical derivation can be performed

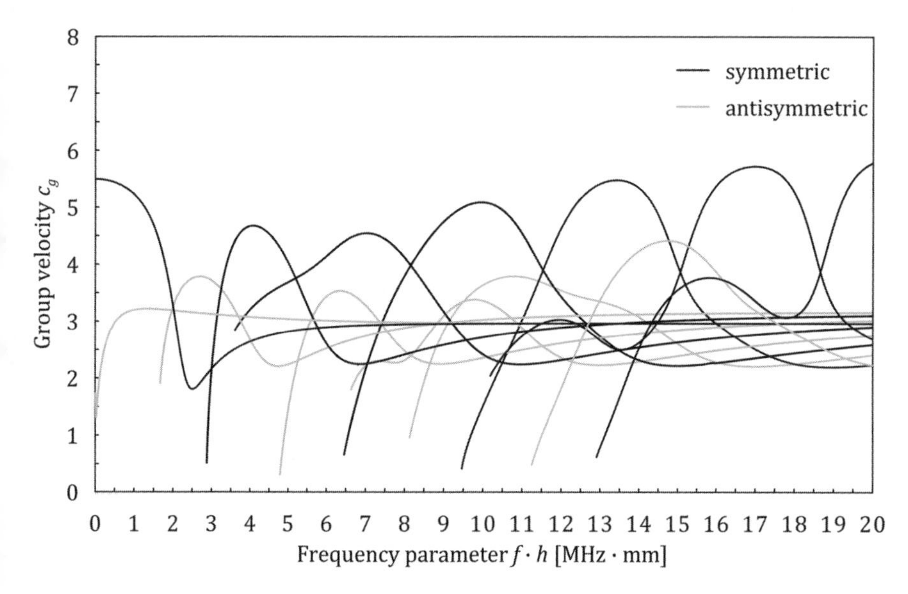

Figure 1.13 Group velocity dispersion curves for symmetric and antisymmetric modes of Lamb waves ($c_L = 6.3$ km/s, $c_S = 3.2$ km/s)

by applying the finite differences formula:

$$\frac{\mathrm{d}c}{\mathrm{d}(fd)} \cong \frac{\Delta c}{\Delta(fd)} \tag{1.97}$$

Graphs of group velocity dispersion curves for an aluminium plate for symmetric and antisymmetric modes of Lamb waves are presented in Figure 1.13.

1.4.4 Distribution of Displacements and Stresses for Various Frequencies of Lamb Waves

After finding the roots of Equations (1.91) and classifying the dispersion curves one can compute the distributions of displacements and stress across the plate thickness according to Equations (1.80), (1.81), (1.88) and (1.89). Sample graphs of displacement and stress distributions across the plate thickness depending on frequency and type of Lamb wave mode are presented in Figures 1.14 and 1.15. It is evident that as the frequency rises, the distribution of displacements and stresses across the plate thickness becomes increasingly

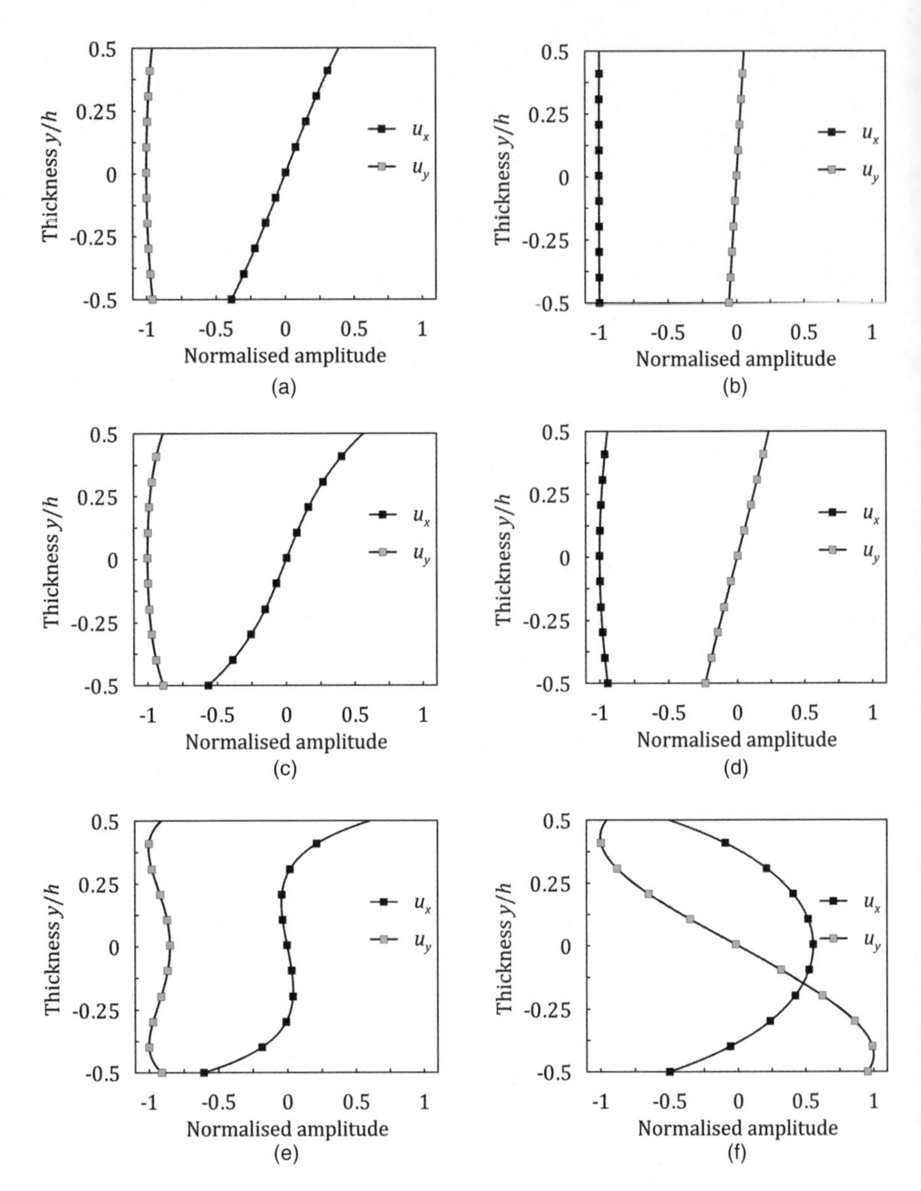

Figure 1.14 Distribution of longitudinal displacements u_x and transverse displacements u_y across the plate thickness for antisymmetric and symmetric modes of Lamb waves for individual frequencies.

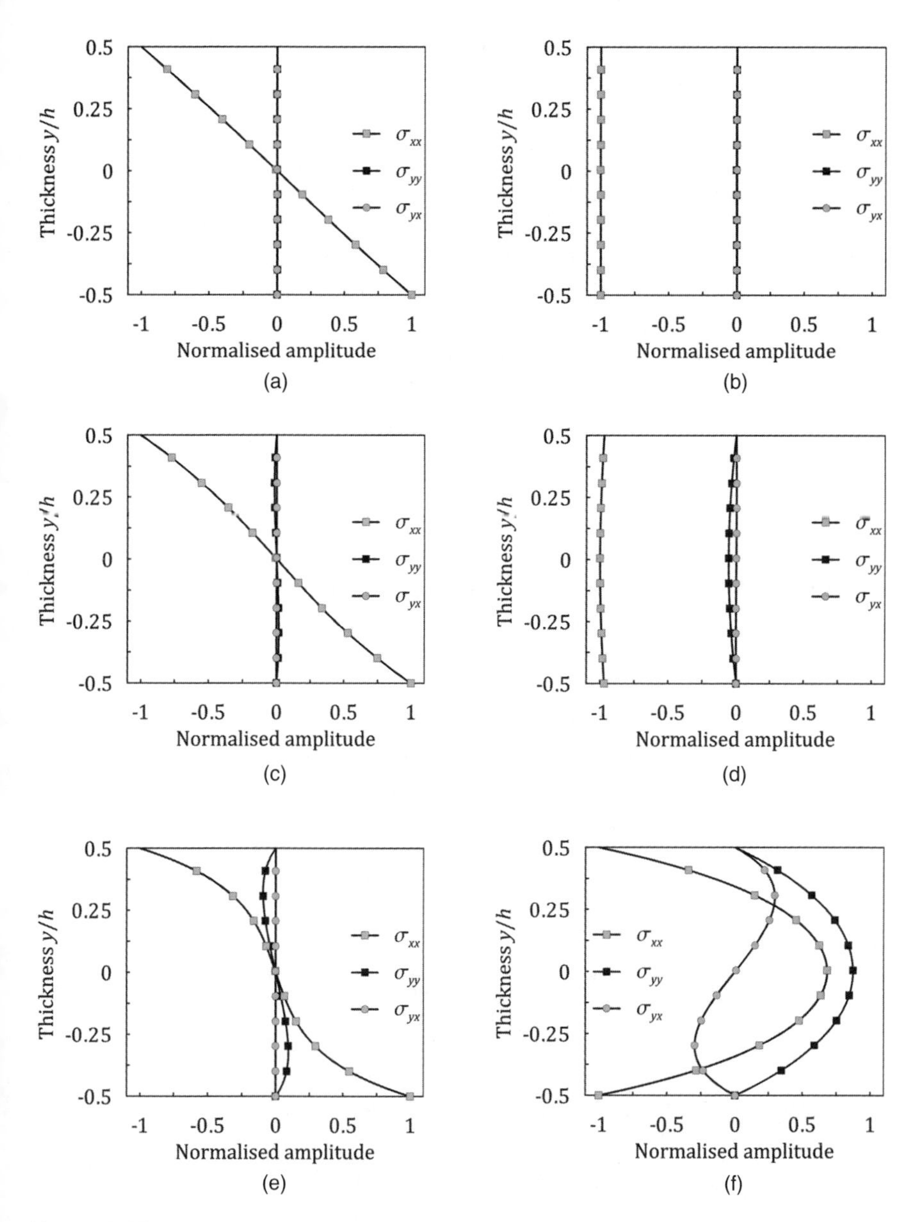

Figure 1.15 Distribution of stresses across plate thickness for antisymmetric and symmetric modes of Lamb waves for individual frequencies

complex. In other words, for higher frequencies one must use polynomials of higher orders for fitting the curves.

1.4.5 Shear Horizontal Waves

Apart from Lamb wave modes that exist in flat plates, there also exist a set of time-harmonic wave motions known as shear horizontal (SH) modes. In case of shear horizontal waves the particle motion (displacements and velocities) are in a plane, that is parallel to the surface of the plate (x, z) (see Figure 1.1). Physically, any mode in the SH family can be considered as the superposition of bulk waves reflecting from the upper and lower surfaces of the plate, polarized horizontally (in the z axis direction). Particle motion has only a u_z component and the wave Equation (1.47) simplifies to:

$$\nabla^2 u_z = \frac{1}{c_S^2}\ddot{u}_z \tag{1.98}$$

It is assumed that the particle motion has the form:

$$u_z(x, y, t) = h(y)\mathrm{e}^{\mathrm{i}(kx-\omega t)} \tag{1.99}$$

The first part of Equation (1.99) represents a standing wave $h(y)$ across the plate thickness. The second part, $\mathrm{e}^{\mathrm{i}(kx-\omega t)}$, represents a wave propagating in the x direction. Substitution of Equation (1.99) into Equation (1.98) and division of both sides by $\mathrm{e}^{\mathrm{i}(kx-\omega t)}$ yields:

$$h''(y) + \eta^2 h(y) = 0 \tag{1.100}$$

where:

$$\eta^2 = \frac{\omega^2}{c_S^2} - k^2 \tag{1.101}$$

The solution of Equation (1.100) has the general form:

$$h(y) = C_1 \sin(\eta y) + C_2 \cos(\eta y) \tag{1.102}$$

Without going into detail, the tractions-free boundary conditions at the upper and lower plate surfaces:

$$\sigma_{yz}(x, \pm d, t) = 0 \tag{1.103}$$

leads to the system of linear homogeneous equations with the determinant:

$$\sin(\eta d)\cos(\eta d) = 0 \tag{1.104}$$

Equation (1.104) is the characteristic equation of SH wave modes and is zero when either:

$$\sin(\eta d) = 0 \tag{1.105}$$

which corresponds to symmetric modes of SH waves, or:

$$\cos(\eta d) = 0 \tag{1.106}$$

which corresponds to antisymmetric modes of SH waves.

Explicit solutions of Equations (1.105) and (1.106) are:

$$\begin{aligned} \eta^S d &= n\pi, \quad n = 0, 1, 2, \ldots \\ \eta^A d &= (2n+1)\frac{\pi}{2}, \quad n = 0, 1, 2, \ldots \end{aligned} \tag{1.107}$$

for symmetric and antisymmetric modes, respectively. Finally, phase velocity dispersion curves of SH wave modes can be obtained from Equations (1.101) and (1.107) by recalling that $k = \omega/c$:

$$c(\omega) = \frac{c_S}{\sqrt{1 - (\eta d)^2 \left(\frac{c_S}{\omega d}\right)^2}} \tag{1.108}$$

Phase velocity dispersion curves calculated according to Equation (1.108) are presented in Figure 1.17. It should be noted that the first symmetric SH wave mode is not dispersive because its eigenvalues is zero ($\eta_0^S d = 0$), and hence Equation (1.108) leads to $c^{S_0}(\omega) = c_S$. Figure 1.16 also indicates the asymptotic behaviour of the SH wave velocity. If $\omega \to \infty$, then $c \to c_S$.

It can be shown that the group velocity of SH waves is inversely proportional to the phase velocity and can be expressed as [5]:

$$c_g(\omega) = c_S \sqrt{1 - (\eta d)^2 \left(\frac{c_S}{\omega d}\right)^2} \tag{1.109}$$

Corresponding to Equation (1.109), the group velocity dispersion curves are presented in Figure 1.17.

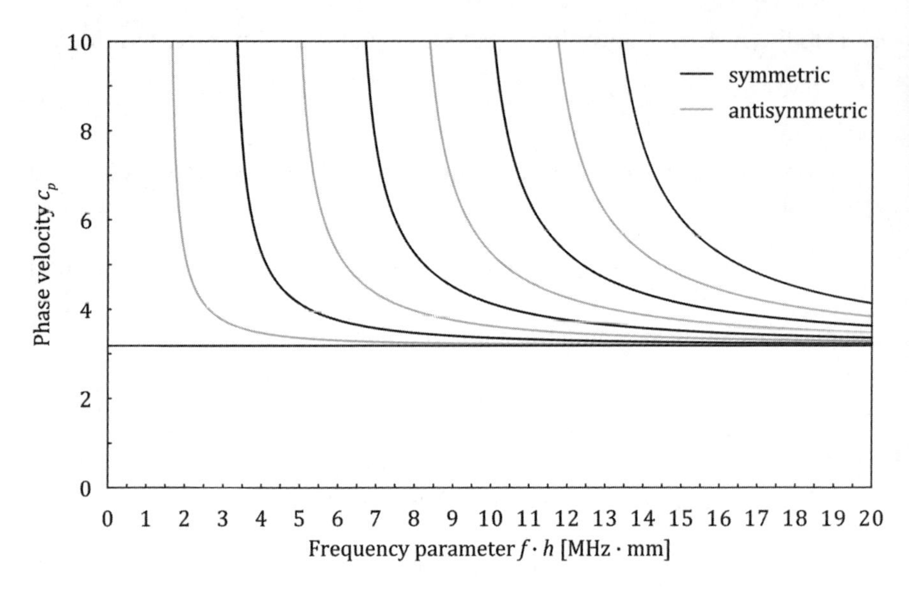

Figure 1.16 Phase velocity dispersion curves for symmetric and antisymmetric modes of SH waves ($c_S = 3.2$ km/s)

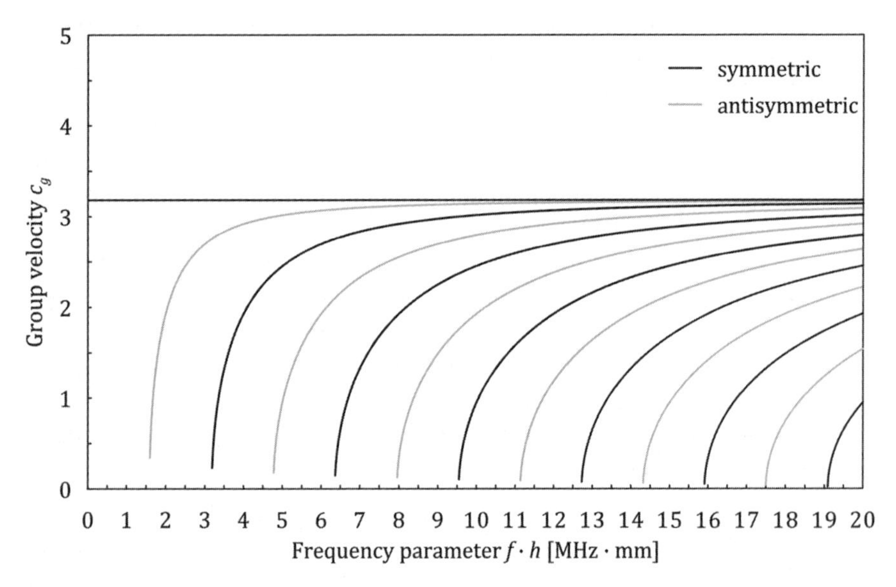

Figure 1.17 Group velocity dispersion curves for symmetric and antisymmetric modes of SH waves ($c_S = 3.2$ km/s)

1.5 Wave Propagation in One-Dimensional Bodies of Circular Cross-Section

1.5.1 Equations of Motion

Propagation of elastic waves in one-dimensional bodies is governed by equations of the linear theory of elasticity, which for isotropic media can be brought to tensor (1.17) or vector form:

$$(\lambda + \mu)\nabla(\nabla \cdot \boldsymbol{u}) + \mu\nabla^2\boldsymbol{u} = \rho\ddot{\boldsymbol{u}} \tag{1.110}$$

The case of one-dimensional bodies of full circular cross-section is analysed here. It is most convenient to analyse this subject using the cylindrical system of coordinates (x, r, θ) instead of the Cartesian system (x, y, z) (see Figure 1.18).

Using Helmholtz decomposition, one can express the displacement field vector $\boldsymbol{u}$ as the sum of the irrotational vector field $\boldsymbol{u}_\phi$ and the solenoidal vector field $\boldsymbol{u}_r$. This can be achieved by assuming that the displacement field vector is generated by a pair of potentials, that is scalar potential ϕ and vector

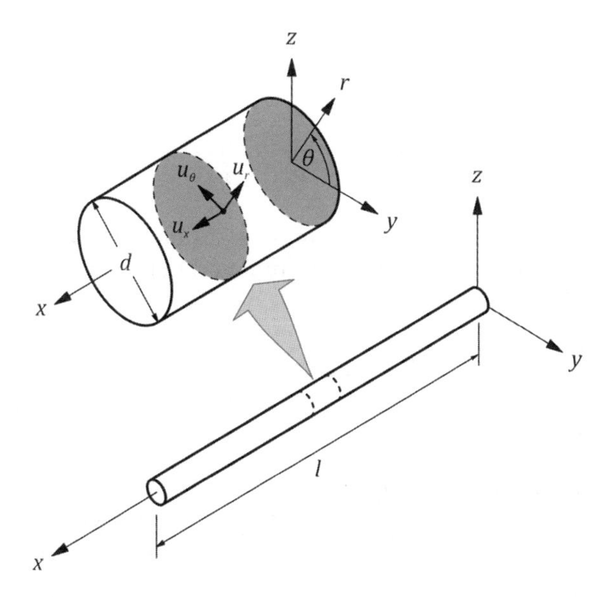

Figure 1.18 Shape of the structural rod element

potential $\boldsymbol{H} = (H_x, H_r, H_\theta)$:

$$\boldsymbol{u} = \boldsymbol{u}_\phi + \boldsymbol{u}_H = \nabla\phi + \nabla \times \boldsymbol{H}, \quad \nabla \cdot \boldsymbol{H} = 0 \tag{1.111}$$

using the following notation:

$$\begin{aligned}
\nabla\phi &= \boldsymbol{i}\frac{\partial\phi}{\partial x} + \boldsymbol{r}\frac{\partial\phi}{\partial r} + \boldsymbol{\theta}\frac{1}{r}\frac{\partial\phi}{\partial\theta} \\
\nabla \cdot \boldsymbol{H} &= \frac{\partial H_x}{\partial x} + \frac{1}{r}\frac{\partial(r H_r)}{\partial r} + \frac{1}{r}\frac{\partial H_\theta}{\partial\theta} \\
\nabla \times \boldsymbol{H} &= \boldsymbol{i}\frac{1}{r}\left[\frac{\partial(r H_\theta)}{\partial r} - \frac{\partial H_r}{\partial\theta}\right] + \boldsymbol{r}\left[\frac{1}{r}\frac{\partial H_x}{\partial\theta} - \frac{\partial H_\theta}{\partial x}\right] \\
&\quad + \boldsymbol{\theta}\left[\frac{\partial H_r}{\partial x} - \frac{\partial H_x}{\partial r}\right] \\
\nabla^2 u &= \frac{\partial^2 u}{\partial x^2} + \frac{\partial^2 u}{\partial r^2} + \frac{1}{r}\frac{\partial u}{\partial r} + \frac{1}{r^2}\frac{\partial^2 u}{\partial\theta^2}
\end{aligned} \tag{1.112}$$

where $\boldsymbol{i}$, $\boldsymbol{r}$ and $\boldsymbol{\theta}$ are unit vectors orientated along axes x, r and θ. Thus, displacement components can be expressed in the following form:

$$\begin{aligned}
u_x &= \frac{\partial\phi}{\partial x} + \frac{1}{r}\frac{\partial(r H_\theta)}{\partial r} - \frac{1}{r}\frac{\partial H_r}{\partial\theta} \\
u_r &= \frac{\partial\phi}{\partial r} + \frac{1}{r}\frac{\partial H_x}{\partial\theta} - \frac{\partial H_\theta}{\partial x} \\
u_\theta &= \frac{1}{r}\frac{\partial\phi}{\partial\theta} + \frac{\partial H_r}{\partial x} - \frac{\partial H_x}{\partial r}
\end{aligned} \tag{1.113}$$

Application of the Helmholtz theorem leads to motion equations identical with Equations (1.42) and (1.43), but formulated in the cylindrical system of coordinates:

$$c_L\nabla^2\phi = \ddot{\phi}, \quad c_S\nabla^2\boldsymbol{H} = \ddot{\boldsymbol{H}} \tag{1.114}$$

1.5.2 Longitudinal Waves

Analysis of longitudinal elastic waves in structural rod elements can be greatly simplified by the assumption of rotational symmetry of the rod with regard to the x axis. Because of this symmetry all displacement and stress components must be independent of the θ angle. In the case of longitudinal waves the u_θ displacement component as well as $\gamma_{x\theta}$ and $\gamma_{r\theta}$ deformation components must be equal to zero, that is $u_\theta = \gamma_{x\theta} = \gamma_{r\theta} = 0$. Moreover, one can demonstrate that a direct consequence of symmetry is that the potential vector $\boldsymbol{H}$ must have only one nonzero component H_θ and the other components H_x and H_r vanish, that is $H_x = H_r = 0$ [4, 7]. Consequently, the nonzero components of the displacement vector in the rod can be expressed as:

$$u_x = \frac{\partial \phi}{\partial x} + \frac{1}{r}\frac{\partial (r H_\theta)}{\partial r}, \quad u_r = \frac{\partial \phi}{\partial r} - \frac{\partial H_\theta}{\partial x} \tag{1.115}$$

After substituting the relationships (1.115) into Equation (1.110) and simplifications, a system of two independent motion equations expressed using scalar potentials ϕ and H_θ is obtained:

$$c_L \nabla^2 \phi = \ddot{\phi}, \quad c_S \left(\nabla^2 H_\theta - \frac{H_\theta}{r^2} \right) = \ddot{H}_\theta \tag{1.116}$$

The second equation of this system can be simplified further, thanks to the fact that:

$$\frac{\partial}{\partial r} \nabla^2 H_\theta = \nabla^2 \frac{\partial H_\theta}{\partial r} - \frac{1}{r^2}\frac{\partial H_\theta}{\partial r}$$

Substituting:

$$H_\theta = -\frac{\partial \psi}{\partial r}$$

leads to:

$$c_L \nabla^2 \phi = \ddot{\phi}, \quad c_S \nabla^2 \psi = \ddot{\psi} \tag{1.117}$$

At the same time, components u_x and u_r of the displacement vector $\boldsymbol{u}$ can be ultimately expressed as:

$$u_x = \frac{\partial \phi}{\partial x} - \frac{\partial^2 \psi}{\partial r^2} - \frac{1}{r}\frac{\partial \psi}{\partial r}, \quad u_r = \frac{\partial \phi}{\partial r} - \frac{\partial^2 \psi}{\partial x \partial r} \tag{1.118}$$

The displacement field in the rod can be easily computed on the basis of Equations (1.118). Nonzero components of the displacement field are as follows:

$$\varepsilon_{xx} = \frac{\partial u_x}{\partial x}, \quad \varepsilon_{rr} = \frac{\partial u_r}{\partial r}, \quad \varepsilon_{\theta\theta} = \frac{u_r}{r}, \quad \gamma_{xr} = \frac{\partial u_r}{\partial x} + \frac{\partial u_x}{\partial r} \tag{1.119}$$

while the stress field can be computed from Hooke's law, recalling the well-known identities:

$$\begin{aligned}
\sigma_{xx} &= 2\mu\varepsilon_{xx} + \lambda(\varepsilon_{xx} + \varepsilon_{rr} + \varepsilon_{\theta\theta}) \\
\sigma_{rr} &= 2\mu\varepsilon_{rr} + \lambda(\varepsilon_{xx} + \varepsilon_{rr} + \varepsilon_{\theta\theta}) \\
\sigma_{\theta\theta} &= 2\mu\varepsilon_{\theta\theta} + \lambda(\varepsilon_{xx} + \varepsilon_{rr} + \varepsilon_{\theta\theta}) \\
\tau_{xr} &= \mu\gamma_{xr}
\end{aligned} \tag{1.120}$$

Harmonic waves propagating in the rod along the x axis can be assumed as the solution of Equations (1.105) in a general complex form:

$$\phi = \hat{\phi}(r)\mathrm{e}^{\mathrm{i}(kx-\omega t)}, \quad \psi = \hat{\psi}(r)\mathrm{e}^{\mathrm{i}(kx-\omega t)} \tag{1.121}$$

where $\hat{\phi}(r)$ and $\hat{\psi}(r)$ are unknown functions. Substitution of the relationships (1.121) into the motion Equations (1.117) leads to a system of Bessel differential equations for functions $\hat{\phi}(r)$ and $\hat{\psi}(r)$:

$$\frac{\mathrm{d}^2\hat{\phi}}{\mathrm{d}r^2} + \frac{1}{r}\frac{\mathrm{d}\hat{\phi}}{\mathrm{d}r} + \alpha^2\hat{\phi} = 0, \quad \frac{\mathrm{d}^2\hat{\psi}}{\mathrm{d}r^2} + \frac{1}{r}\frac{\mathrm{d}\hat{\psi}}{\mathrm{d}r} + \beta^2\hat{\psi} = 0 \tag{1.122}$$

where:

$$\alpha^2 = \frac{\omega^2}{c_L^2} - k^2, \quad \beta^2 = \frac{\omega^2}{c_S^2} - k^2$$

which have solutions in the form of Bessel functions of the first type: $J_0(\alpha r)$ and $J_0(\beta r)$, as well as of the second type: $Y_0(\alpha r)$ and $Y_0(\beta r)$. As Bessel functions of the second type exhibit a singularity in the origin $r = 0$, this branch of solutions is discarded, leading to the following form of solutions of the problem being analysed:

$$\hat{\phi} = AJ_0(\alpha r), \quad \hat{\psi} = BJ_0(\beta r) \tag{1.123}$$

where A and B are some constants.

Taking into account the general form of solutions given as Equation (1.121), one can finally write that:

$$\phi = AJ_0(\alpha r)e^{i(kx-\omega t)}, \quad \psi = BJ_0(\beta r)e^{i(kx-\omega t)} \tag{1.124}$$

Propagation of longitudinal elastic waves in a rod requires meeting the boundary conditions of stresses vanishing on the external rod surface, which accompany the motion equation system (1.117):

$$\sigma_{rr}(x,r) = \tau_{xr}(x,r) = 0, \quad dla \quad 0 \le x \le l, \quad r = a = \frac{d}{2} \tag{1.125}$$

where l is the length and d is the rod diameter.

After substituting Equations (1.124) into Equations (1.119), using the identities of Equations (1.120) again and some simplifications, the boundary conditions of the vanishing stress components σ_{rr} and τ_{xr} lead to a system of two uniform equations expressed through the solutions from formulas (1.124).

A given system of equations has a nontrivial solution only if its determinant vanishes. In the analysed case this condition leads directly to a certain nonlinear equation known in literature as the Pochhammer frequency equation for longitudinal modes propagating in rods; this equation relates the angular frequency ω with wavenumber k. The Pochhammer frequency equation has the following form:

$$\begin{aligned} &\frac{2\alpha}{a}\left(\beta^2+k^2\right)J_1(\alpha a)J_1(\beta a) - \left(\beta^2-k^2\right)J_0(\alpha a)J_1(\beta a) \\ &-4k^2\alpha\beta J_1(\alpha a)J_0(\beta a) = 0 \end{aligned} \tag{1.126}$$

It is worth noting that this equation was derived for the first time in 1876 by a Prussian mathematician Leo Pochhammer [8], who studied vibrations of circular cylinders. This equation was also studied by many other researchers (e.g. Chree [9], Love [10], Davis [11], Pao and Mindlin [12] and Graff [13]), but due to its complexity its roots remained unknown for many years.

1.5.3 Solution of Pochhammer Frequency Equation

In the analysed case Pochhammer frequency equation was solved using original dedicated software developed by M. Krawczuk and A. Żak for the MATLAB® environment [14]. Phase velocity and group velocity values for waves propagating in the rod were calculated under the assumption that the rod was made of aluminium alloy of Young's modulus $E = 72.7$ GPa, Poisson's coefficient $\upsilon = 0.33$, material density $\rho = 2700$ kg/m^3 and diameter

$d = 0.01$ m. Characteristic velocities were $c_L = 6.3$ km/s and $c_S = 3.2$ km/s, respectively.

The calculation range was set out by a frequency range from 0.1 Hz to 20 MHz and a phase frequency range c from 2 km/s to 50 km/s. Roots of the Pochhammer frequency equation were sought in nodes of a regular grid of 400×2000 nodes with an assumed accuracy of $\delta \leq 0.001\%$.

Solving the equation involves applying the conjugate bisection method [15]. In the first step, roots were located as a function of the phase velocity $c_p = \omega/k$ for the given frequency, which was treated like a parameter in Equation (1.126). In the second step, the phase velocity c was considered as a parameter and the roots were located as a function of frequency f. In this way the second step of calculations improved the solution obtained in the first step for those of the analysed areas where phase velocity changes were very large. Group velocity values were also computed numerically by derivation of the wavenumber curves $k = k(\omega)$ with regard to the angular frequency ω.

Results obtained for changes of the group velocity to phase velocity ratio c_g/c_p as a function of the parameter fd, where f is frequency and d is rod diameter, are shown in Figure 1.19. As can be seen in Figure 1.20, the phase

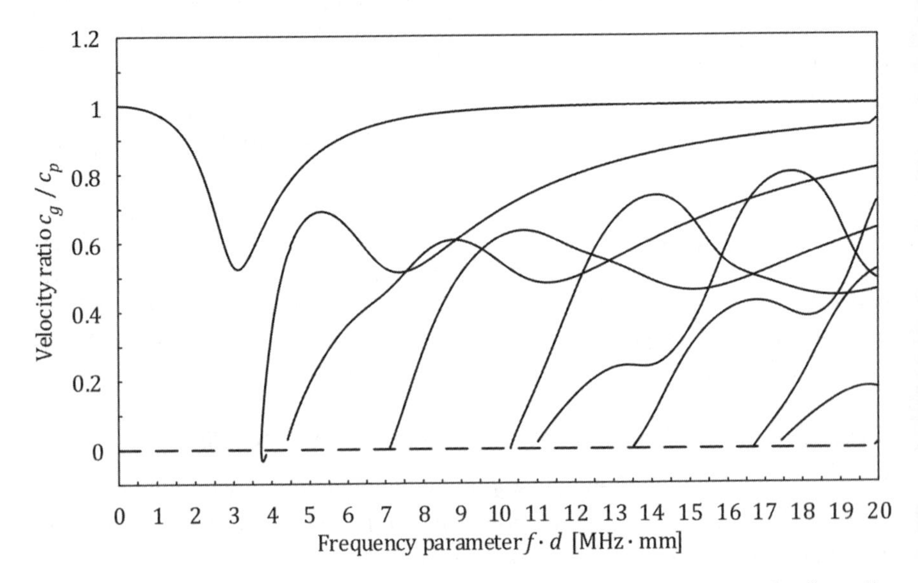

Figure 1.19 Dispersion curves of the group velocity to phase velocity ratio for the case of longitudinal modes in an aluminium rod ($c_L = 6.3$ km/s, $c_S = 3.2$ km/s)

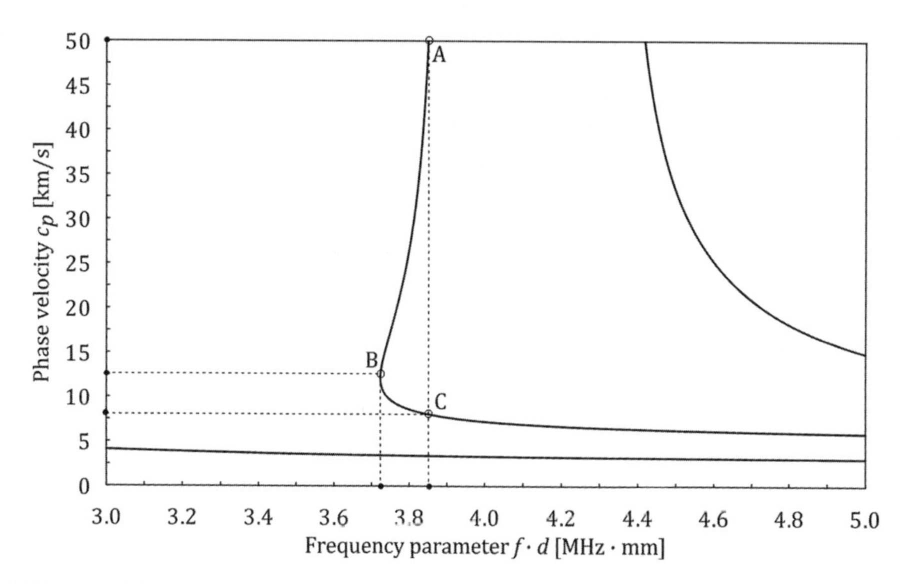

Figure 1.20 Increase of the second mode of the phase velocity dispersion curve for an aluminium rod

velocity dispersion curve for the second mode exhibits very unusual behaviour near the cut-off frequency for this mode between points A, B and C. One can see that between points A and B the group velocity and the phase velocity have opposite signs. This suggests that the direction of energy transfer in the rod may be opposite to the wave propagation direction. In other words, wave motion carries energy in one direction, but wave propagation seems to occur in the other direction. This phenomenon, called wave backpropagation, was investigated and documented in the literature by many researchers (e.g. Meeker and Meitzler [16], Meitzler [17], Alippi *et al.* [18] and Marston [19]) and still is the subject of research, especially in the case of electromagnetic waves.

In the frequency parameter fd range from a cut-off frequency of 3.72 MHz·mm (point B) to 3.85 MHz·mm (points A and C) the phase velocity curve c_p takes double values, which indicates two different zones of group velocity c_g values. The first branch of the phase velocity curve $c_p = c_p(fd)$ between points A and B is an area of high phase velocities, where phase velocity c_p and group velocity c_g have opposite signs. The second branch between points B and C is an area of low phase velocities, where phase velocity c_p and group velocity c_g have the same signs.

1.5.4 Torsional Waves

Torsional waves are a consequence of vanishing displacements u_r and u_z. Because of rotational symmetry, displacement u_θ must be independent of θ. For torsional waves the motion equation is as follows [13]:

$$\frac{\partial^2 u_\theta}{\partial r^2} + \frac{1}{r}\frac{\partial u_\theta}{\partial r} - \frac{u_\theta}{r} + \frac{\partial u_\theta}{\partial x^2} = \frac{1}{c_S^2}\frac{\partial^2 u_\theta}{\partial t^2} \tag{1.127}$$

Harmonic waves of the following form are assumed:

$$u_\theta = V(r)\mathrm{e}^{\mathrm{i}(kx-\omega t)} \tag{1.128}$$

Substituting Equation (1.128) into Equation (1.127) and solving the differential equation for the unknown function $V(r)$ leads to:

$$u_\theta = \frac{1}{\beta} B J_1(\beta r)\mathrm{e}^{\mathrm{i}(kx-\omega t)} \tag{1.129}$$

where B is any constant.

From the boundary conditions:

$$\sigma_{rr}(x,r) = \tau_{xr}(x,r) = \tau_{r\theta}(x,r) = 0, \quad \text{for} \quad 0 \le x \le l, \quad r = a = \frac{d}{2} \tag{1.130}$$

the only nontrivial condition is the following one:

$$\tau_{r\theta}(x,r) = 0, \quad \text{for} \quad 0 \le x \le l, \quad r = a = \frac{d}{2} \tag{1.131}$$

This condition leads to a dispersion equation for torsional modes propagating in rods that relates the angular frequency ω with wavenumber k:

$$(\beta a)J_0(\beta a) - 2J_1(\beta a) = 0 \tag{1.132}$$

The first three roots of Equation (1.132) are:

$$\beta_1 = 0, \quad \beta_2 a = 5.136, \quad \beta_3 a = 8.417$$

One should note that $\beta = 0$ is also a solution of the dispersion equation. The limit transition $\beta \to 0$ in Equation (1.129) leads to the following equation:

$$u_\theta = \frac{1}{2} Br e^{i(kx-\omega t)} \tag{1.133}$$

This displacement represents the lowest torsional mode. In the lowest mode displacement the amplitude u_θ is proportional to the radius. The motion corresponding to the solution is rotation of each rod cross-section as a whole around its centre. One should note that $\beta = 0$, which implies that the phase velocity is equal to the shear wave velocity c_S:

$$\beta^2 - \frac{\omega^2}{c_S^2} - k^2, \quad \beta = 0, \quad \to \quad c_S = \frac{\omega}{k} = c$$

Thus the lowest torsional mode is nondispersive. Higher modes are dispersive. For the given frequency, solutions of Equation (1.132) in the form of wavenumbers k_n can take real or imaginary values. For real values the $k_n(\omega)$ branches are hyperbolically shaped, while for imaginary values they are circles.

1.5.5 Flexural Waves

Flexural waves are dependent on the circumferential angle θ and in the displacement vector all three components are nonzero and change according to simple trigonometry-based relationships:

$$\begin{aligned} u_x &= U_x(r)\cos\theta e^{i(kx-\omega t)} \\ u_r &= U_r(r)\sin\theta e^{i(kx-\omega t)} \\ u_\theta &= U_\theta(r)\cos\theta e^{i(kx-\omega t)} \end{aligned} \tag{1.134}$$

After substituting the displacement components (1.134) into the system of displacement Equations (1.114) a system of three ordinary differential equations containing the functions U_x, U_y and $U_\theta$is obtained. Without going into the details of solving these equations, their ultimate form is as follows:

$$
\begin{aligned}
U_x(r) &= \mathrm{i}kAJ_1(\alpha r) - \frac{C}{r}\frac{\partial}{\partial r}\left[rJ_2(\beta r)\right] - \frac{C}{r}J_2(\beta r) \\
U_r(r) &= A\frac{\partial}{\partial r}J_1(\alpha r) + \frac{B}{r}J_1(\beta r) + \mathrm{i}kCJ_2(\beta r) \\
U_\theta(r) &= -\frac{A}{r}J_1(\alpha r) + \mathrm{i}kCJ_2(\beta r) - B\frac{\partial}{\partial r}J_1(\beta r)
\end{aligned}
\tag{1.135}
$$

Particular integrals of (1.135) are chosen in such fashion that they do not have singularities on the rod axis. From the conditions of zero stresses on the cylinder surface:

$$
\sigma_{rr}(x,r) = \tau_{xr}(x,r) = \tau_{r\theta}(x,r) = 0, \quad \text{for} \quad 0 \le x \le l, \quad r = a = \frac{d}{2} \tag{1.136}
$$

after applying the relationships (1.135) a system of three equations [20] is obtained. This is a system of equations uniform with regard to constants A, B and C. The condition of the vanishing system determinant leads to a frequency equation [13], from which one can determine subsequent roots k.

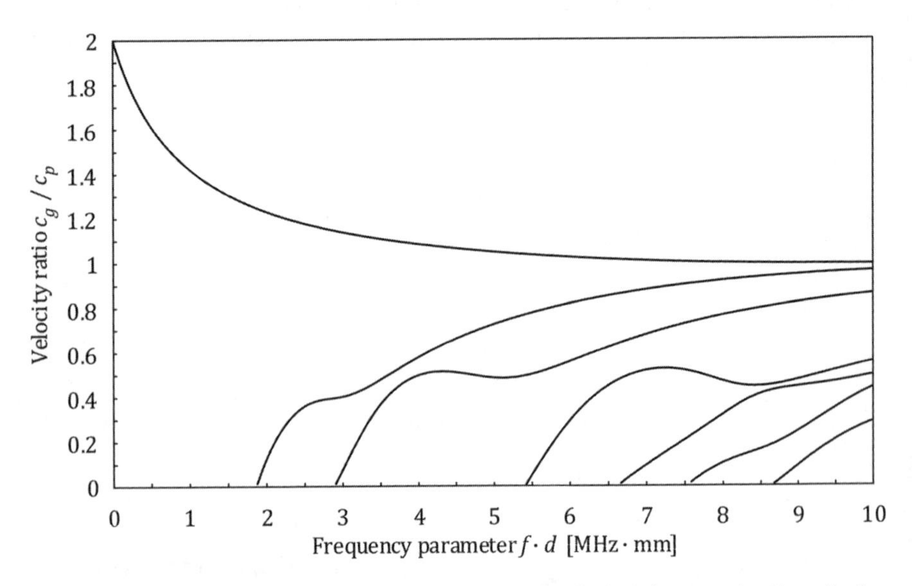

Figure 1.21 Dispersion curves of the group velocity to phase velocity ratio for the case of flexural modes in an aluminium rod ($c_L = 6.3$ km/s, $c_S = 3.2$ km/s)

The frequency equation for flexural waves in rods of circular cross-section was investigated by Pao and Mindlin [12].

Results obtained for changes of the group velocity to phase velocity ratio c_g/c_p as a function of the parameter fd, where f is frequency and d is rod diameter, are shown in Figure 1.21.

References

1. Lamb, H. (1917) On waves in an elastic plate. *Proceedings of the Royal Society of London*, **93**, 293–312.
2. Worlton, D.C. (1961) Experimental confirmation of Lamb waves at megacycle frequencies, *Journal of Applied Physics*, **32**, 967–971.
3. Kolsky, H. (1963) *Stress Waves in Solids*, Dover Publications, Inc., New York.
4. Rose, J.L. (1999) *Ultrasonic Waves in Solid Media*, Cambridge University Press, Cambridge.
5. Giurgiutiu, V. (2007) *Structural Health Monitoring with Piezoelectric Wafer Active Sensors*, Academic Press.
6. Doyle, J.F. (1997) *Wave Propagation in Structures*, Springer-Verlag New York, Inc., New York.
7. Achenbach, J.D. (1973) *Wave Propagation in Elastic Solids*, North-Holland Publishing Company, Amsterdam.
8. Pochhammer, L. (1876) Biegung des Kreiscylinders – Fortpflanzungs-Geschwindigkeit Kleiner Schwingungen in einem Kreiscylinder. *Journal für die reine und angewandte Mathematik*, **81**, 33–61.
9. Chree, C. (1889) The equations of an isotropic elastic solid in polar and cylindrical coordinates, their solutions and applications. *Proceedings of the Cambridge Philosophical Society. Mathematical and Physical Sciences*, **14**, 250–369.
10. Love, A.E. (1927) *A Treatise on the Mathematical Theory of Elasticity*, 4th edn, Dover Publications, New York, Dover.
11. Davis, R.M. (1948) A critical study of the Hopkinson pressure bar. *Philosophical Transactions of the Royal Society of London. Series A, Mathematical and Physical Sciences*, **240**, 375–457.
12. Pao, Y.H. and Mindlin, R.D. (1960) Dispersion of flexural waves in an elastic, circular cylinder. *Journal of Applied Mechanics*, **27**, 513–520.
13. Graff, K.F. (1991) *Wave Motion in Elastic Solids*, Dover Publications, New York, Dover.
14. Żak, A. and Krawczuk M. (2010) Assessment of rod behaviour theories used in spectral finite element modelling. *Journal of Sound and Vibration*, **329**(11), 2099–2113.
15. Ralston, A. (1965) *A First Course in Numerical Analysis*, McGraw-Hill Book Company, New York.

16. Meeker, T.R. and Meitzler A.H. (1964) Guided wave propagation in elongated cylinders and plates, Chapter 2, in *Physical Acoustics*, vol. 1, Part A, Academic Press, New York.
17. Meitzler, A.H. (1965) Backward wave transmission of stress pulses in elastic cylinders and plates. *The Journal of Acoustical Society of America*, **38**, 835–842.
18. Alippi, A., Bettucci, A. and Germano, M. (2000) Anomalous propagation characteristics of evanescent waves. *Ultrasonics*, **38**, 817–820.
19. Marston, P.L. (2003) Negative group velocity Lamb waves on plates and applications to the scattering of sound by shells. *The Journal of the Acoustical Society of America*, **113**, 2659–2662.
20. Bancroft, D. (1941) The velocity of longitudinal waves in cylindrical bars. *Physical Review*, **59**, 588–593.

2

Spectral Finite Element Method

The spectral finite element method is a relatively new computational technique that basically combines two different numerical techniques, that is spectral methods [1] and the finite element method [2].

Spectral methods are a special class of techniques employed for solving problems described by partial differential equations numerically. Such problems are most often associated with such phenomena as wave propagation, interference and diffraction in continuous media of various types (gases, liquids, solids), gas or liquid flows, multiphase flows, diffusion and many others. In the case of spectral methods, solutions are sought in the investigated area using Fourier series or approximating polynomials possessing special properties. Such polynomials are usually orthogonal Chebyshev polynomials or very-high-order Lobatto polynomials over nonuniformly spaced nodes, in which solutions are sought. Thanks to this approach, in spectral methods the computation error ε decreases exponentially as a function of the approximating polynomial n, that is $\varepsilon \approx O[(1/n)^n]$, therefore guaranteeing very fast (spectral) convergence of the solution to exact solutions. Despite unquestioned benefits, spectral methods are very difficult to translate to algorithms, require much computational power and are not suited to solving problems in geometrically complex areas.

The finite element method is employed to solve complex problems from various disciplines of physical sciences described by partial differential

Guided Waves in Structures for SHM: The Time-Domain Spectral Element Method, First Edition.
Wieslaw Ostachowicz, Pawel Kudela, Marek Krawczuk and Arkadiusz Zak.

equations or integral equations, such as solid mechanics, fluid mechanics, thermodynamics, electro- and magneto-statics and -dynamics, conjugate field problems and many others. A characteristic property of the finite element method is discretisation (division) of the analysed area into a certain number of smaller subareas called finite elements, within which one seeks solutions described by approximating polynomials over uniformly spaced nodes. Because of the so-called Runge phenomenon [1] (the emergence of large errors of solutions, resulting from oscillation of approximating polynomial values near finite element edges with an increase in the approximating polynomial order n), usually polynomials of the first and second order (i.e. linear or quadratic), and less often of the third order, are used as approximating polynomials. Advantages of the finite element method include ease of algorithmisation and capability of solving problems in geometrically complex areas, while its disadvantage, resulting from using low-order approximation polynomials, is the need for dense discretisation of the analysed area.

The spectral finite element method is essentially a combination of both mentioned methods; it combines the properties of approximating polynomials of spectral methods and the approach to discretising the analysed area particular to the finite element method. Thanks to this combination, very fast (spectral) convergence of solutions with an increase in the order of approximating polynomials is achieved together with the absence of limitations regarding the geometry of analysed areas and with significantly lower requirements concerning discretisation density.

Lobatto polynomials, Chebyshev polynomials and Laguerre polynomials [3] are employed as approximating polynomials in the spectral finite element method. These polynomials are defined by a system of appropriately selected nodes. Unlike for the finite element method, in the case of the spectral finite element method the nodes are nonuniformly spaced and their locations correspond to zeros of certain polynomials. This allows the Runge phenomenon [1] to be avoided, that is large oscillations of approximating polynomials near the area edges. The fact that zero points of these polynomials are more dense near area edges allows the edge oscillations that are natural for high-order polynomials to be avoided and opens up a chance to employ high-order approximating polynomials, which is virtually impossible with the finite element method.

The spectral finite element method, like the finite element method, uses the finite element idea. A finite element is a simple geometrical object (one-, two- or three-dimensional) for which particular points called nodes as well as particular approximating functions called shape functions or node functions have been specified. These are used to describe the distribution of the

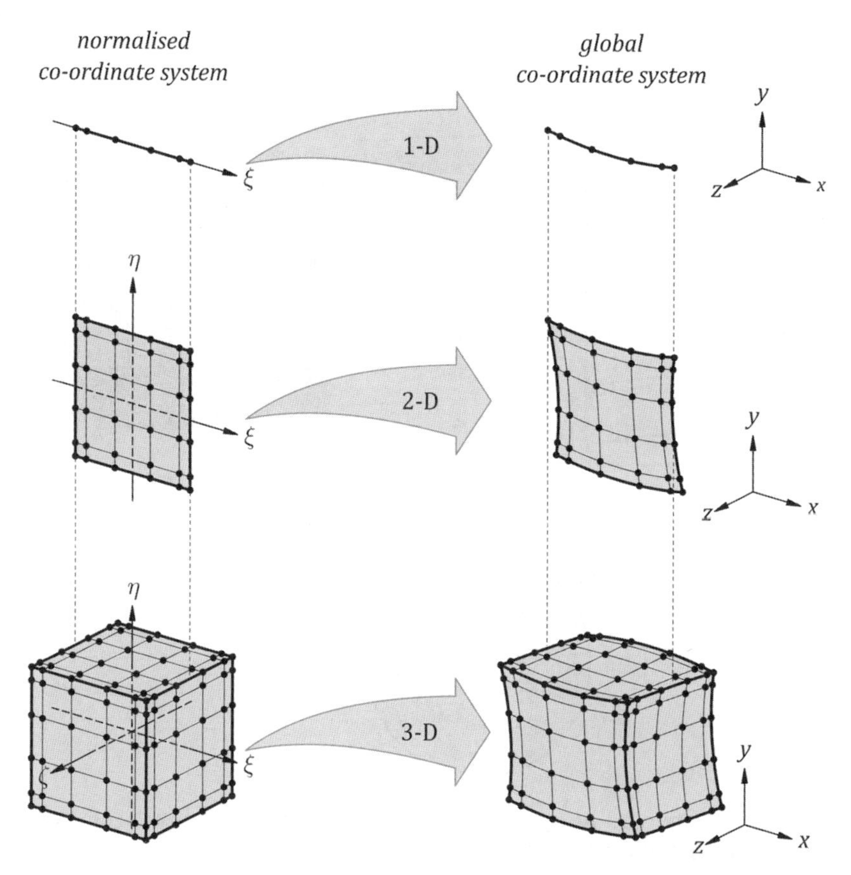

Figure 2.1 One-, two- and three-dimensional spectral finite elements of order 5 in normalised coordinate systems and their representations in the global system

analysed physical properties inside the element and on its boundaries. Figure 2.1 presents one-, two- and three-dimensional spectral finite elements of the fifth order in a normalised coordinate system and their representations in the global coordinate system.

Individual steps of defining the mathematical model and solving it in order to obtain the solution for the spectral finite element method are analogous to the case of the finite element method. They can be divided into the subsequent steps:

1. The analysed object is conceptually divided into a finite number of geometrically simple elements, henceforth called spectral finite elements.

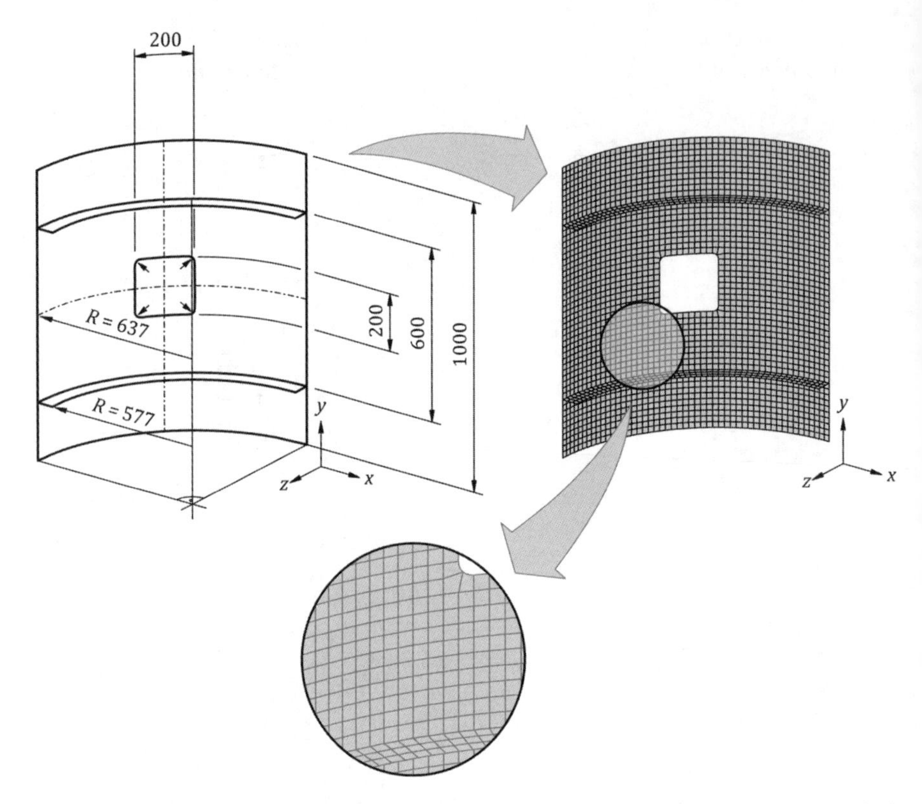

Figure 2.2 Section of aeroplane fuselage discretised with shell-type spectral finite elements of the 5th order

In individual elements characteristic points called nodes are chosen. Figure 2.2 presents a section of aeroplane fuselage discretised with shell-type spectral finite elements.

2. It is assumed that spectral finite elements employed for discretisation are only connected together in a finite number of nodes located on their edges.
3. Certain mathematical functions are chosen, with their form depending on the number of nodes in elements. These are used to describe the distribution of the analysed physical properties inside the spectral finite elements, depending on their node values. These functions are called node functions or shape functions. In the case of spectral finite elements these are Lobatto polynomials, Chebyshev polynomials or Laguerre polynomials [1, 3].
4. Ordinary or partial differential equations describing the physical phenomenon being analysed are transformed by applying the weighed

residual method or the method of minimising the variation functional of the phenomenon [4] to equations of the spectral finite element method. The first approach is called the weak formulation of the method, while the second one is called the strong formulation. Typical equations of the spectral finite element method are ordinary differential equations of the first and second order, and also, for some phenomena, algebraic equations. These equations are composed at the level of individual elements and called local equations. The transformations mentioned above are closely related to the notion of characteristic matrices of elements that are derived thanks to these methods and that can be identified with matrices of inertia, damping and stiffness of individual spectral finite elements.

5. One aggregates the respective characteristic matrices of spectral finite elements used for discretising the object, as well as aggregates of the load vectors. As a result, global characteristic matrices are obtained and the equations of the spectral finite element method are expressed in their terms. This aggregation takes into account the topology of elements used for discretisation, that is their location within the object and orientation in space, as presented in Figure 2.3. In the case of nonstationary problems the load vectors must additionally account for the initial conditions. The total number of equations generated in this way, as discussed in item 4, is equal to the total number of nodes multiplied by the number of degrees of freedom in every one of them.

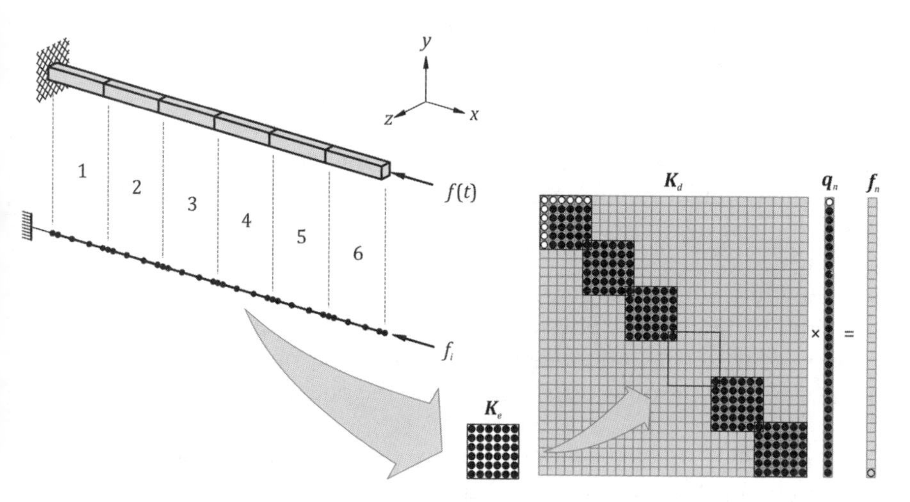

Figure 2.3 General outline of discretisation using spectral finite elements

6. The created system of equations, and thus the global characteristic matrices, are complemented with appropriate boundary conditions. This is implemented by means of introducing appropriate modifications to matrices of coefficients of the global system of equations and to the global vector of the right-hand equation sides.
7. The created system of equations is solved by employing appropriate numerical methods, which leads to obtaining values of the sought physical properties in nodes of individual elements.
8. Depending on needs, additional physical properties are calculated. Such properties are functions of the determined node values. For example, when investigating typical solid mechanics phenomena, one usually chooses displacements as unknown node values, and strains and stresses are then calculated on the basis of computed fields of displacements inside the individual spectral finite elements.
9. If the analysed phenomenon is nonstationary, operations presented in items 5 to 8 are repeated until the relevant condition that indicates the completion of computation is met. Such a condition can be, for example, reaching a predetermined value of the computed property in one of the nodes, phenomenon duration or some other parameter.

One should remember that with the spectral finite element method, as with the finite element method, numerical solutions are obtained in a finite number of points of a numerical model, that is in the nodes of individual elements. The goal of obtaining numerical solutions convergent to exact solutions imposes a number of requirements on the shape functions of spectral finite elements, also called node functions:

- Shape functions describing the field of the sought physical properties should guarantee their continuity within the elements and their conformity at boundaries of elements up to the order one less than the order of the highest derivative appearing in the differential equation of the phenomenon.
- Shape functions must guarantee their capability to describe the constant values of the sought physical properties or their derivatives within the element up to the order one less than the order of the highest derivative appearing in the differential equation of the phenomenon.

Meeting the above conditions usually provides monotonic convergence of the obtained solutions to the exact solution with an increasing number of elements and simultaneous shrinking of their geometrical size. Thus, the correct choice of shape functions is of key importance with both the spectral finite element method and the finite element method.

2.1 Shape Functions in the Spectral Finite Element Method

The fundamental issue of many numerical problems is approximation of the unknown function $f(x)$, usually with a linear combination of some base functions possessing specific properties. A typical example of such a procedure may be the approximation of a continuous and differentiable function $f(x)$ in the range $x \in [a, b]$ represented by n initial terms of its expansion into a Taylor series in the neighbourhood of point a:

$$f(x) \approx f(a) + \frac{x-a}{1!} f^{(1)}(a) + \frac{(x-a)^2}{2!} f^{(2)}(a) + \cdots + \frac{(x-a)^n}{n!} f^{(n)}(a) \quad (2.1)$$

Accuracy or error of this approximation $R_n(x, a)$ can be expressed in the form of the Lagrange residual:

$$R_n(x, a) = \frac{(x-a)^{n+1}}{(n+1)!} f^{(n+1)}(\xi), \quad \xi \in [a, b] \quad (2.2)$$

when an additional condition is met:

$$\lim_{x \to a} \frac{R_n(x, a)}{(x-a)^n} = 0 \quad (2.3)$$

In the general case of an approximating function $f(x)$ one can consider a class of certain base functions $\{p_n(x)\}(n = 0, 1, 2, \ldots)$, which are polynomials, or a class of periodic functions, which can be trigonometric functions $\{\sin(nx), \cos(nx)\}(n = 0, 1, 2, \ldots)$. There are also other classes of base functions, which can be useful in specific approximation cases, for example rational functions, exponential functions [1,5] and so on. This fact can be expressed as follows:

$$f(x) \approx a_0 g_0(x) + a_1 g_1(x) + \cdots + a_m g_m(x) \quad (2.4)$$

where $a_i (i = 0, 1, \ldots, m)$ are constant coefficients and $\{g_n(x)\}(n = 0, 1, 2, \ldots)$ represent the general class of base functions.

Among many possible base functions that can be used to construct shape functions in the finite element method, both in the classical and spectral approach, polynomials of various types are used most often. This results from the fact that polynomials naturally meet the conditions imposed on the shape functions listed in the previous chapter. The finite element method utilises low-order approximating polynomials over uniformly spaced nodes, while the spectral finite element method uses nonuniformly spaced nodes, which allows polynomials of very high order to be employed. These can be Lobatto, Chebyshev or Laguerre polynomials, among others.

2.1.1 Lobatto Polynomials

The general form of Lobatto polynomial of the nth order $L_n(\xi)$ can be expressed through the first derivative of the Legendre polynomial of the $(n+1)$th order $P_{n+1}(\xi)$:

$$L_n(\xi) = \frac{d}{d\xi} P_{n+1}(\xi), \quad n = 0, 1, 2, \ldots \tag{2.5}$$

while Legendre polynomials of the nth order are defined using Rodrigues's formula [6]:

$$P_n(\xi) = \frac{1}{2^n n!} \frac{d^n}{d\xi^n} \left(\xi^2 - 1\right)^n, \quad n = 0, 1, 2, \ldots \tag{2.6}$$

The forms of several initial Lobatto polynomials $L_n(\xi)$ together with the general form of the polynomial of the kth order are listed in Table 2.1 and their plots in the $\xi \in [-1, +1]$ range are presented in Figure 2.4.

An important property of Lobatto polynomials is their orthogonality with weight $(1 - \xi^2)$ in the range $\xi \in [-1, +1]$:

$$\int_{-1}^{+1} L_i(\xi) L_j(\xi) \left(1 - \xi^2\right) d\xi = \frac{2(i+1)(i+2)}{2i+3} \delta_{ij}, \quad i, j = 0, 1, 2, \ldots \tag{2.7}$$

where δ_{ij} is the Kronecker delta.

In the spectral finite element method nodes of an nth-order element in a normalised coordinate system are taken as [3]:

$$\xi_1 = -1, \quad \xi_2 = r_1, \quad \xi_3 = r_2, \ldots, \xi_{n-1} = r_{n-2}, \quad \xi_n = +1 \tag{2.8}$$

where $r_1, r_2, \ldots, r_{n-2}$ are roots of the Lobatto polynomial $L_{n-2}(\xi)$ of the order $n-2$. Thus, coordinates of nodes ξ_i of a spectral finite element of the nth

Table 2.1 Forms of several initial Lobatto polynomials $L_n(\xi)$ together with the general form of polynomial of the kth order

Polynomial order	Polynomial form
$n = 0$	$L_0(\xi) = 1$
$n = 1$	$L_1(\xi) = 3\xi$
$n = 2$	$L_2(\xi) = \frac{3}{2}(5\xi^2 - 1)$
$n = 3$	$L_3(\xi) = \frac{5}{2}(7\xi^2 - 3)\xi$
$n = 4$	$L_4(\xi) = \frac{15}{8}(21\xi^4 - 14\xi^2 + 1)$
$n = 5$	$L_5(\xi) = \frac{21}{8}(33\xi^4 - 30\xi^2 + 5)\xi$
...	...
$n = k$	$L_k(\xi) = \frac{1}{2^{k+1}(k+1)!}\frac{d^{k+2}}{d\xi^{k+2}}(\xi^2 - 1)^{k+1}$

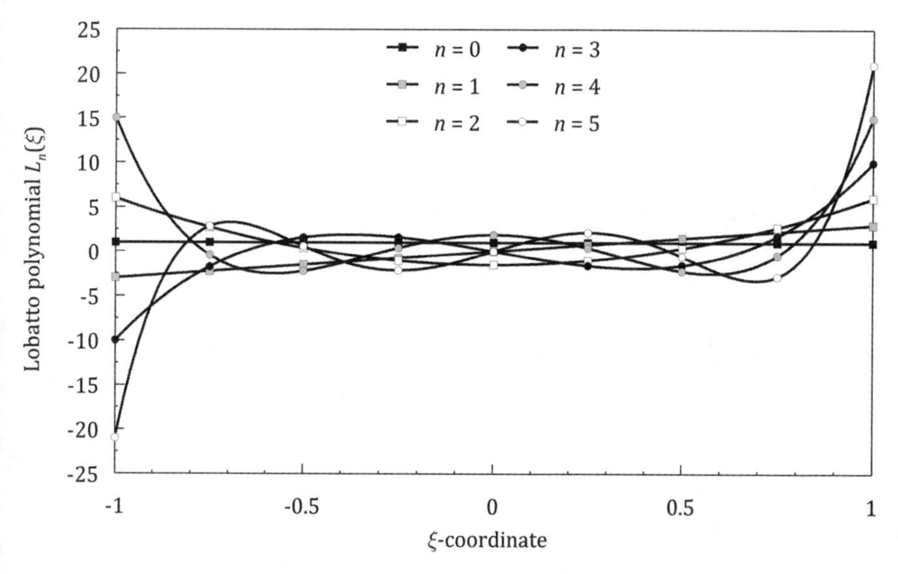

Figure 2.4 Plots of several initial Lobatto polynomials $L_n(\xi)$ in the range $\xi \in (-1, +1)$

order in the normalised coordinate system can be derived from the following relation:

$$L_n^c(\xi) \equiv \left(1 - \xi^2\right) L_{n-2}(\xi) = 0, \quad n = 2, 3, \ldots \tag{2.9}$$

In the subject literature the polynomial $L_n^c(\xi)$ expressed by relation (2.9) is also called a full (complete) Lobatto polynomial of the nth order, while its roots r_i are identified with coordinates of nodes ξ_i of the spectral finite element of the nth order that was expressed in the normalised coordinate system [3]. Values of coordinates of Lobatto nodes in the normalised coordinate system for spectral finite elements of several initial orders are listed in Table 2.2.

2.1.2 Chebyshev Polynomials

Chebyshev polynomials of the second type $U_n(\xi)$ can be defined using a recursive formula:

$$\begin{cases} U_0(\xi) = 1 \\ U_1(\xi) = 2\xi \\ \ldots \\ U_{n+1}(\xi) = 2\xi U_n(\xi) - U_{n-1}(\xi), \quad n = 0, 1, 2, \ldots \end{cases} \tag{2.10}$$

The forms of several initial Chebyshev polynomials $U_n(\xi)$ together with the general form of polynomial of the kth order are specified in Table 2.3 and their plots in the $\xi \in [-1, +1]$ range are presented in Figure 2.5.

An important property of Chebyshev polynomials of the second type is their orthogonality with weight $\sqrt{1 - \xi^2}$ in the range $\xi \in [-1, +1]$:

$$\int_{-1}^{+1} U_i(\xi) U_j(\xi) \sqrt{1 - \xi^2} d\xi = \frac{\pi}{2} \delta_{ij}, \quad i, j = 0, 1, 2, \ldots \tag{2.11}$$

where δ_{ij} is Kronecker delta.

In the spectral finite element method nodes of an nth-order element in a normalised coordinate system are taken as [2, 3]:

$$\xi_1 = -1, \quad \xi_2 = r_1, \quad \xi_3 = r_2, \ldots, \xi_{n-1} = r_{n-2}, \quad \xi_n = +1 \tag{2.12}$$

where $r_1, r_2, \ldots, r_{n-2}$ are roots of the Chebyshev polynomial of the second type $U_{n-2}(\xi)$ of the order $n - 2$.

Table 2.2 Values of coordinates of Lobatto nodes in the normalised coordinate system for spectral finite elements of several initial orders

Element order	Node coordinates
$n = 1$	$\xi_1 = -1.0000000000000000$
	$\xi_2 = +1.0000000000000000$
$n = 2$	$\xi_1 = -1.0000000000000000$
	$\xi_2 = 0.0000000000000000$
	$\xi_3 = +1.0000000000000000$
$n = 3$	$\xi_1 = -1.0000000000000000$
	$\xi_2 = -0.4472135954999579$
	$\xi_3 = +0.4472135954999579$
	$\xi_4 = +1.0000000000000000$
$n = 4$	$\xi_1 = -1.0000000000000000$
	$\xi_2 = -0.6546536707079771$
	$\xi_3 = 0.0000000000000000$
	$\xi_4 = +0.6546536707079771$
	$\xi_5 = +1.0000000000000000$
$n = 5$	$\xi_1 = -1.0000000000000000$
	$\xi_2 = -0.7650553239294647$
	$\xi_3 = -0.2852315164806451$
	$\xi_4 = +0.2852315164806451$
	$\xi_5 = +0.7650553239294647$
	$\xi_6 = +1.0000000000000000$
$n = 6$	$\xi_1 = -1.0000000000000000$
	$\xi_2 = -0.8302238962785669$
	$\xi_3 = -0.4688487934707142$
	$\xi_4 = 0.0000000000000000$
	$\xi_5 = +0.4688487934707142$
	$\xi_6 = +0.8302238962785669$
	$\xi_7 = +1.0000000000000000$

Thus, coordinates of nodes ξ_i of a spectral finite element of the nth order in the normalised coordinate system can be derived from the following relation:

$$T_n^c(\xi) \equiv \left(1 - \xi^2\right) U_{n-2}(\xi) = 0, \quad n = 2, 3, \ldots \tag{2.13}$$

In the subject literature the polynomial $T_n^c(\xi)$ described by relation (2.13) is also called a full Chebyshev polynomial of the nth order, while its roots r_i are identified with coordinates of nodes ξ_i of the spectral finite element of the nth

Table 2.3 Forms of several initial Chebyshev polynomials $U_n(\xi)$ together with the general form of polynomial of the kth order

Polynomial order	Polynomial form
$n = 0$	$U_0(\xi) = 1$
$n = 1$	$U_1(\xi) = 2\xi$
$n = 2$	$U_2(\xi) = 4\xi^2 - 1$
$n = 3$	$U_3(\xi) = 8\xi^3 - 4\xi$
$n = 4$	$U_4(\xi) = 16\xi^4 - 12\xi^2 + 1$
$n = 5$	$U_5(\xi) = 32\xi^5 - 32\xi^3 + 6\xi$
...	...
$n = k$	$U_k(\xi) = \dfrac{(\xi + \sqrt{\xi^2 - 1})^{k+1} - (\xi - \sqrt{\xi^2 - 1})^{k+1}}{2\sqrt{\xi^2 - 1}}$

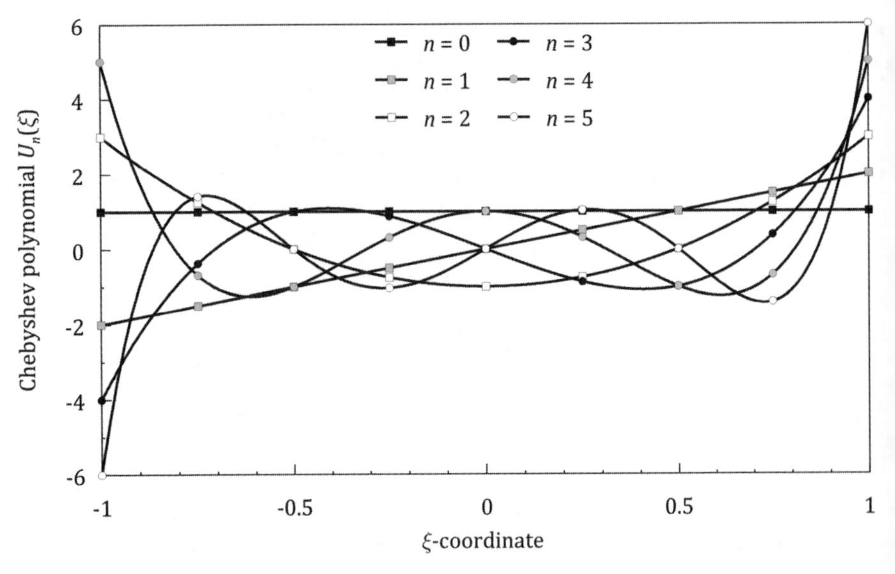

Figure 2.5 Plots of several initial Chebyshev polynomials $U_n(\xi)$ in the range $\xi \in (-1, +1)$

Table 2.4 Values of coordinates of Chebyshev nodes in the normalised coordinate system for spectral finite elements of several initial orders

Element order	Node coordinates
$n = 1$	$\xi_1 = -1.0000000000000000$
	$\xi_2 = +1.0000000000000000$
$n = 2$	$\xi_1 = -1.0000000000000000$
	$\xi_2 = 0.0000000000000000$
	$\xi_3 = +1.0000000000000000$
$n = 3$	$\xi_1 = -1.0000000000000000$
	$\xi_2 = -0.5000000000000000$
	$\xi_3 = +0.5000000000000000$
	$\xi_4 = +1.0000000000000000$
$n = 4$	$\xi_1 = -1.0000000000000000$
	$\xi_2 = -0.7071067811865475$
	$\xi_3 = 0.0000000000000000$
	$\xi_4 = +0.7071067811865475$
	$\xi_5 = +1.0000000000000000$
$n = 5$	$\xi_1 = -1.0000000000000000$
	$\xi_2 = -0.8090169943749474$
	$\xi_3 = -0.3090169943749474$
	$\xi_4 = +0.3090169943749474$
	$\xi_5 = +0.8090169943749474$
	$\xi_6 = +1.0000000000000000$
$n = 6$	$\xi_1 = -1.0000000000000000$
	$\xi_2 = -0.8660254037844386$
	$\xi_3 = -0.5000000000000000$
	$\xi_4 = 0.0000000000000000$
	$\xi_5 = +0.5000000000000000$
	$\xi_6 = +0.8660254037844386$
	$\xi_7 = +1.0000000000000000$

order that was expressed in the normalised coordinate system [2, 3]. Values of node coordinates in the normalised coordinate system can be calculated from the relationship:

$$\xi_i = \cos\left(\frac{\pi\, i}{n-1}\right), \quad i = 0, 1, 2, \ldots, n-1 \tag{2.14}$$

Table 2.4 lists values of Chebyshev node coordinates derived from relationship (2.14) for spectral finite elements of several initial orders.

2.1.3 Laguerre Polynomials

Laguerre polynomials $G_n(\xi)$ can be derived on the basis of the Rodrigues formula [6]:

$$G_n(\xi) = \frac{e^{\xi}}{n!}\frac{d^n}{d\xi^n}\left(e^{-\xi}\xi^n\right), \quad n = 0, 1, 2, \ldots \tag{2.15}$$

The forms of several initial Laguerre polynomials $G_n(\xi)$ together with the general form of the polynomial of the kth order are specified in Table 2.5 and their plots in the $\xi \in [0, +\infty]$ range are presented in Figure 2.6.

An important property of Laguerre polynomials is their orthogonality with weight $e^{-\xi}$ in the range $\xi \in [0, +\infty]$:

$$\int_0^{+\infty} G_i(\xi)G_j(\xi)e^{-\xi}d\xi = \delta_{ij}, \quad i, j = 0, 1, 2, \ldots \tag{2.16}$$

where δ_{ij} is Kronecker delta.

Table 2.5 Forms of several initial Laguerre polynomials $G_n(\xi)$ together with the general form of polynomial of the kth order

Polynomial order	Polynomial form
$n = 0$	$G_0(\xi) = 1$
$n = 1$	$G_1(\xi) = 1 - \xi$
$n = 2$	$G_2(\xi) = \frac{1}{2}(\xi^2 - 4\xi + 2)$
$n = 3$	$G_3(\xi) = \frac{1}{6}(-\xi^3 + 9\xi^2 - 18\xi + 6)$
$n = 4$	$G_4(\xi) = \frac{1}{24}(\xi^4 - 16\xi^3 + 72\xi^2 - 96\xi + 24)$
$n = 5$	$G_5(\xi) = \frac{1}{120}(-\xi^5 + 25\xi^4 - 200\xi^3 + 600\xi^2 - 600\xi + 120)$
...	...
$n = k$	$G_k(\xi) = \frac{1}{k+1}((2k + 1 - \xi) + L_k(\xi) - kL_{k-1}(\xi))$

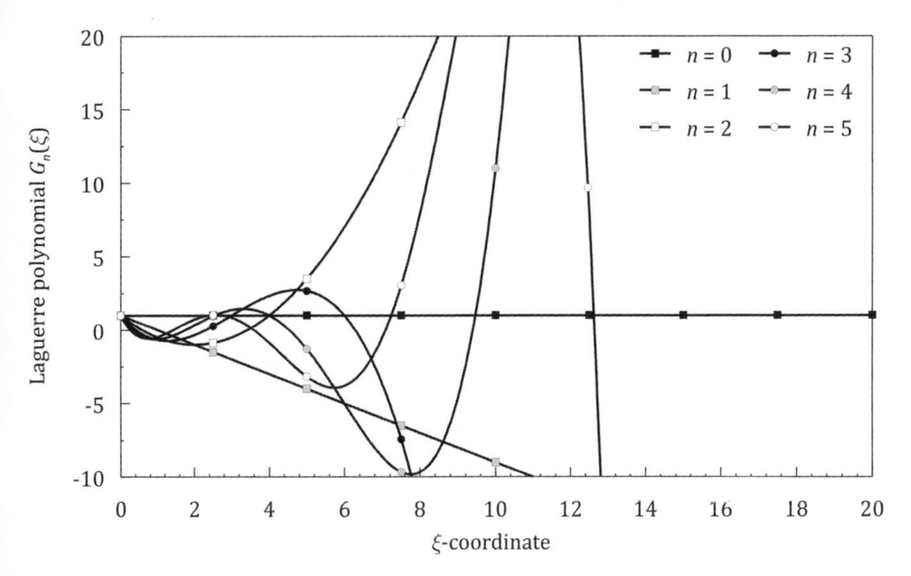

Figure 2.6 Plots of several initial Laguerre polynomials $G_n(\xi)$ in the range $\xi \in (0, +\infty)$

In the spectral finite element method nodes of an nth-order element in the normalised coordinate system are taken as [1]:

$$\xi_1 = 0, \quad \xi_2 = r_1, \quad \xi_3 = r_2, \ldots, \xi_{n-1} = r_{n-1} \tag{2.17}$$

where $r_1, r_2, \ldots, r_{n-1}$ are roots of the Laguerre polynomial $G_{n-1}(\xi)$ of the order $n - 1$.

Thus, coordinates of nodes ξ_i of a spectral finite element of the nth order in the normalised coordinate system can be derived from the following relation:

$$G_n^c(\xi) \equiv \xi G_{n-1}(\xi) = 0, \quad n = 1, 2, \ldots \tag{2.18}$$

In the subject literature the polynomial $G_n^c(\xi)$ expressed with relation (2.18) is also called a full Laguerre polynomial of the nth order, while its roots r_i are identified with coordinates of nodes ξ_i of the spectral finite element of the nth order in the normalised coordinate system [1]. Table 2.6 lists values of Laguerre node coordinates derived from relationship (2.18) for spectral finite elements of several initial orders.

Table 2.6 Values of coordinates of Laguerre nodes in the normalised coordinate system for spectral finite elements of several initial orders

Element order	Node coordinates
$n = 1$	$\xi_1 = 0.0000000000000000$
	$\xi_2 = +1.0000000000000000$
$n = 2$	$\xi_1 = 0.0000000000000000$
	$\xi_2 = +0.5857864376269050$
	$\xi_3 = +3.4142135623730950$
$n = 3$	$\xi_1 = 0.0000000000000000$
	$\xi_2 = +0.4157745567834791$
	$\xi_3 = +2.2942803602790417$
	$\xi_4 = +6.2899450829374792$
$n = 4$	$\xi_1 = 0.0000000000000000$
	$\xi_2 = +0.3225476896193923$
	$\xi_3 = +1.7457611011583466$
	$\xi_4 = +4.5366202969211280$
	$\xi_5 = +9.3950709123011331$
$n = 5$	$\xi_1 = 0.0000000000000000$
	$\xi_2 = +0.2635603197181409$
	$\xi_3 = +1.4134030591065168$
	$\xi_4 = +3.5964257710407221$
	$\xi_5 = +7.0858100058588376$
	$\xi_6 = +12.640800844275783$
$n = 6$	$\xi_1 = 0.0000000000000000$
	$\xi_2 = +0.2228466041792607$
	$\xi_3 = +1.1889321016726230$
	$\xi_4 = +2.9927363260593141$
	$\xi_5 = +5.7751435691045105$
	$\xi_6 = +9.8374674183825899$
	$\xi_7 = +15.982873980601702$

2.2 Approximating Displacement, Strain and Stress Fields

In a general case, approximating polynomials discussed in the previous item allow for changes to be made of the unknown physical properties within spectral finite elements depending on their node values. Such node values can be, for example, constituents of the displacement field, constituents of the stress or strain field, temperature or many other parameters. In the displacement formulation [2, 7, 8] of the spectral finite element method the approximating polynomials are employed to describe changes of the displacements u,v and w within the spectral finite elements as functions of the node displacements q_i^u, q_i^v, q_i^w. This formulation can be expressed in the global coordinate system as:

$$\begin{cases} u(x,y,z) = \sum_{i=1}^{m} N_i(x,y,z) q_i^u \\ v(x,y,z) = \sum_{i=1}^{m} N_i(x,y,z) q_i^v \\ w(x,y,z) = \sum_{i=1}^{m} N_i(x,y,z) q_i^w \end{cases} \tag{2.19}$$

where m denotes the number of nodes of a spectral finite element and $N_i (i = 1, \ldots, m)$ are shape functions. Shape functions are also called node functions. Relationship (2.19) can be expressed in the following simplified form:

$$\boldsymbol{q}^e = \mathbf{N}^e \boldsymbol{q}_n^e \tag{2.20}$$

where $\boldsymbol{q}^e = [u, v, w]^{\mathrm{T}}$ denotes the displacement vector of the spectral finite element, $\mathbf{N}^e$ is the shape function matrix and $\boldsymbol{q}_n^e = [q_1^u, q_1^v, q_1^w, \ldots, q_m^u, q_m^v, q_m^w]^{\mathrm{T}}$ is a vector of the element's node displacements. One should note that the shape functions $N_i (i = 1, \ldots, m)$ do not depend on time and therefore analogous approximation methods are applicable to vectors of the velocity fields $\dot{\boldsymbol{q}}^e$ and vectors of the acceleration fields $\ddot{\boldsymbol{q}}^e$:

$$\dot{\boldsymbol{q}}^e = \mathbf{N}^e \dot{\boldsymbol{q}}_n^e, \quad \ddot{\boldsymbol{q}}^e = \mathbf{N}^e \ddot{\boldsymbol{q}}_n^e \tag{2.21}$$

where $\dot{\boldsymbol{q}}_n^e$ and $\ddot{\boldsymbol{q}}_n^e$ denote the vectors of node velocities and of node accelerations of the element, respectively.

The strain vector $\boldsymbol{\varepsilon}^e$ and stress vector $\boldsymbol{\sigma}^e$ in any point of the spectral finite element can be expressed in terms of the element's node displacements $\boldsymbol{q}_n^e$. To this end one needs to approximate the displacement field with the given relationship (2.20), as well as to use the relations between strains and displacements and between stress and displacements known from the theory of elasticity [9].

The theory of elasticity was being developed particularly vigorously in the nineteenth century and in the beginning of the twentieth century by such renowned mathematicians and physicists as Cauchy, Lagrange, Navier and

Poisson. The mathematical theory of elasticity describes the parameters that characterise the geometry of displacements, as well as internal forces called stresses that appear when an object is deformed. The theory of elasticity used a theoretical, idealised and simplified model of a solid as a *material continuum*. This model neglects the molecular structure of the solid, thus ignoring a number of real properties of solids, assuming a model of continuous distribution of the material in space instead. The *material continuum* is considered as a continuous medium in the mathematical sense. Thus, it is assumed that points of the medium that were close to each other before deformation remain close to each other after deformation as well. The option of discontinuities appearing in the *material continuum* during deformation is thus excluded.

The linear theory of elasticity assumes that solids obey Hooke's law [9]. These solids may be either isotropic, that is exhibiting identical properties in every direction, or anisotropic, that is with properties depending on direction. Distributions of displacements, strains and stresses in a solid under load from external forces are described for the given boundary conditions by the fundamental relationships of the theory of elasticity. These relationships include the relations between strain and displacement and between stress and strain or displacement.

The relationship between the strain state and displacement state assumes the following general form:

$$\boldsymbol{\varepsilon}^e = \boldsymbol{\Gamma}_l \boldsymbol{q}^e + \boldsymbol{\Gamma}_n \boldsymbol{q}^e \tag{2.22}$$

where $\boldsymbol{\Gamma}_l$ and $\boldsymbol{\Gamma}_n$ denote, respectively, the linear and nonlinear differentiation operators, expressed as:

$$\boldsymbol{\Gamma}_l = \begin{bmatrix} \frac{\partial}{\partial x} & 0 & 0 \\ 0 & \frac{\partial}{\partial y} & 0 \\ 0 & 0 & \frac{\partial}{\partial z} \\ \frac{\partial}{\partial y} & \frac{\partial}{\partial x} & 0 \\ 0 & \frac{\partial}{\partial z} & \frac{\partial}{\partial y} \\ \frac{\partial}{\partial z} & 0 & \frac{\partial}{\partial x} \end{bmatrix}, \quad \boldsymbol{\Gamma}_n = \boldsymbol{\Gamma}_n(\boldsymbol{q}) = \frac{1}{2} \begin{bmatrix} (\partial_x \boldsymbol{q})^{\mathrm{T}} \partial_x \\ (\partial_y \boldsymbol{q})^{\mathrm{T}} \partial_y \\ (\partial_z \boldsymbol{q})^{\mathrm{T}} \partial_z \\ (\partial_x \boldsymbol{q})^{\mathrm{T}} \partial_y + (\partial_y \boldsymbol{q})^{\mathrm{T}} \partial_x \\ (\partial_y \boldsymbol{q})^{\mathrm{T}} \partial_z + (\partial_z \boldsymbol{q})^{\mathrm{T}} \partial_y \\ (\partial_z \boldsymbol{q})^{\mathrm{T}} \partial_x + (\partial_x \boldsymbol{q})^{\mathrm{T}} \partial_z \end{bmatrix} \tag{2.23}$$

in which the differentiation operator ∂_α ($\alpha = x, y, z$) is defined as follows:

$$\partial_\alpha = \begin{bmatrix} \dfrac{\partial}{\partial \alpha} & 0 & 0 \\ 0 & \dfrac{\partial}{\partial \alpha} & 0 \\ 0 & 0 & \dfrac{\partial}{\partial \alpha} \end{bmatrix} \tag{2.24}$$

As the above relationships describe operators of differentiation performed on the element displacement vector $\boldsymbol{q}^e$, approximating functions of an appropriate class must be chosen. This requirement results from the necessity of guaranteeing that the assumption of continuity is observed. By the use of relationships (2.20) and (2.22), one can finally express the vector of strains ε^e inside the spectral finite element through the vector of node displacements q_n^e as:

$$\boldsymbol{\varepsilon}^e = \boldsymbol{\Gamma}_l \boldsymbol{N}^e \boldsymbol{q}_n^e + \boldsymbol{\Gamma}_n \boldsymbol{N}^e \boldsymbol{q}_n^e = \boldsymbol{B}_l^e \boldsymbol{q}_n^e + \boldsymbol{B}_n^e \boldsymbol{q}_n^e = \boldsymbol{B}^e \boldsymbol{q}_n^e \tag{2.25}$$

where $\boldsymbol{B}_l^e$ and $\boldsymbol{B}_n^e$ denote matrices of, respectively, linear and nonlinear relationships between strains $\boldsymbol{\varepsilon}^e$ and node displacements of element $\boldsymbol{q}_n^e$.

Relationships between the states of stresses and strains can in general be written as:

$$\boldsymbol{\sigma} = \boldsymbol{D}\boldsymbol{\varepsilon} \tag{2.26}$$

where the matrix $\boldsymbol{D}$ is called the matrix of elasticity coefficients.

In the most general case of a *generalised* or *anisotropic material*, the matrix of elasticity coefficients $\boldsymbol{D}$ is a full matrix and its construction requires 21 material constants. In the case of materials exhibiting symmetry with regard to chosen directions the number of material constants is smaller.

In the case of a material exhibiting a single plane of symmetry the number of material constants is reduced from 21 to 13, while in the case of a material exhibiting two planes of symmetry, also known as an orthotropic material (from *orthogonally anisotropic*), the number of material constants needed to construct matrix $\boldsymbol{D}$ further reduces from 13 to 9. At this point one should note that the vast majority of laminar composite materials exhibit properties of orthotropic materials.

In the case of an orthotropic and linearly elastic material the matrix of elasticity coefficients $\boldsymbol{D}$ has the following form:

$$\boldsymbol{D} = \begin{bmatrix} Q_{11} & Q_{12} & Q_{13} & 0 & 0 & 0 \\ Q_{12} & Q_{22} & Q_{23} & 0 & 0 & 0 \\ Q_{13} & Q_{23} & Q_{33} & 0 & 0 & 0 \\ 0 & 0 & 0 & Q_{44} & 0 & 0 \\ 0 & 0 & 0 & 0 & Q_{55} & 0 \\ 0 & 0 & 0 & 0 & 0 & Q_{66} \end{bmatrix} \tag{2.27}$$

where the way coefficients $Q_{ij}(i, j = 1, 2, 3)$ and $Q_{kk}(k = 4, 5, 6)$ of matrix $\boldsymbol{D}$ are defined is presented in [10].

For an isotropic and linearly elastic material whose elastic properties are independent of the direction, matrix $\boldsymbol{D}$ assumes the following, much simpler form:

$$\boldsymbol{D} = \begin{bmatrix} \lambda + 2\mu & \lambda & \lambda & 0 & 0 & 0 \\ \lambda & \lambda + 2\mu & \lambda & 0 & 0 & 0 \\ \lambda & \lambda & \lambda + 2\mu & 0 & 0 & 0 \\ 0 & 0 & 0 & \mu & 0 & 0 \\ 0 & 0 & 0 & 0 & \mu & 0 \\ 0 & 0 & 0 & 0 & 0 & \mu \end{bmatrix} \tag{2.28}$$

where λ and μ are Lame's constants. These can be expressed with the well-known relationships:

$$\lambda = \frac{\nu E}{(1+\nu)(1-2\nu)}, \quad \mu = \frac{E}{2(1+\nu)} \tag{2.29}$$

where E denotes Young's modulus and ν is Poisson's ratio. Typical values of Young's modulus E and Poisson's ratio ν for selected structural materials are listed in Table 2.7.

To summarise, one can say that relationships (2.20), (2.25) and (2.26) allow the stress vector $\boldsymbol{\sigma}^e$ to be expressed inside the spectral finite element through the vector of node displacements $\boldsymbol{q}_n^e$ as:

$$\boldsymbol{\sigma}^e = \boldsymbol{D}^e \boldsymbol{\varepsilon}^e = \boldsymbol{D}^e \boldsymbol{B}_l^e \boldsymbol{q}_n^e + \boldsymbol{D}^e \boldsymbol{B}_n^e \boldsymbol{q}_n^e = \boldsymbol{D}^e \boldsymbol{B}^e \boldsymbol{q}_n^e \tag{2.30}$$

where $\boldsymbol{D}^e$ denotes the matrix of elasticity coefficients associated with a linear elastic material assigned to the given spectral finite element.

Table 2.7 Typical values of Young's modulus E and Poisson's ratio ν for selected structural materials

Material	Young's modulus (GPa)	Poisson's constant	Density (kg/m^3)
Epoxide resin	3.43	0,35	1250
Thermosetting resin	3.5	0,35	1200
Polyamide (nylon)	3.5	0,35	1100
Polyamide (DuPont)	2.5	0,34	1420
Aluminium: 2024-T3, 6061-T6	73.1	0,33	2700
Aluminium: 7079-T6	71.7	0,33	2740
Magnesium: AZ31B-H24	44.83	0,35	1700
Magnesium: HK31A-H24	44.14	0,35	1790
Steel	210	0,3	7800
Steel: ASI 304	193.1	0,29	8030
Steel: ASI C1020	203.4	0,29	7850
Steel: 4340	210	0,29	7800
Alloy steel	210	0,28	7800
Titanium: 11	115	0,31	4500
Titanium: B 120 VCA	102	0,3	4850
Glass fibre	65.5	0,23	2250
Kevlar fibre	130	0,22	1450
Carbon fibre	275.6	0,2	1900
Boron fibre	399.6	0,21	2580

2.3 Equations of Motion of a Body Discretised Using Spectral Finite Elements

Equations of motion of a body discretised using spectral finite elements can be derived from Euler–Lagrange equations, also called Lagrange equations. These were introduced by a Swiss mathematician Leonhard Euler and an Italian mathematician Joseph Louis Lagrange in 1750, and nowadays they constitute the base form of variation calculus. They result from the principle of least action and their solutions are functions for which a certain functional $\mathtt{L}$ is stationary. They can be expressed as follows:

$$\frac{\mathrm{d}}{\mathrm{d}t}\left\{\frac{\partial \mathrm{L}}{\partial \dot{q}}\right\} - \left\{\frac{\partial \mathrm{L}}{\partial q}\right\} + \left\{\frac{\partial \mathrm{R}}{\partial \dot{q}}\right\} = 0 \tag{2.31}$$

where $\mathrm{L} = \mathrm{T} - \Pi_p$ denotes the Lagrangian function of a dynamic system and $\mathtt{T}$, Π_p and $\mathtt{R}$ denote its kinetic energy, potential energy and dissipation function, respectively. Vectors $\boldsymbol{q}$ and $\dot{\boldsymbol{q}}$ are the displacement vector and velocity vector, respectively.

The kinetic energy T of a dynamic system, its potential energy Π_p and dissipation function R can in a general case be expressed by the following relationships [8]:

$$\begin{cases} \mathrm{T} = \dfrac{1}{2}\displaystyle\int_V \rho\,\dot{\boldsymbol{q}}^{\mathrm{T}}\dot{\boldsymbol{q}}\,\mathrm{d}V \\ \Pi_p = \dfrac{1}{2}\displaystyle\int_V \boldsymbol{\varepsilon}^{\mathrm{T}}\boldsymbol{\sigma}\,\mathrm{d}V - \int_V \boldsymbol{q}^{\mathrm{T}}\boldsymbol{\psi}_V\,\mathrm{d}V - \int_A \boldsymbol{q}^{\mathrm{T}}\boldsymbol{\psi}_A\,\mathrm{d}A \\ \mathrm{R} = \dfrac{1}{2}\displaystyle\int_V \mu\,\dot{\boldsymbol{q}}^{\mathrm{T}}\dot{\boldsymbol{q}}\,\mathrm{d}V \end{cases} \tag{2.32}$$

where ρ and μ denote, respectively, the density and damping coefficient and $\boldsymbol{\psi}_V$ and $\boldsymbol{\psi}_A$ represent the volume and surface force vectors. One should note that the vectors $\boldsymbol{\psi}_V$ and $\boldsymbol{\psi}_A$ depend on time t, that is $\boldsymbol{\psi}_V = \boldsymbol{\psi}_V(t)$ and $\boldsymbol{\psi}_A = \boldsymbol{\psi}_A(t)$.

The kinetic energy T^e, potential energy Π_p^e and dissipation function R^e of a spectral finite element can be expressed in a form that takes into account the chosen method of approximating the fields of displacement (2.20), velocity (2.21), strain (2.25) and stress (2.30), as:

$$\begin{cases} \mathrm{T}^e = \dfrac{1}{2}(\dot{\boldsymbol{q}}_n^e)^{\mathrm{T}}\left[\displaystyle\int_{V^e} \rho\,(\mathbf{N}^e)^{\mathrm{T}}\,\mathbf{N}^e\,\mathrm{d}V^e\right]\dot{\boldsymbol{q}}_n^e \\ \Pi_p^e = \dfrac{1}{2}(\boldsymbol{q}_n^e)^{\mathrm{T}}\left[\displaystyle\int_{V^e} (\boldsymbol{B}_l^e + \boldsymbol{B}_n^e)^{\mathrm{T}}\,\boldsymbol{D}^e\,(\boldsymbol{B}_l^e + \boldsymbol{B}_n^e)\,\mathrm{d}V^e\right]\boldsymbol{q}_n^e \\ \qquad - (\boldsymbol{q}_n^e)^{\mathrm{T}}\left[\displaystyle\int_{V^e} (\mathbf{N}^e)^{\mathrm{T}}\,\boldsymbol{\psi}_V\,\mathrm{d}V^e + \int_{A^e} (\mathbf{N}^e)^{\mathrm{T}}\,\boldsymbol{\psi}_A\,\mathrm{d}A^e + \boldsymbol{f}_c^e\right] \\ \mathrm{R}^e = \dfrac{1}{2}(\dot{\boldsymbol{q}}_n^e)^{\mathrm{T}}\left[\displaystyle\int_{V^e} \mu\,(\mathbf{N}^e)^{\mathrm{T}}\,\mathbf{N}^e\,\mathrm{d}V^e\right]\dot{\boldsymbol{q}}_n^e \end{cases} \tag{2.33}$$

where the potential energy Π_p^e was complemented by work of focused forces associated with the vector of focused node forces $\boldsymbol{f}_c^e$ of the element. This vector is also dependent on time t, that is $\boldsymbol{f}_c^e = \boldsymbol{f}_c^e(t)$.

At this point one should note that the expressions inside the integrals in relationships (2.33) denote so-called characteristic matrices and vectors of the spectral finite element. These are the inertia matrix $\boldsymbol{M}^e$, stiffness matrix $\boldsymbol{K}^e$ and damping matrix $\boldsymbol{C}^e$, as well as vectors of its volume loads $\boldsymbol{f}_V^e(t)$ and surface loads $\boldsymbol{f}_A^e(t)$:

$$\begin{cases} \boldsymbol{M}^e = \displaystyle\int_{V^e} \rho\,(\mathbf{N}^e)^{\mathrm{T}}\,\mathbf{N}^e \mathrm{d}V^e \\ \boldsymbol{K}^e = \displaystyle\int_{V^e} (\boldsymbol{B}_l^e + \boldsymbol{B}_n^e)^{\mathrm{T}}\,\mathbf{D}^e\,(\boldsymbol{B}_l^e + \boldsymbol{B}_n^e)\,\mathrm{d}V^e \\ \boldsymbol{C}^e = \displaystyle\int_{V^e} \mu\,(\mathbf{N}^e)^{\mathrm{T}}\,\mathbf{N}^e \mathrm{d}V^e \\ \boldsymbol{f}_V^e = \displaystyle\int_{V^e} (\mathbf{N}^e)^{\mathrm{T}}\,\boldsymbol{\psi}_V \mathrm{d}V^e \\ \boldsymbol{f}_A^e = \displaystyle\int_{A^e} (\mathbf{N}^e)^{\mathrm{T}}\,\boldsymbol{\psi}_A \mathrm{d}A^e \end{cases} \tag{2.34}$$

After employing relationships (2.34), one can express the kinetic energy T^e, potential energy Π_p^e and dissipation function R^e of a spectral finite element as:

$$\begin{cases} \mathrm{T}^e = \dfrac{1}{2}\,(\dot{\boldsymbol{q}}_n^e)^{\mathrm{T}}\,\boldsymbol{M}^e\,\dot{\boldsymbol{q}}_n^e \\ \Pi_p^e = \dfrac{1}{2}\,(\boldsymbol{q}_n^e)^{\mathrm{T}}\,\boldsymbol{K}^e\,\boldsymbol{q}_n^e - (\boldsymbol{q}_n^e)^{\mathrm{T}}\left[\boldsymbol{f}_V^e + \boldsymbol{f}_A^e + \boldsymbol{f}_c^e\right] \\ \mathrm{R}^e = \dfrac{1}{2}\,(\dot{\boldsymbol{q}}_n^e)^{\mathrm{T}}\,\boldsymbol{C}^e\,\dot{\boldsymbol{q}}_n^e \end{cases} \tag{2.35}$$

which, after substitution into Lagrange's formula (2.31), leads to the equation of motion on the single spectral finite element level of the following form:

$$\boldsymbol{M}^e\,\ddot{\boldsymbol{q}}_n^e + \boldsymbol{C}^e\,\dot{\boldsymbol{q}}_n^e + \boldsymbol{K}^e\,\boldsymbol{q}_n^e = \boldsymbol{f}_n^e(t) \tag{2.36}$$

where the load vector $f_n^e(t)$ is the sum of the volume loads $f_V^e(t)$, surface loads $f_A^e(t)$ and focused loads $f_c^e(t)$ acting upon the spectral finite element:

$$f_n^e(t) = f_V^e(t) + f_A^e(t) + f_c^e(t) \tag{2.37}$$

Equations of motion of a body discretised with spectral finite elements can be obtained in a formal manner by generalising Equation (2.36) from the single element level to the whole object level:

$$\mathbf{M}\,\ddot{\mathbf{q}}_n + \mathbf{C}\,\dot{\mathbf{q}}_n + \mathbf{K}\,\mathbf{q}_n = \mathbf{f}_n(t) \tag{2.38}$$

where $\mathbf{M}$, $\mathbf{K}$ and $\mathbf{C}$ this time denote the global characteristic matrices of inertia, stiffness and damping, respectively, and $\mathbf{f}_n(t)$ is the global vector of node loads. The following relationships are valid:

$$\begin{cases} \mathbf{M} = \sum_{e=1}^{E} (\mathbf{T}^e)^{\mathrm{T}} \mathbf{M}^e \mathbf{T}^e, \quad \mathbf{C} = \sum_{e=1}^{E} (\mathbf{T}^e)^{\mathrm{T}} \mathbf{C}^e \mathbf{T}^e, \quad \mathbf{K} = \sum_{e=1}^{E} (\mathbf{T}^e)^{\mathrm{T}} \mathbf{K}^e \mathbf{T}^e \\ \ddot{\mathbf{q}}_n = \sum_{e=1}^{E} (\mathbf{T}^e)^{\mathrm{T}} \ddot{\mathbf{q}}_n^e, \quad \dot{\mathbf{q}}_n = \sum_{e=1}^{E} (\mathbf{T}^e)^{\mathrm{T}} \dot{\mathbf{q}}_n^e \\ \mathbf{q}_n = \sum_{e=1}^{E} (\mathbf{T}^e)^{\mathrm{T}} \mathbf{q}_n^e, \quad \mathbf{f}_n = \sum_{e=1}^{E} (\mathbf{T}^e)^{\mathrm{T}} \mathbf{f}_n^e \end{cases} \tag{2.39}$$

where e denotes the number of spectral finite elements and E is the total number of elements employed for discretisation. The symbol $\mathbf{T}^e$ denotes a zero–one matrix reflecting the topology and fashion of object discretisation, which allows for movement from the single element level to the whole object level, that is from local equations written down for individual elements (2.36) to global Equations (2.38).

At this point one should note that equations of motion at the level of the whole object, expressed by relationship (2.38), need to be complemented with relevant boundary conditions resulting from bounds that restrict free motion of the object and initial conditions resulting from the character of the analysed problem.

In practice, boundary conditions are implemented by means of appropriate modifications to elements of characteristic matrices and characteristic vectors or of modifications of the structure of characteristic matrices and characteristic vectors that form the equations of motion.

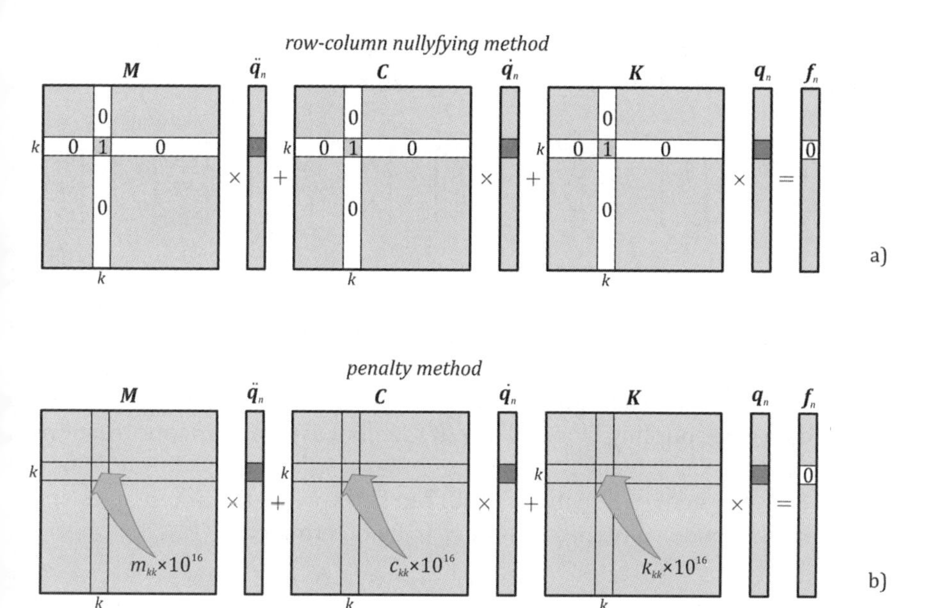

Figure 2.7 Schematic outline of two methods of imposing *zero boundary conditions*: (a) zeroing rows and columns of characteristic matrices, (b) introducing the penalty factor

The simplest situation is encountered in the case of so-called *zero boundary conditions*. Zero boundary conditions specify such boundary conditions, for which displacements, velocities and accelerations corresponding to degrees of freedom in node locations take values of zero. Accounting for this fact requires zeroing the rows and columns of characteristic matrices corresponding to degrees of freedom of the imposed zero values, except for elements on diagonals of characteristic matrices, which should be fixed as having values of one. Similarly, relevant elements of the forcing vector must be zeroed as well. This process has been outlined in Figure 2.7(a).

A different approach involves multiplying the terms on diagonals of the characteristic matrix that correspond to degrees of freedom of zero displacements, velocities and accelerations by a very large number, which is sometimes called the *penalty factor* in the literature. This process requires no additional modifications of nondiagonal terms in the rows and columns associated with these degrees of freedom, as shown in Figure 2.7(b).

If displacements, velocities or accelerations in bound locations are nonzero, the above method can be utilised as well. To this end, characteristic matrices and vectors need to be modified by being divided into blocks corresponding to the unknown node values and the imposed node values:

$$\begin{bmatrix} \boldsymbol{M}_{11} & \boldsymbol{M}_{12} \\ \boldsymbol{M}_{12}^{\mathrm{T}} & \boldsymbol{M}_{22} \end{bmatrix} \begin{Bmatrix} \ddot{\boldsymbol{q}}_n \\ \ddot{\boldsymbol{z}}_n \end{Bmatrix} + \begin{bmatrix} \boldsymbol{C}_{11} & \boldsymbol{C}_{12} \\ \boldsymbol{C}_{12}^{\mathrm{T}} & \boldsymbol{C}_{22} \end{bmatrix} \begin{Bmatrix} \dot{\boldsymbol{q}}_n \\ \dot{\boldsymbol{z}}_n \end{Bmatrix} + \begin{bmatrix} \boldsymbol{K}_{11} & \boldsymbol{K}_{12} \\ \boldsymbol{K}_{12}^{\mathrm{T}} & \boldsymbol{K}_{22} \end{bmatrix} \begin{Bmatrix} \boldsymbol{q}_n \\ \boldsymbol{z}_n \end{Bmatrix} = \begin{Bmatrix} \boldsymbol{f}_n(t) \\ \boldsymbol{r}_n(t) \end{Bmatrix} \tag{2.40}$$

where blocks $\boldsymbol{M}_{11}$, $\boldsymbol{C}_{11}$ and $\boldsymbol{K}_{11}$ of characteristic matrices correspond to unknown node values $\ddot{\boldsymbol{q}}_n$, $\dot{\boldsymbol{q}}_n$ and $\boldsymbol{q}_n$, blocks $\boldsymbol{M}_{22}$, $\boldsymbol{C}_{22}$ and $\boldsymbol{K}_{22}$ of characteristic matrices correspond to imposed node values $\ddot{\boldsymbol{z}}_n$, $\dot{\boldsymbol{z}}_n$ and $\boldsymbol{z}_n$, and blocks $\boldsymbol{M}_{12}$, $\boldsymbol{C}_{12}$ and $\boldsymbol{K}_{12}$ are coupling blocks. The $\boldsymbol{r}_n(t)$ symbol denotes dynamic reactions of nodes.

The above system of Equations (2.40) can be expressed in a form allowing the unknown node values $\ddot{\boldsymbol{q}}_n$, $\dot{\boldsymbol{q}}_n$ and $\boldsymbol{q}_n$ to be determined:

$$\boldsymbol{M}_{11}\ddot{\boldsymbol{q}}_n + \boldsymbol{C}_{11}\dot{\boldsymbol{q}}_n + \boldsymbol{K}_{11}\boldsymbol{q}_n = \boldsymbol{f}_n(t) - (\boldsymbol{M}_{22}\ddot{\boldsymbol{z}}_n + \boldsymbol{C}_{22}\dot{\boldsymbol{z}}_n + \boldsymbol{K}_{22}\boldsymbol{z}_n) \tag{2.41}$$

which in turn yields the capability to compute the dynamic reactions of bounds $\boldsymbol{r}_n(t)$ from the following equation:

$$\boldsymbol{r}_n(t) = \boldsymbol{M}_{12}^{\mathrm{T}}\ddot{\boldsymbol{q}}_n + \boldsymbol{C}_{12}^{\mathrm{T}}\dot{\boldsymbol{q}}_n + \boldsymbol{K}_{12}^{\mathrm{T}}\boldsymbol{q}_n + \boldsymbol{M}_{22}\ddot{\boldsymbol{z}}_n + \boldsymbol{C}_{22}\dot{\boldsymbol{z}}_n + \boldsymbol{K}_{22}\boldsymbol{z}_n \tag{2.42}$$

One should note that the described approach allows both time-independent bounds (*scleronomic bounds*) and time-dependent bounds (*renomic bounds*) to be modelled.

Methods of imposing initial conditions at the whole object level are intimately connected with employed methods of solving the global equations of motion (2.38), which will be discussed in detail in a later part of the present chapter.

2.4 Computing Characteristic Matrices of Spectral Finite Elements

In order to solve the equations of motion (2.38) of an object discretised with spectral finite elements, one has first to compute the values of characteristic matrices of individual elements according to relationships (2.34). In a general case, these computations utilise integrals of the following form:

$$A^e = \int_{V^e} \boldsymbol{F}(x, y, z)\mathrm{d}V^e \tag{2.43}$$

where the function being integrated, $\boldsymbol{F}(x, y, z)$, depends on the shape functions $\boldsymbol{N}^e = \boldsymbol{N}^e(x, y, z)$ (2.20) or their derivatives $\boldsymbol{B}^e = \boldsymbol{B}^e(x, y, z)$ (2.25) with regard to the global coordinates x, y and z.

Direct integration of relationship (2.43) with regard to the global coordinates x, y and z is usually ineffective and not desirable due to the fact that the areas of integration V^e can take on arbitrary shapes. Because of this, in practical computations the above integration is performed in certain areas that are regular with regard to the normalised coordinates ξ, η and ζ, such that $(\xi, \eta, \zeta) \in [-1, +1]$. Depending on the dimensions, such areas can be segments, equilateral triangles or squares, and also three-dimensional regular tetrahedrons or cubes. To this end, one needs to make shape functions $\boldsymbol{N}^e$ depend on the same normalised coordinates $\boldsymbol{N}^e = \boldsymbol{N}^e(\xi, \eta, \zeta)$.

Switching from the global coordinate system to a normalised coordinate system allows the integral in Equation (2.43) to be supplanted by the following expression:

$$A^e = \int_{-1}^{+1}\int_{-1}^{+1}\int_{-1}^{+1} \boldsymbol{F}(\xi, \eta, \zeta)\det(\mathbf{J})\mathrm{d}\xi\, \mathrm{d}\eta\, \mathrm{d}\zeta \tag{2.44}$$

where the symbol $\mathbf{J}$ denotes the Jacobian determinant of the mapping of area V^e on to the normalised area. In relationship (2.44) the function $\boldsymbol{F}(\xi, \eta, \zeta)$ being integrated depends on the normalised coordinates ξ, η and ζ. In case the integrated function contains derivatives of shape functions $\boldsymbol{B}^e = \boldsymbol{B}^e(x, y, z)$ (2.25) with regard to the global coordinates x, y and z, additional transformation of these derivatives to the system of the normalised coordinates ξ, η and ζ is required, as follows:

$$\begin{Bmatrix} \dfrac{\partial N_i}{\partial x} \\ \dfrac{\partial N_i}{\partial y} \\ \dfrac{\partial N_i}{\partial z} \end{Bmatrix} = \mathbf{J}^{-1} \begin{Bmatrix} \dfrac{\partial N_i}{\partial \xi} \\ \dfrac{\partial N_i}{\partial \eta} \\ \dfrac{\partial N_i}{\partial \zeta} \end{Bmatrix}, \qquad i = 1, \ldots, m \tag{2.45}$$

where the symbol **J** denotes the Jacobian determinant of the mapping of the area of the spectral finite element V^e from the system of global coordinates x, y and z to the system of normalised coordinates ξ, η and ζ. This Jacobian determinant can be expressed by the following relationship:

$$\mathbf{J} = \begin{bmatrix} \frac{\partial x}{\partial \xi} & \frac{\partial y}{\partial \xi} & \frac{\partial z}{\partial \xi} \\ \frac{\partial x}{\partial \eta} & \frac{\partial y}{\partial \eta} & \frac{\partial z}{\partial \eta} \\ \frac{\partial x}{\partial \zeta} & \frac{\partial y}{\partial \zeta} & \frac{\partial z}{\partial \zeta} \end{bmatrix} \tag{2.46}$$

The Jacobian determinant of this mapping **J** can be calculated on the basis of values of coordinates of nodes of the spectral finite element in the global coordinate system and of certain mapping functions $\boldsymbol{T}^e = \boldsymbol{T}^e(\xi, \eta, \zeta)$ that describe the element shape in this system:

$$\mathbf{J} = \begin{bmatrix} \sum_{i=1}^{p} \frac{\partial T_i}{\partial \xi} x_i & \sum_{i=1}^{p} \frac{\partial T_i}{\partial \xi} y_i & \sum_{i=1}^{p} \frac{\partial T_i}{\partial \xi} z_i \\ \sum_{i=1}^{p} \frac{\partial T_i}{\partial \eta} x_i & \sum_{i=1}^{p} \frac{\partial T_i}{\partial \eta} y_i & \sum_{i=1}^{p} \frac{\partial T_i}{\partial \eta} z_i \\ \sum_{i=1}^{p} \frac{\partial T_i}{\partial \zeta} x_i & \sum_{i=1}^{p} \frac{\partial T_i}{\partial \zeta} y_i & \sum_{i=1}^{p} \frac{\partial T_i}{\partial \zeta} z_i \end{bmatrix} = \begin{bmatrix} \frac{\partial T_1}{\partial \xi} & \frac{\partial T_2}{\partial \xi} & \cdots \\ \frac{\partial T_1}{\partial \eta} & \frac{\partial T_2}{\partial \eta} & \cdots \\ \frac{\partial T_1}{\partial \zeta} & \frac{\partial T_2}{\partial \zeta} & \cdots \end{bmatrix} \begin{bmatrix} x_1 & y_1 & z_1 \\ x_2 & y_2 & z_2 \\ \vdots & \vdots & \vdots \end{bmatrix} \tag{2.47}$$

where $T_i (i = 1, \ldots, p)$ and $(x_i, y_i, z_i)(i = 1, \ldots, p)$ are coordinates of selected points of the element.

At this point one should note that if the functions describing the geometry of the spectral finite element $T_i (i = 1, \ldots, p)$ are shape functions of the element $N_i (i = 1, \ldots, p)$, equalities $p = m$ and $(T_i = N_i)(i = 1, \ldots, m)$ take place and such an element is called an *isoparametric element*. If the element geometry is described using more parameters than are employed for describing variabilities of displacement, strain or stress fields, the inequality $p > m$ takes place and such an element is called a *superparametric element*. In the opposite case, that is when $p < m$, the element is called a *subparametric* one.

As mentioned above, except for uncommon exceptions, direct integration of relationship (2.43) is ineffective and not desirable. In practice, this integration is usually performed using numerical integration methods [5, 7]. In the spectral finite element method the choice of a suitable numerical integration procedure, called *quadrature* in the subject literature, is a very important factor affecting method effectiveness and directly depends on the type of approximating polynomials used: Lobatto, Chebyshev or Laguerre ones.

Numerical integration of characteristic matrices of spectral finite elements using Lobatto polynomials as approximating polynomials employs Lobatto quadrature, also called Gauss–Lobatto quadrature or Gauss–Lobatto–Legendre quadrature [3, 5]. In the case of elements using Chebyshev polynomials, the best results are obtained using Gauss quadrature, also called Gauss–Legendre quadrature [3, 5]. In the case of spectral finite elements constructed on the basis of Laguerre polynomials, numerical integration is performed using Gauss–Laguerre quadrature.

In the general case, relationship (2.44) can be written as:

$$\begin{aligned} A^e &= \int_{-1}^{+1}\int_{-1}^{+1}\int_{-1}^{+1} \boldsymbol{F}(\xi,\eta,\zeta)\det(\mathbf{J})\mathrm{d}\xi\,\mathrm{d}\eta\,\mathrm{d}\zeta \\ &= \sum_{i=1}^{q_1}\sum_{j=1}^{q_2}\sum_{k=1}^{q_3} w_i w_j w_k \boldsymbol{F}(a_i, a_j, a_k)\det(\mathbf{J}) \end{aligned} \tag{2.48}$$

where the symbols w_i, w_j and w_k denote quadrature weights and the symbols a_i, a_j and a_k denote the respective abscissae. Values of weights and abscissae depend on quadrature type and the number of quadrature points q_1, q_2 and q_3 necessary for correctly computing integral (2.48) depends on the order n of the employed approximating polynomials.

2.4.1 Lobatto Quadrature

Numerical integration using Lobatto quadrature is based on the following formula:

$$\int_{-1}^{+1} f(\xi)\mathrm{d}\xi = \frac{2}{q(q-1)}[f(-1)+f(+1)] + \sum_{i=1}^{q-2} w_i f(a_i) + \mathrm{E}_1 \tag{2.49}$$

where q is the number of quadrature points and E_1 is the quadrature error.

Lobatto quadrature weights $w_i(i = 1, \ldots, q)$ can be computed according to the following relationship:

$$w_i = \frac{2}{q(q-1)P_{q-1}(a_i)^2}, \quad i = 1, 2, \ldots, q \tag{2.50}$$

where $P_{q-1}(\xi)$ is the Legendre polynomial of order $q-1$ and coordinates of abscissae $a_i(i = 1, \ldots, q)$ are computed as roots of the equation:

$$\left(1 - a_i^2\right) \frac{\mathrm{d}}{\mathrm{d}\xi} P_{q-1}(a_i) = 0, \quad i = 1, 2, \ldots, q \tag{2.51}$$

The error of Lobatto quadrature E_1 (2.49) can be expressed as:

$$\mathrm{E}_1 = -\frac{q(q-1)^3 2^{2p-1}((q-2)!)^4}{(2q-1)((2q-2)!)^3} f^{(2q-2)}(\eta), \quad \eta \in [-1, +1] \tag{2.52}$$

Relationship (2.52) allows one to state that the Lobatto quadrature error E_1 vanishes for functions $f(\xi)$ being polynomials of order n not higher than $2q-3$.

In the case of full Lobatto polynomials $L_n^c(\xi)$, coordinates of abscissae $a_i(i = 1, \ldots, n)$ of quadrature (2.49) are at the same time coordinates of nodes $\xi_i(i = 1, \ldots, n)$ of the spectral finite element of nth order and equality $(\xi_i = a_i)(i = 1, \ldots, n)$ takes place. Values of weights $w_i(i = 1, \ldots, n)$ and abscissae $a_i(i = 1, \ldots, n)$ of Lobatto quadrature, computed using relationships (2.50) and (2.51) for full Lobatto polynomials $L_n^c(\xi)$ of several initial orders, are listed in Table 2.8.

2.4.2 Gauss Quadrature

Numerical integration using Gauss quadrature is based on the following formula:

$$\int_{-1}^{+1} f(\xi)\mathrm{d}\xi = \sum_{i=1}^{q} w_i f(a_i) + \mathrm{E}_2 \tag{2.53}$$

where q is the number of quadrature points and E_2 is the quadrature error.

Table 2.8 Values of weights and abscissae of Lobatto quadrature for full Lobatto polynomials $L_n^c(\xi)$ of several initial orders

Element order	Abscissa coordinates	Weights
$n = 1$	$a_1 = -1.0000000000000000$	$w_1 = 1.0000000000000000$
	$a_2 = +1.0000000000000000$	$w_2 = 1.0000000000000000$
$n = 2$	$a_1 = -1.0000000000000000$	$w_1 = 0.3333333333333333$
	$a_2 = 0.0000000000000000$	$w_2 = 1.3333333333333333$
	$a_3 = +1.0000000000000000$	$w_3 = 0.3333333333333333$
$n = 3$	$a_1 = -1.0000000000000000$	$w_1 = 0.1666666666666667$
	$a_2 = -0.4472135954999579$	$w_2 = 0.8333333333333333$
	$a_3 = +0.4472135954999579$	$w_3 = 0.8333333333333333$
	$a_4 = +1.0000000000000000$	$w_4 = 0.1666666666666667$
$n = 4$	$a_1 = -1.0000000000000000$	$w_1 = 0.1000000000000000$
	$a_2 = -0.6546536707079771$	$w_2 = 0.5444444444444444$
	$a_3 = 0.0000000000000000$	$w_3 = 0.7111111111111111$
	$a_4 = +0.6546536707079771$	$w_4 = 0.5444444444444444$
	$a_5 = +1.0000000000000000$	$w_5 = 0.1000000000000000$
$n = 5$	$a_1 = -1.0000000000000000$	$w_1 = 0.0666666666666667$
	$a_2 = -0.7650553239294647$	$w_2 = 0.3784749562978470$
	$a_3 = -0.2852315164806451$	$w_3 = 0.5548583770354864$
	$a_4 = +0.2852315164806451$	$w_4 = 0.5548583770354864$
	$a_5 = +0.7650553239294647$	$w_5 = 0.3784749562978470$
	$a_6 = +1.0000000000000000$	$w_6 = 0.0666666666666667$
$n = 6$	$a_1 = -1.0000000000000000$	$w_1 = 0.0476190476190476$
	$a_2 = -0.8302238962785669$	$w_2 = 0.2768260473615659$
	$a_3 = -0.4688487934707142$	$w_3 = 0.4317453812098626$
	$a_4 = 0.0000000000000000$	$w_4 = 0.4876190476190476$
	$a_5 = +0.4688487934707142$	$w_5 = 0.4317453812098626$
	$a_6 = +0.8302238962785669$	$w_6 = 0.2768260473615659$
	$a_7 = +1.0000000000000000$	$w_7 = 0.0476190476190476$

Gauss quadrature weights $w_i (i = 1, \ldots, q)$ can be computed according to the following relationship:

$$w_i = \frac{2}{(1 - a_i^2)\left(\frac{\mathrm{d}}{\mathrm{d}\xi} P_q(a_i)\right)^2}, \quad i = 1, 2, \ldots, q \tag{2.54}$$

where $P_q(\xi)$ is the Legendre polynomial of order $q-$ and coordinates of abscissae $a_i(i = 1, \ldots, q)$ are its roots:

$$P_q(a_i) = 0, \quad i = 1, 2, \ldots, q \tag{2.55}$$

The error of Gauss quadrature E_2 (2.53) can be expressed as:

$$\text{E}_2 = \frac{2^{2q+1}(q!)^4}{(2q+1)((2q)!)^3} f^{(2q)}(\eta), \quad \eta \in [-1, +1] \tag{2.56}$$

Relationship (2.56) allows one to state that the Gauss quadrature error E_2 vanishes for functions $f(\xi)$ being polynomials of order n not higher than $2q-1$.

In the case of full Chebyshev polynomials $T_n^c(\xi)$, coordinates of abscissae $a_i(i = 1, \ldots, n)$ of quadrature (2.53) are not coordinates of nodes $\xi_i(i = 1, \ldots, n)$ of the spectral finite element of order n. Values of weights $w_i(i = 1, \ldots, n)$ and abscissae $a_i(i = 1, \ldots, n)$ of Gauss quadrature, computed using relationships (2.54) and (2.55) for full Chebyshev polynomials $T_n^c(\xi)$ of several initial orders, are listed in Table 2.9.

2.4.3 Gauss–Laguerre Quadrature

Numerical integration using Laguerre quadrature is based on the following formula:

$$\int_0^{+\infty} f(\xi)\mathrm{d}\xi = \sum_{i=1}^{q} w_i f(a_i) + \text{E}_3 \tag{2.57}$$

where q is the number of quadrature points and E_3 is the quadrature error.

Laguerre quadrature weights $w_i(i = 1, \ldots, q)$ can be computed according to the following relationship:

$$w_i = \frac{a_i}{(q+1)^2 \left[G_{q+1}(a_i)\right]^2}, \quad i = 1, 2, \ldots, q \tag{2.58}$$

where $G_{q+1}(\xi)$ is the Laguerre polynomial of order $q+1$ and coordinates of abscissae $a_i(i = 1, \ldots, q)$ are roots of the Laguerre polynomial of qth order:

$$G_q(a_i) = 0, \quad i = 1, 2, \ldots, q \tag{2.59}$$

Table 2.9 Values of weights and abscissae of Gauss quadrature for full Chebyshev polynomials $T_n^c(\xi)$ of several initial orders

Element order	Abscissa coordinates	Weights
$n = 1$	$a_1 = -0.5773502691896258$	$w_1 = 1.0000000000000000$
	$a_2 = +0.5773502691896258$	$w_2 = 1.0000000000000000$
$n = 2$	$a_1 = -0.7745966692414834$	$w_1 = 0.5555555555555556$
	$a_2 = 0.0000000000000000$	$w_2 = 0.8888888888888889$
	$a_3 = +0.7745966692414834$	$w_3 = 0.5555555555555556$
$n = 3$	$a_1 = -0.8611363115940526$	$w_1 = 0.3478548451374539$
	$a_2 = -0.3399810435848563$	$w_2 = 0.6521451548625461$
	$a_3 = +0.3399810435848563$	$w_3 = 0.6521451548625461$
	$a_4 = +0.8611363115940526$	$w_4 = 0.3478548451374539$
$n = 4$	$a_1 = -0.9061798459386640$	$w_1 = 0.2369268850561891$
	$a_2 = -0.5384693101056831$	$w_2 = 0.4786286704993665$
	$a_3 = 0.0000000000000000$	$w_3 = 0.5688888888888889$
	$a_4 = +0.5384693101056831$	$w_4 = 0.4786286704993665$
	$a_5 = +0.9061798459386640$	$w_5 = 0.2369268850561891$
$n = 5$	$a_1 = -0.9324695142031520$	$w_1 = 0.1713244923791703$
	$a_2 = -0.6612093864662645$	$w_2 = 0.3607615730481386$
	$a_3 = -0.2386191860831969$	$w_3 = 0.4679139345726910$
	$a_4 = +0.2386191860831969$	$w_4 = 0.4679139345726910$
	$a_5 = +0.6612093864662645$	$w_5 = 0.3607615730481386$
	$a_6 = +0.9324695142031520$	$w_6 = 0.1713244923791703$
$n = 6$	$a_1 = -0.9491079123427585$	$w_1 = 0.1294849661688697$
	$a_2 = -0.7415311855993944$	$w_2 = 0.2797053914892767$
	$a_3 = -0.4058451513773972$	$w_3 = 0.3818300505051189$
	$a_4 = 0.0000000000000000$	$w_4 = 0.4179591836734694$
	$a_5 = +0.4058451513773972$	$w_5 = 0.3818300505051189$
	$a_6 = +0.7415311855993944$	$w_6 = 0.2797053914892767$
	$a_7 = +0.9491079123427585$	$w_7 = 0.1294849661688697$

The error of Laguerre quadrature E_3 (2.57) can be expressed as:

$$\mathrm{E}_3 = \frac{(q!)^2}{(2q)!} f^{(2q)}(\eta), \quad \eta \in [0, +\infty] \tag{2.60}$$

Relationship (2.60) allows one to state that the Laguerre quadrature error E_3 vanishes for functions $f(\xi)$ being polynomials of order n not higher than $2q - 1$.

Table 2.10 Values of weights and abscissae of Gauss quadrature for full Laguerre polynomials $G_n^c(\xi)$ of several initial orders

Element order	Abscissa coordinates	Weights
$n = 1$	$a_1 = 0.5857864376269050$	$w_1 = 0.8535533905932738$
	$a_2 = 3.4142135623730950$	$w_2 = 0.1464466094067262$
$n = 2$	$a_1 = 0.4157745567834791$	$w_1 = 0.7110930099291730$
	$a_2 = 2.2942803602790417$	$w_2 = 0.2785177335692408$
	$a_3 = 6.2899450829374792$	$w_3 = 0.0103892565015861$
$n = 3$	$a_1 = 0.3225476896193923$	$w_1 = 0.6031541043416336$
	$a_2 = 1.7457611011583466$	$w_2 = 0.3574186924377997$
	$a_3 = 4.5366202969211280$	$w_3 = 0.0388879085150054$
	$a_4 = 9.3950709123011331$	$w_4 = 0.0005392947055613$
$n = 4$	$a_1 = 0.2635603197181409$	$w_1 = 0.5217556105828087$
	$a_2 = 1.4134030591065168$	$w_2 = 0.3986668110831759$
	$a_3 = 3.5964257710407221$	$w_3 = 0.0759424496817076$
	$a_4 = 7.0858100058588376$	$w_4 = 0.0036117586799220$
	$a_5 = 12.640800844275783$	$w_5 = 0.0000233699723858$
$n = 5$	$a_1 = 0.2228466041792607$	$w_1 = 0.4589646739499636$
	$a_2 = 1.1889321016726230$	$w_2 = 0.4170008307721210$
	$a_3 = 2.9927363260593141$	$w_3 = 0.1133733820740450$
	$a_4 = 5.7751435691045105$	$w_4 = 0.0103991974531491$
	$a_5 = 9.8374674183825899$	$w_5 = 0.0002610172028149$
	$a_6 = 15.982873980601702$	$w_6 = 0.0000008985479064$
$n = 6$	$a_1 = 0.1930436765603624$	$w_1 = 0.4093189517012739$
	$a_2 = 1.0266648953391920$	$w_2 = 0.4218312778617198$
	$a_3 = 2.5678767449507462$	$w_3 = 0.1471263486575053$
	$a_4 = 4.9003530845264846$	$w_4 = 0.0206335144687169$
	$a_5 = 8.1821534445628608$	$w_5 = 0.0010740101432807$
	$a_6 = 12.734180291797814$	$w_6 = 0.0000158654643486$
	$a_7 = 19.395727862262540$	$w_7 = 0.0000000317031548$

In the case of full Laguerre polynomials $G_n^c(\xi)$, coordinates of abscissae $a_i (i = 1, \ldots, n)$ of quadrature (2.57) are not coordinates of nodes $\xi_i (i = 1, \ldots, n)$ of the spectral finite element of nth order. Values of weights $w_i (i = 1, \ldots, n)$ and of abscissae $a_i (i = 1, \ldots, n)$ of Laguerre quadrature, computed using relationships (2.58) and (2.59) for full Laguerre polynomials $G_n^c(\xi)$ of several initial orders, are listed in Table 2.10.

Summarising, one needs to underline again that the choice of approximating polynomials and the associated choice of quadratures employed to determine numerically the values of characteristic matrices of spectral finite elements directly affects the form of these matrices.

Using Lobatto approximating polynomials (2.9) and Lobatto quadratures (2.49) results in a diagonal form of the matrix of inertia of element $\boldsymbol{M}^e$, while the stiffness matrix $\boldsymbol{K}^e$ is a full matrix. In the case of using Chebyshev (2.13) or Laguerre (2.18) approximating polynomials and employing the respective Gauss (2.53) and Gauss–Laguerre (2.57) quadratures, the full (consistent) forms of inertia matrix $\boldsymbol{M}^e$ and stiffness matrix $\boldsymbol{K}^e$ are obtained. On the one hand, diagonality of the inertia matrix $\boldsymbol{M}^e$ produced by Lobatto quadrature (2.49) greatly reduces complexity and shortens the time needed for numerical computations performed later. This has profound consequences for using the spectral finite element method. On the other hand, its disadvantage can be reduced accuracy (2.52) $\mathrm{E}_1 \sim f^{(2q-3)}(\eta)(\eta \in [-1, +1])$ when compared with Gauss (2.56) or Gauss–Laguerre quadratures (2.60) $(\mathrm{E}_2, \mathrm{E}_3) \sim f^{(2q-1)}(\eta)(\eta \in [-1, +1])$ or with approximating polynomials $f(\xi)(\xi \in [-1, +1])$ over q nodes, which can result in, for example, underestimating values of computed natural vibration frequencies of the analysed systems. The more accurate Gauss (2.56) and Gauss–Laguerre (2.60) quadratures, which yield inertia matrices $\boldsymbol{M}^e$ of spectral finite elements in full form, require much more computational power. For this reason, in the case of very big and complex discrete models, inertia matrix diagonalisation is attempted using indirect methods. These include diagonal scaling and row summing – methods well known in the literature [11]. Unfortunately, the process of diagonalising full matrices of inertia $\boldsymbol{M}^e$ introduces additional numerical errors, which can be significant compared to errors resulting from decreased accuracy of Lobatto quadrature. For these reasons the choice of the type of approximating polynomials (Lobatto, Chebyshev, Laguerre, etc.) and the type of employed quadrature (Lobatto, Gauss, Gauss–Laguerre, etc.) should always be considered thoroughly.

2.5 Solving Equations of Motion of a Body Discretised Using Spectral Finite Elements

Equations of motion of a body discretised using spectral finite elements essentially form a system of differential equations of the second order. At this point one should note that in the particular case:

$$\boldsymbol{B}^e = \boldsymbol{B}_l^e, \quad \boldsymbol{B}_n^e = 0 \tag{2.61}$$

when the matrix of dependencies between strain and displacements is linear, the equations of motion (2.38) form a system of differential equations of the second order with constant coefficients.

In the following part of the present section it is assumed that the nonlinear part of the matrix $\boldsymbol{B}^e$ from relationship (2.25) is neglected, that is $\boldsymbol{B}^e_n = 0$. Such an assumption may be justified by the fact that for wave propagation in most media, mutual interaction of waves or interaction of waves with edges and discontinuities of various kinds assumes linearity of the occurring phenomena. This is the case for interference or diffraction of propagating elastic waves [12].

Depending on the form of the right-hand-side vector $\boldsymbol{f}_n(t)$ of the system of motion Equations (2.38), three cases can be distinguished:

- forcing with an harmonic signal,
- forcing with a periodic signal,
- forcing with a nonperiodical (stochastic) signal.

While exact solutions do exist for certain specific forcing cases, with the present state of development of numerical methods and techniques they are not useful in practice.

2.5.1 Forcing with an Harmonic Signal

Differential equations of motion for forcing with an harmonic signal assume the following form:

$$\boldsymbol{M}\,\ddot{\boldsymbol{q}}_n + \boldsymbol{C}\,\dot{\boldsymbol{q}}_n + \boldsymbol{K}\,\boldsymbol{q}_n = [A_i \sin(\omega t + \varphi_i)]^{\mathrm{T}}, \quad i = 1, 2, \ldots, n \tag{2.62}$$

where the right-hand-side vector is composed of harmonic signals of amplitudes $A_i (i = 1, 2, \ldots, n)$ and phases $\varphi_i (i = 1, 2, \ldots, n)$ and identical frequencies ω.

Equations (2.62) for fixed conditions can be solved in the same way as one for a system with a single degree of freedom, using the Fourier integral transform [1] and assuming zero initial conditions in the equation. After such a transform the equations of motion (2.62) assume the form of a system of algebraic equations with complex coefficients:

$$\left(\boldsymbol{K} - \omega^2 \boldsymbol{M} + \mathrm{j}\omega \boldsymbol{C}\right) \boldsymbol{q}_n(\mathrm{j}\omega) = \boldsymbol{f}_n(\mathrm{j}\omega) \tag{2.63}$$

where $\mathrm{j} = \sqrt{-1}$ is the imaginary unit.

The only unknown quantity in this system is the complex vector of displacements $q_n(j\omega)$. This system of equations can be solved in many ways; the most common ones include Gauss's method and its modifications [13].

After calculating the values of the complex displacement vector $q_n(j\omega)$ in the frequency domain ω, one can derive the displacement vector in the time domain $q_n(t)$ by employing the following relationships:

$$q_n(t) = [q_i \sin(\omega t + \psi_i)]^{\mathrm{T}}, \quad i = 1, 2, \ldots, n \tag{2.64}$$

where:

$$q_i = \sqrt{[\mathrm{Re}\, q_i(j\omega)]^2 + [\mathrm{Im}\, q_i(j\omega)]^2}, \quad i = 1, 2, \ldots, n \tag{2.65}$$

is the amplitude of vibrations of the ith degree of freedom of the discrete model, while:

$$\psi_i = \tan^{-1}\left(\frac{\mathrm{Im}\, q_i(j\omega)}{\mathrm{Re}\, q_i(j\omega)}\right), \quad i = 1, 2, \ldots, n \tag{2.66}$$

is its vibration phase.

2.5.2 Forcing with a Periodic Signal

In the case of forcing with a periodic signal, the right-hand-side vector $f_n(t)$ of the system of motion Equations (2.38) has the following property:

$$f_n(t) = f_n(t + T) \tag{2.67}$$

where T denotes the period of the signal.

Every periodic forcing signal can be represented as an infinite Fourier series [1]. In practice, such a series usually contains several initial terms of the expansion. In such a case the right-hand-side vector $f_n(t)$ can be conveniently expressed as a series of cases of harmonic forcing:

$$f_n(t) = \left[\sum_{\nu=0}^{\mu} f_{\nu i} \sin(\omega_\nu t + \psi_{\nu i})\right]^{\mathrm{T}}, \quad i = 1, 2, \ldots, n \tag{2.68}$$

where μ denotes the number of terms of Fourier series included in the expansion of the forcing signal, also called its harmonic components, $f_{\nu i}(\nu = 0, 1, \ldots, \mu)(i = 1, 2, \ldots, n)$ and $\psi_{\nu i}(\nu = 0, 1, \ldots, \mu)(i = 1, 2, \ldots, n)$ are the amplitude and phase of νth harmonic component of the signal, respectively, and $\omega_\nu(\nu = 0, 1, \ldots, \mu)$ is its frequency.

Thanks to the assumed linearity of the analysed system of Equations (2.62) one can apply the superposition principle and compute forced vibrations for individual harmonic components, and then sum the results:

$$q_n(t) = \sum_{\nu=0}^{\mu} q_\nu(t) = \left[\sum_{\nu=0}^{\mu} q_{\nu i} \sin(\omega_\nu t + \psi_{\nu i})\right]^{\mathrm{T}}, \quad i = 1, 2, \ldots, n \tag{2.69}$$

2.5.3 Forcing with a Nonperiodic Signal

In cases of nonperiodic forcing many diverse methods of numerical integration of the equations of motion (2.38) are used [7, 8]. One of the most effective ones is the *Newmark method*, also known as the *Newmark β−method*. It is an implicit method of direct integration of the equations of motion, for which the relationships between the vectors of displacements q_n, velocities $\dot{q}_n$ and accelerations $\ddot{q}_n$ in two subsequent time steps are as follows:

$$\begin{cases} q_n(t + \Delta t) = q_n(t) + \dot{q}_n(t)\Delta t + \left(\dfrac{1}{2} - \beta\right)\ddot{q}_n(t)\Delta t^2 + \beta\,\ddot{q}_n(t + \Delta t)\Delta t^2 \\ \dot{q}_n(t + \Delta t) = \dot{q}_n(t)\Delta t + (1 - \gamma)\ddot{q}_n(t)\Delta t + \gamma\,\ddot{q}_n(t + \Delta t)\Delta t \end{cases} \tag{2.70}$$

where γ and β are parameters affecting the accuracy and stability of the method. If $\gamma = 1/2$, no artificial damping is introduced into the method. The parameter β is associated with the method of approximating acceleration along time step Δt. If $\beta = 1/6$, acceleration changes linearly during the time step. For $\beta = 1/4$ acceleration is constant during a time step, while for $\beta = 1/8$ it changes in a stepwise fashion [8].

The Newmark method is unconditionally stable for $\beta \geq 1/4$ and $\gamma \geq 1/2$, but in computational practice usually the values of $\beta = 1/4$ and $\gamma = 1/2$ are chosen. The sequence of numerical computations associated with integrating equations of motion using the Newmark method has been listed in Table 2.11.

Another of the methods of numerical integration of equations of motion (2.38) used is the *central difference method*, classified among the explicit methods, in which the vectors of velocities $\dot{q}_n$ and accelerations $\ddot{q}_n$ at a given time

Table 2.11 Algorithm of direct integration of the equations of motion using the Newmark method

I. Preliminary computations

1. Constructing matrices of inertia $\boldsymbol{M}$, damping $\boldsymbol{C}$ and stiffness $\boldsymbol{K}$.
2. Initialising vectors of displacement $\boldsymbol{q}_n$, velocity $\dot{\boldsymbol{q}}_n$ and acceleration $\ddot{\boldsymbol{q}}_n$.
3. Choosing the length of the integration step Δt and values of parameters β and γ:

$$\gamma \geq \frac{1}{2}, \quad \beta \geq \frac{1}{4}\left(\frac{1}{2}+\gamma\right)^2$$

4. Choosing values of auxiliary parameters:

$$a_0 = \frac{1}{\beta \Delta t^2}, \quad a_1 = \frac{1}{\beta \Delta t}, \quad a_2 = \frac{1}{2\beta} - 1, \quad a_3 = \frac{\gamma}{\beta \Delta t}$$

$$a_4 = \frac{\gamma}{\beta} - 1, \quad a_5 = \left(\frac{\gamma}{2\beta} - 1\right)\Delta t, \quad a_6 = (1-\gamma)\Delta t, \quad a_7 = \gamma \Delta t$$

5. Constructing the effective stiffness matrix $\tilde{\boldsymbol{K}}$:

$$\tilde{\boldsymbol{K}} = \boldsymbol{K} + a_0 \boldsymbol{M} + a_1 \boldsymbol{C}$$

6. Triangulating the effective stiffness matrix $\tilde{\boldsymbol{K}}$:

$$\tilde{\boldsymbol{K}} = \boldsymbol{L}\boldsymbol{D}\boldsymbol{L}^{\mathrm{T}}$$

II. Computations during a typical step Δt

1. Computing the effective loading vector $\tilde{\boldsymbol{f}}_n(t+\Delta t)$:

$$\tilde{\boldsymbol{f}}_n(t+\Delta t) = \boldsymbol{f}_n(t+\Delta t) + \boldsymbol{M}\left(a_0 \boldsymbol{q}_n(t) + a_1 \dot{\boldsymbol{q}}_n(t) + a_2 \ddot{\boldsymbol{q}}_n(t)\right) + \boldsymbol{C}\left(a_3 \boldsymbol{q}_n(t) + a_4 \dot{\boldsymbol{q}}_n(t) + a_5 \ddot{\boldsymbol{q}}_n(t)\right)$$

2. Solving the system with regard to the displacement vector $\boldsymbol{q}_n(t+\Delta t)$:

$$\boldsymbol{L}\boldsymbol{D}\boldsymbol{L}^{\mathrm{T}} \boldsymbol{q}_n(t+\Delta t) = \boldsymbol{f}_n(t+\Delta t)$$

3. Computing vectors of velocity $\dot{\boldsymbol{q}}_n(t+\Delta t)$ and acceleration $\ddot{\boldsymbol{q}}_n(t+\Delta t)$:

$$\ddot{\boldsymbol{q}}_n(t+\Delta t) = a_0(\boldsymbol{q}_n(t+\Delta t) - \boldsymbol{q}_n(t)) - a_1 \dot{\boldsymbol{q}}_n(t) - a_2 \ddot{\boldsymbol{q}}_n(t)$$

$$\dot{\boldsymbol{q}}_n(t+\Delta t) = \dot{\boldsymbol{q}}_n(t+\Delta t) + a_6 \ddot{\boldsymbol{q}}_n(t) + a_7 \ddot{\boldsymbol{q}}_n(t+\Delta t)$$

moment t are usually expressed as:

$$\begin{cases} \dot{q}_n(t) = \dfrac{q_n(t+\Delta t) - q_n(t-\Delta t)}{2\Delta t} \\ \ddot{q}_n(t) = \dfrac{q_n(t+\Delta t) - 2q_n(t) + q_n(t-\Delta t)}{\Delta t^2} \end{cases} \tag{2.71}$$

After using the above relationships, the equations of motion (2.38) reduce to a system of algebraic equations with constant coefficients for each given time step Δt:

$$\begin{cases} \left(\dfrac{1}{\Delta t^2}\boldsymbol{M} + \dfrac{1}{2\Delta t}\boldsymbol{C}\right) \boldsymbol{q}_n(t+\Delta t) = \boldsymbol{r}_n(t) \\ \boldsymbol{r}_n(t) = \boldsymbol{f}_n(t) - \left(\boldsymbol{K} - \dfrac{2}{\Delta t^2}\boldsymbol{M}\right) \boldsymbol{q}_n(t) - \left(\dfrac{1}{\Delta t^2}\boldsymbol{M} - \dfrac{1}{2\Delta t}\boldsymbol{C}\right) \boldsymbol{q}_n(t-\Delta t) \end{cases} \tag{2.72}$$

Because solving the system (2.72) for time $t + \Delta t$ requires knowing the solutions for earlier time steps t and $t - \Delta t$, for time $t = 0$ it is necessary to declare the values of solutions for the moment $t = -\Delta t$. For this aim one employs the initial conditions for displacements $\boldsymbol{q}_n(0)$, velocities $\dot{\boldsymbol{q}}_n(0)$ and accelerations $\ddot{\boldsymbol{q}}_n(0)$ together with the following equation:

$$\boldsymbol{q}_n(-\Delta t) = \boldsymbol{q}_n(0) - \dot{\boldsymbol{q}}_n(0)\Delta t + \frac{1}{2}\ddot{\boldsymbol{q}}_n(0)\Delta t^2 \tag{2.73}$$

The scheme of numerical computations associated with integrating the equations of motion using the central differences method for arbitrary forms of the inertia matrix $\boldsymbol{M}$ and damping matrix $\boldsymbol{C}$ are presented in Table 2.12.

A disadvantage of the central differences method can be its conditional stability, which requires that the length of the time step Δt be smaller than some critical value that is closely related to the dynamic properties of the discretised system:

$$\Delta t \leq \Delta t_{cr} = \frac{T_n}{\pi} \tag{2.74}$$

where T_n is the shortest period of natural vibrations of the discrete model. In practical computations this condition is actually met very rarely, because it is known that the dynamics of a discrete model modelled with spectral finite elements can be approximated *sufficiently well* with a superposition of several

Table 2.12 Algorithm of direct integration of the equations of motion using the central differences method for arbitrary forms of the inertia matrix *M* and damping matrix *C*

I. Preliminary computations

1. Constructing matrices of inertia $\boldsymbol{M}$, damping $\boldsymbol{C}$ and stiffness $\boldsymbol{K}$.
2. Initialising vectors of displacement $\boldsymbol{q}_n$, velocity $\dot{\boldsymbol{q}}_n$ and acceleration $\ddot{\boldsymbol{q}}_n$.
3. Choosing the length of the integration step Δt and values of auxiliary parameters:

$$a_0 = \frac{1}{\Delta t^2}, \quad a_1 = \frac{1}{2\Delta t}, \quad a_2 = 2a_0, \quad a_3 = \frac{1}{a_2}$$

4. Computing values of the displacement vector $\boldsymbol{q}_n(-\Delta t)$:

$$\boldsymbol{q}_n(-\Delta t) = \boldsymbol{q}_n(0) - \dot{\boldsymbol{q}}_n(0)\Delta t + a_3\ddot{\boldsymbol{q}}_n(0)$$

5. Constructing the effective inertia matrix $\tilde{\boldsymbol{M}}$:

$$\tilde{\boldsymbol{K}} = a_0\boldsymbol{M} + a_1\boldsymbol{C}$$

6. Triangulating the effective inertia matrix $\tilde{\boldsymbol{M}}$:

$$\tilde{\boldsymbol{M}} = \boldsymbol{L}\boldsymbol{D}\boldsymbol{L}^{\mathrm{T}}$$

II. Computations during a typical step Δt

1. Computing the effective loading vector $\tilde{\boldsymbol{f}}_n(t)$:

$$\tilde{\boldsymbol{f}}_n(t) = \boldsymbol{f}_n(t) - (\boldsymbol{K} - a_2\boldsymbol{M})\boldsymbol{q}_n(t) - (a_0\boldsymbol{M} - a_1\boldsymbol{C})\boldsymbol{q}_n(t - \Delta t)$$

2. Solving the system with regard to the displacement vector $\boldsymbol{q}_n(t + \Delta t)$:

$$\boldsymbol{L}\boldsymbol{D}\boldsymbol{L}^{\mathrm{T}}\boldsymbol{q}_n(t + \Delta t) = \boldsymbol{f}_n(t)$$

3. Computing vectors of velocity $\dot{\boldsymbol{q}}_n(t)$ and acceleration $\ddot{\boldsymbol{q}}_n(t)$, if required:

$$\ddot{\boldsymbol{q}}_n(t) = a_0\,(\boldsymbol{q}_n(t + \Delta t) - 2\,\boldsymbol{q}_n(t) + \boldsymbol{q}_n(t - \Delta t))$$
$$\dot{\boldsymbol{q}}_n(t) = a_1\,(\boldsymbol{q}_n(t + \Delta t) - \boldsymbol{q}_n(t - \Delta t))$$

modes of natural vibrations. Thus, choosing a time step Δt many times longer than dictated by condition (2.74) usually leads to good results.

At this point one should note that in the case of diagonal characteristic forms of matrices of inertia $\boldsymbol{M}$ and damping $\boldsymbol{C}$, solving system (2.72) requires no triangulation or inversion of the matrix $\boldsymbol{M}/\Delta t^2 + \boldsymbol{C}/2/\Delta t$ in each subsequent time step Δt and any required computations can be performed at an individual element level. Such a procedure allows systems of equations to be solved using very large numbers of degrees of freedom (very complex discrete models) quickly and effectively, which is characteristic of the spectral finite element method.

It is noteworthy that the cost of performed numerical computations associated with analysing very complex discrete models using the methods of direct integration of equations of motion (2.38) mentioned above or other similar methods is directly proportional to the total number of time steps of the analysis N. For this reason numerical computations employing direct integration methods are most effective when the required analysis time is relatively short and the number of integration steps N is not large. In the case of numerical computations involving a large number of integration steps N transforming the equations of motion (2.38) to so-called modal coordinates may be justified.

The following transform then becomes useful:

$$q_n(t) = \tilde{\boldsymbol{Y}} \boldsymbol{x}_n(t) \tag{2.75}$$

where $\tilde{\boldsymbol{Y}}$ is a certain rectangular matrix of the order $n \times r$ and $\boldsymbol{x}_n(t)$ is a vector dependent on time t and of the order $r(r \leq n)$.

After substituting relationship (2.75) into the system of equations of motion (2.38) a new system of equations is obtained:

$$\tilde{\boldsymbol{M}} \ddot{\boldsymbol{x}}_n + \tilde{\boldsymbol{C}} \dot{\boldsymbol{x}}_n + \tilde{\boldsymbol{K}} \boldsymbol{x}_n = \boldsymbol{v}_n(t) \tag{2.76}$$

where:

$$\begin{cases} \tilde{\boldsymbol{M}} = \boldsymbol{Y}^{\mathrm{T}} \boldsymbol{M} \boldsymbol{Y}, & \tilde{\boldsymbol{C}} = \boldsymbol{Y}^{\mathrm{T}} \boldsymbol{C} \boldsymbol{Y} \\ \tilde{\boldsymbol{K}} = \boldsymbol{Y}^{\mathrm{T}} \boldsymbol{K} \boldsymbol{Y}, & \boldsymbol{v}_n = \boldsymbol{Y}^{\mathrm{T}} \boldsymbol{f}_n \end{cases} \tag{2.77}$$

and where the characteristic matrices $\tilde{\boldsymbol{M}}$, $\tilde{\boldsymbol{C}}$ and $\tilde{\boldsymbol{K}}$ are of the order $r(r \leq n)$.

One can show that for specific cases of transformation matrices $\tilde{\boldsymbol{Y}}$ the resulting matrices $\tilde{\boldsymbol{M}}$, $\tilde{\boldsymbol{C}}$ and $\tilde{\boldsymbol{K}}$ may be diagonal, which leads to decoupling the system of equations of motion in the system of coordinates associated with the new vector $\boldsymbol{x}_n(t)$.

Most often a matrix with columns being eigenvectors associated with the analysed system is used as the transformation matrix $\tilde{\boldsymbol{Y}}$. For this end one has to solve a so-called natural vibration problem of the following form:

$$\boldsymbol{M}\ddot{\boldsymbol{q}}_n + \boldsymbol{K}\boldsymbol{q}_n = 0 \tag{2.78}$$

for which a solution is postulated:

$$\boldsymbol{q}_n(t) = [\varphi_i \sin(\omega t + \psi_i)]^{\mathrm{T}}, \quad i = 1, 2, \ldots, n \tag{2.79}$$

where $\boldsymbol{\varphi}$ denotes the vibration amplitude vector of coordinates $\varphi_i (i = 1, 2, \ldots, n)$. Substitution of the postulated solution (2.79) into relationship (2.78) leads to a new system of equations:

$$\boldsymbol{K}\boldsymbol{\varphi} = \omega^2 \boldsymbol{M}\boldsymbol{\varphi} \tag{2.80}$$

As the stiffness matrix $\boldsymbol{K}$ is nonnegatively definite and the inertia matrix $\boldsymbol{M}$ is positively definite, solutions to the system of Equations (2.80) are real pairs $(\omega_i, \boldsymbol{\varphi}_i)(i = 1, \ldots, n)$ for which the following inequality takes place:

$$0 \leq \omega_1 \leq \omega_2 \leq \ldots \leq \omega_n, \quad i = 1, 2, \ldots, n \tag{2.81}$$

The vector $\boldsymbol{\varphi}_i = [\varphi_{1\alpha}, \varphi_{2\alpha}, \ldots, \varphi_{n\alpha}](\alpha = 1, 2, \ldots, n)$ is called the αth form of natural vibrations of system (2.80) and ω_α is the corresponding natural vibration frequency. Assuming the following notation:

$$\boldsymbol{\varphi} = [\boldsymbol{\varphi}_1, \boldsymbol{\varphi}_2, \ldots, \boldsymbol{\varphi}_n] = \begin{bmatrix} \varphi_{11} & \varphi_{12} & \cdots & \varphi_{1n} \\ \varphi_{21} & \varphi_{22} & \cdots & \varphi_{2n} \\ \vdots & \vdots & \ddots & \vdots \\ \varphi_{n1} & \varphi_{n2} & \cdots & \varphi_{nn} \end{bmatrix}, \quad \boldsymbol{\omega}^2 = \begin{bmatrix} \omega_1^2 & & & \\ & \omega_2^2 & & \\ & & \ddots & \\ & & & \omega_n^2 \end{bmatrix} \tag{2.82}$$

it can be seen that:

$$\boldsymbol{\varphi}^{\mathrm{T}}\boldsymbol{K}\boldsymbol{\varphi} = \boldsymbol{\omega}^2, \quad \boldsymbol{\varphi}^{\mathrm{T}}\boldsymbol{M}\boldsymbol{\varphi} = \boldsymbol{I} \tag{2.83}$$

where $\boldsymbol{I}$ is the unitary matrix of dimensions $n \times n$.

Thanks to using the matrix of eigenvectors $\tilde{\boldsymbol{Y}}$ (2.82) as the transformation matrix $\boldsymbol{\varphi}$ (2.75), the system of equations of motion (2.76) can be represented in a system of modal coordinates associated with the vector $\boldsymbol{x}_n(t)$ as n independent (decoupled) differential equations with constant coefficients:

$$\ddot{x}_\alpha(t) + 2\omega_\alpha \xi_\alpha \dot{x}_\alpha(t) + \omega_\alpha^2 x_\alpha(t) = \upsilon_\alpha(t), \quad \alpha = 1, 2, \ldots, n \tag{2.84}$$

with the assumption of proportional damping, that is one for which the following relationship takes place:

$$\boldsymbol{\varphi}_\alpha{}^{\mathrm{T}} \boldsymbol{C} \boldsymbol{\varphi}_\beta = 2\omega_\alpha \xi_\alpha \delta_{\alpha\beta}, \quad \alpha, \beta = 1, 2, \ldots, n \tag{2.85}$$

The solution of Equation (2.84) can be found numerically, using one of many direct integration methods [7], or analytically, employing Duhamel's integral:

$$\begin{aligned} x_\alpha(t) = {} & \frac{1}{\omega_\alpha} \int_{t_0}^{t} \upsilon_\alpha(\tau) \mathrm{e}^{-\xi_\alpha \omega_\alpha (t-\tau)} \sin(\tilde{\omega}_\alpha (t-\tau))\, \mathrm{d}\tau \\ & + \mathrm{e}^{-\xi_\alpha \omega_\alpha t} \left(A_\alpha \sin(\tilde{\omega}_\alpha t) + B_\alpha \cos(\tilde{\omega}_\alpha t) \right) \end{aligned} \tag{2.86}$$

where $\tilde{\omega}_\alpha = \omega_\alpha \sqrt{1 - \xi_\alpha^2} (\alpha = 1, 2, \ldots, n)$ and where $A_\alpha (\alpha = 1, 2, \ldots, n)$ and $B_\alpha (\alpha = 1, 2, \ldots, n)$ are integration constants resulting from the assumed initial conditions. Solutions $x_\alpha(t) (\alpha = 1, 2, \ldots, n)$ obtained in this fashion can be expressed in a global coordinate system without further difficulties, by using the transform (2.75).

In both the finite element method and the spectral finite element method the biggest problem is finding the values of elements of the damping matrix $\boldsymbol{C}^e$. The seemingly simple relationship:

$$\boldsymbol{C}^e = \int_{V^e} \mu \left(\mathbf{N}^e \right)^{\mathrm{T}} \mathbf{N}^e \mathrm{d}V^e \tag{2.87}$$

in practice requires knowing the damping coefficient μ, which in computing poses significant problems. For this reason the damping matrix of a spectral finite element $\boldsymbol{C}^e$ is often assumed to constitute a linear combination of its inertia matrix $\boldsymbol{M}^e$ and stiffness matrix $\boldsymbol{K}^e$:

$$\boldsymbol{C}^e = \alpha\, \boldsymbol{M}^e + \beta\, \boldsymbol{K}^e \tag{2.88}$$

Values of elements of the damping matrix $\boldsymbol{C}^e$ are most often computed according to the Rayleigh model [8], which uses the following relationship:

$$\xi_i = \frac{\alpha + \beta\,\omega_i^2}{2\omega_i}, \qquad i = 1, 2, \ldots, n \tag{2.89}$$

where $\xi_i (i = 1, 2, \ldots, n)$ is a modal damping associated with the ith form of natural vibration.

Finding values of the coefficients α and β requires experimental determination of critical damping values ($\xi_i = 1$) for two different natural vibration frequencies. In the case of modal analysis this poses no particular difficulties, but for problems associated with elastic wave propagation in high-frequency ranges the measurement is not simple any more.

One can also use simplified models in which damping of a spectral finite element $\boldsymbol{C}^e$ is proportional only to the element's inertia matrix $\boldsymbol{M}^e$ or stiffness matrix $\boldsymbol{K}^e$. In the first case the coefficient α can be determined for a given frequency ω from the following relationship:

$$\alpha = 2\omega\xi \tag{2.90}$$

where ξ denotes critical damping for the given frequency ω. It should be underlined that in this case waves of lower frequencies are attenuated more strongly.

In the second case, on the other hand, the coefficient β can be determined for a given frequency ω from the following relationship:

$$\beta = \frac{2\xi}{\omega} \tag{2.91}$$

where, like before, ξ denotes critical damping for the given frequency ω. Unlike for relationship (2.90), in this case waves of higher frequencies are attenuated more strongly.

From the numerical point of view it is more advantageous to use the model of the damping matrix with the spectral finite element $\boldsymbol{C}^e$ being proportional to the inertia matrix $\boldsymbol{M}^e$, because for the inertia matrix $\boldsymbol{M}^e$ in the diagonal form the element's damping matrix $\boldsymbol{C}^e$ also becomes diagonal. This allows the obtained equations of motion (2.38) to be solved efficiently and quickly using the finite difference method (2.71).

Using the element's damping $\boldsymbol{C}^e$ proportional to the stiffness matrix $\boldsymbol{K}^e$ is definitely more correct in the case of problems associated with elastic wave propagation. Unfortunately, this approach leads to the full form of the element's damping matrix $\boldsymbol{C}^e$ and significantly reduces the computational benefits compared with the case of the spectral finite element's damping matrix $\boldsymbol{C}^e$ being diagonal together with its inertia matrix $\boldsymbol{M}^e$. In such cases the equations of motion (2.38) are solved using the Newmark method (2.70).

References

1. Boyd, J.P. (2000) *Chebyshev and Fourier Spectral Methods*, Dover Publications, Inc., New York.
2. Zienkiewicz, O.C. (1989) *The Finite Element Method*, McGraw-Hill Book Company, London.
3. Pozrikidis, C. (2005) *Introduction to Finite and Spectral Element Methods Using MATLAB*, Chapman & Hall/CRC.
4. Finlayson, B.A. (1972) *The Method of Weighted Residuals and Variational Principles, with Application in Fluid Mechanics, Heat and Mass Transfer*, Academic Press.
5. Ralston, A. (1965) *A First Course of Numerical Analysis*, McGraw-Hill Book Company, London.
6. Askey, R. (2005) The 1839 paper on permutations: its relation to the Rodrigues formula and further developments, in *Mathematics and Social Utopias in France: Olinde Rodrigues and His Times* (eds S.L. Altmann and E.L. Ortiz), History and Mathematics 28, American Mathematical Society, Providence, RI, pp. 105–118.
7. Kleiber, M. (1989) *Incremental Finite Element Modelling in Non-linear Solid Mechanics*, E. Horwood Ltd.
8. Rao, S.S. (1981) *The Finite Element Method in Engineering*, Pergamon Press.
9. Timoshenko, S. and Goodier, J.N. (1951) *Theory of Elasticity*, McGraw Hill Book Company, London.
10. Vinson, J.R. and Sierakowski, R.L. (1989) *Behavior of Structures Composed of Composite Materials*, Martinus-Nijhoff Inc., Dordrecht.
11. Dauksher, W. and Emery, A.F. (1997) Accuracy in modelling the acoustic wave equation with Chebyshev spectral finite elements. *Finite Elements in Analysis and Design*, **26**, 115–128.
12. Rayleigh, J.W.S. (1945) *The Theory of Sound*, Dover Publications, Inc., New York.
13. Kiełbasiński, A. and Schwetlick, H. (1992) *Numeryczna Algebra Liniowa*, Wydawnictwo Naukowo Techniczne, Warszawa.

3

Three-Dimensional Laser Vibrometry

Laser vibrometry is a modern technology enabling contactless measurement of vibration velocity of object surfaces. Laser vibrometry allows the recording of not only vibrations (standing waves) but also propagating elastic waves. Throughout the present book laser vibrometry is employed for recording elastic waves propagating in structural elements. Laser vibrometry allows for visualising and analysing the phenomenon of elastic wave propagation.

Even though laser vibrometry is becoming increasingly popular for registering elastic wave propagation, several other methods are still being used. In order to demonstrate the advantages of this method, also other elastic wave registration techniques are discussed (Section 3.2). Laser vibrometry is a modern technology enabling contactless measurement of object surface vibration velocity. For this reason correct generation of elastic waves in the analysed structure is very important. Elastic wave generation becomes particularly important in the context of their application for damage detection and location. For this reason the techniques of elastic wave generation are also presented in Section 3.1.

Guided Waves in Structures for SHM: The Time-Domain Spectral Element Method, First Edition.
Wieslaw Ostachowicz, Pawel Kudela, Marek Krawczuk and Arkadiusz Zak.

3.1 Review of Elastic Wave Generation Methods

Multiple methods of elastic wave generation are known; however, each of them has advantages and disadvantages that make it suitable for certain conditions.

3.1.1 Force Impulse Methods

The simplest method of elastic wave generation involves applying an impulse of force (impact) to an analysed structure. Such a loading method is usually realised by impacting the analysed structure with a modal hammer or steel ball [1]. In the case of composite materials a method based on breaking pencil lead [2], called the Hsu-Nielsen method, is used [3, 4]. This method is very simple and does not require using any transducers. It employs forcing in the form of an impulse. In the case of Lamb waves, impact-type forcing causes wave dispersion. Moreover, forcing of this type generates multiple modes of Lamb waves simultaneously. This approach to elastic wave generation is most often used in the acoustic emission method. The acoustic emission method involves locating the source of elastic waves [1]. An object impacting a structure (e.g. a bird hitting a plane in flight) is an example of impulse-type forcing.

The elastic wave generation method utilising impact-type forcing is not used for damage detection and location. The reason is the need to generate specific wave modes in these circumstances. In damage detection and location, methods utilising the propagation of Lamb waves generating at most two main wave modes (S_0 and A_0) is desirable. However, using two wave modes propagating at different velocities makes identification of wave reflections associated with individual modes more difficult. For this reason methods allowing for generating chosen wave modes are used. In the used solutions the amplitude of the chosen wave mode is many times larger than that of the other mode.

Because of these restrictions, methods that allow for generating elastic waves of narrow frequency spectra are employed for solving problems of damage detection and location. In addition, these methods should allow for controlling such forcing parameters as frequency, duration, amplitude, modulation, and so on.

3.1.2 Ultrasonic Methods

For the reasons specified above conventional plane bulk longitudinal wave transducers used for ultrasonic analyses are well suited for generating elastic

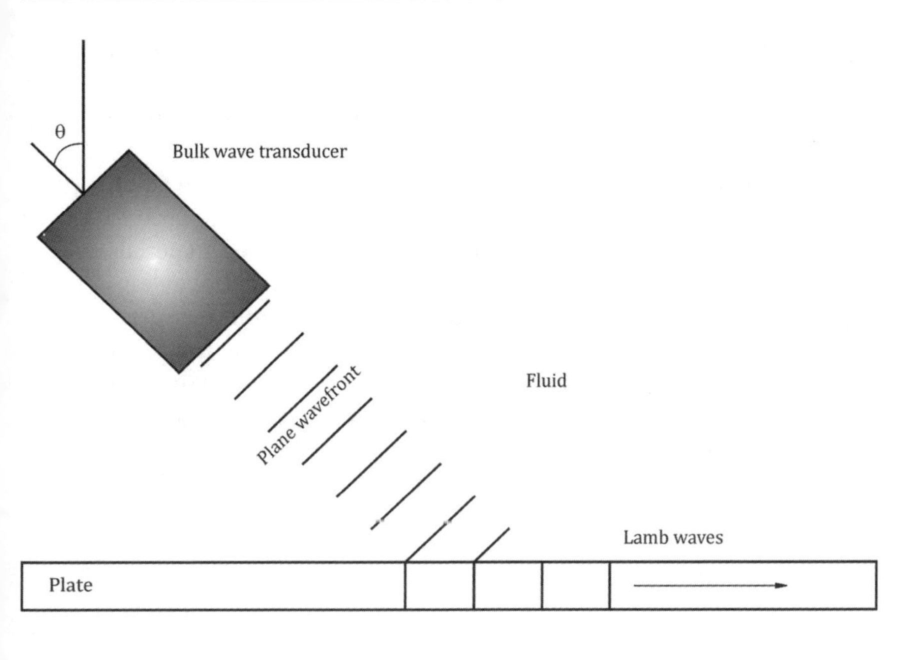

Figure 3.1 Idea of elastic wave generation in a plate using a conventional ultrasonic transducer (5)

waves [5–7]. An ultrasonic transducer is immersed in a fluid (e.g. water) together with the object in which elastic waves are to be generated. Figure 3.1 presents the idea of Lamb elastic wave generation in the case of the analysed object being a plate. A transducer orientated at a certain angle generates a longitudinal wave. The longitudinal wave propagates in the fluid until it reaches the object, in which a Lamb elastic wave is generated by means of internal reflections. The longitudinal wave falling on to the mentioned object at a certain angle generates normal stress on its surface. Normal stress varies harmonically with the wavelength λ [5]:

$$\lambda = \frac{c_L}{f \sin \Theta} \tag{3.1}$$

where c_L denotes the velocity of longitudinal wave propagation in the fluid, f denotes the frequency of the generated wave and Θ denotes the transducer orientation angle.

A disadvantage of this method is the necessity of using a fluid, in which both the transducer and analysed object (in which waves are generated) must be immersed. This condition precludes this method from application in many practical situations.

The method of elastic wave generation using air as the intermediate medium exhibits poor efficiency because of the large difference in acoustic impedance between the air and most of the analysed materials. In the discussed case the elastic wave is converted into a Lamb wave to a small degree only, while to a much larger degree it reflects from the surface of the object in which waves are to be generated [5]. Ultrasonic transducers integrated with coupling material (usually *Perspex*, better known as *Plexiglas*), called wedge-coupled angle-adjustable ultrasonic probes (Figure 3.2), allow the application of fluid as the intermediate medium to be bypassed [5]. Such heads are universally used for Lamb wave generation and registration and can be actively tuned in order to generate the chosen elastic wave mode.

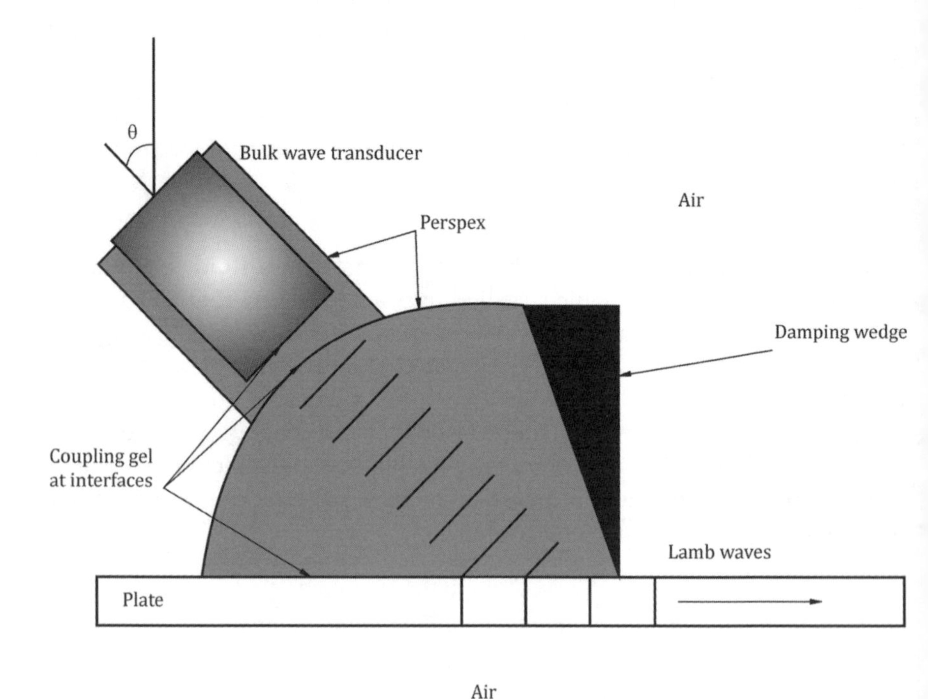

Figure 3.2 Idea of elastic wave generation in a plate using wedge-shaped angle heads (5)

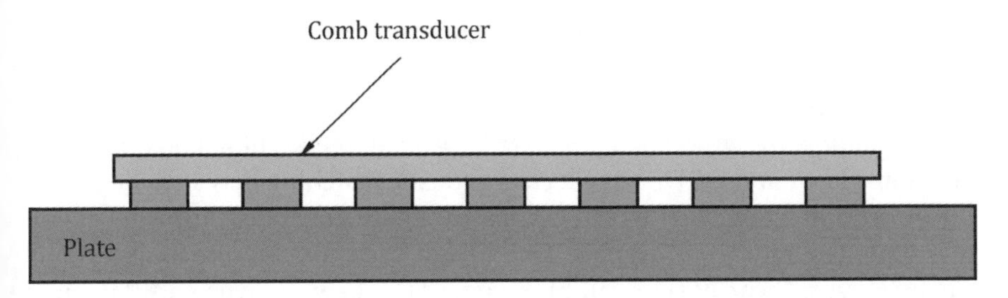

Figure 3.3 Scheme of a comb-type ultrasonic transducer (5)

Tuning is performed by choosing the angle of ultrasonic transducer orientation. In this case Snell's law governing the wave incidence angle and refraction angle is used [8].

Elastic waves are also generated using comb-type ultrasonic transducers, consisting of a number of periodically spaced tips (Figure 3.3). Tips of a comb-type transducer deform the object surface. The spacing of transducer tips corresponds to the wavelength of the generated wave. Comb-type transducers can also be made in the form of a matrix of ultrasonic transducers spaced at uniform distances equal to the wavelength of a generated elastic wave. Widths of all transducer tips as well as gaps between tips must have constant values. The sum of the transducer tip width and the gap between tips matches the wavelength of the surface wave generated by transducers [9]. A disadvantage of using transducers of this type is the need to tune them to a specific frequency of generated waves (wavelength) in a permanent fashion. This is accomplished by means of altering the transducer tip spacing. Transducers of this type allow Lamb waves and Rayleigh waves to be generated [5].

Ultrasonic methods of elastic wave generation are losing importance nowadays. No paper involving the application of wedge-type transducers was published in 2011; there were three such papers published in 2010 and two in 2009. As far as wedge-type coupled angle-adjustable ultrasonic transducers are concerned, no such paper was published in 2011 or 2010; there was one in 2009 and three in 2008 [10].

3.1.3 Methods Based on the Electromagnetic Effect

Elastic wave generation and registration can also be achieved using so-called electro-magnetic acoustic transducers (EMATs) [11–13]. Transducers of this type are used in cases involving metal structures. This results from the requirement for the element, in which elastic waves are to be generated, to be constructed of ferromagnetic and conductive material. These transducers operate on the grounds of Lorentz's law. They are composed of a fixed magnet and a solenoid, through which current flows. Eddy currents are induced in the element in which elastic waves are to be generated. These eddy currents generate elastic waves. Due to a lack of acoustic matching between the transducer, air and element in which elastic waves are generated, transducers of this type generate weak signals. EMATs are used to generate longitudinal elastic waves, shear horizontal (SH) waves [14], Lamb waves and Rayleigh waves [5]. Disadvantages of transducers of this type include their significant weight and size.

3.1.4 Methods Based on the Piezoelectric Effect

If the analysis is aimed at damage detection and location, elastic waves are usually generated using piezoceramic transducers. The ubiquity of solutions based on piezoceramic transducers results from their properties. Transducers of this type are very thin and light. Such transducers can be easily attached to analysed structures or embedded in composite elements during their construction. The most frequently used piezoelectric material is a solid-state solution of two perovskites: $PbTiO_3$ (lead titanate) and $PbZrO_3$ (lead zirconate). This solid-state solution is more commonly known as PZT. Among the disadvantages of piezoceramic transducers their fragility is most prominent. PZT is easily damaged when handled without due care. Piezoelectric transducers use the reverse piezoelectric effect for elastic wave generation. The reverse piezoelectric effect causes deformations of piezoelectric elements under the influence of an external electric field. Elastic wave generation using piezoelectric transducers requires equipment that meets specific requirements. Figure 3.4 presents the equipment used in the Laboratory of Department of Mechanics of Intelligent Structures of IFFM for measuring elastic wave propagation. The equipment consists of a programmable signal generator (1), an amplifier for signals exciting the piezoelectric transducer (2) and an oscilloscope (3) used for registration of elastic waves. A disadvantage of the presented equipment is the fact that it involves many measuring devices of large sizes and weights. The presented system is well suited for performing laboratory measurements, but one can hardly imagine using it to perform measurements on real structures. Another disadvantage of this equipment set is that there is only a single channel applicable for elastic wave generation. The oscilloscope supports a limited number of channels only. If more piezoelectric transducers are used, they must be connected sequentially. For these reasons compact devices including an

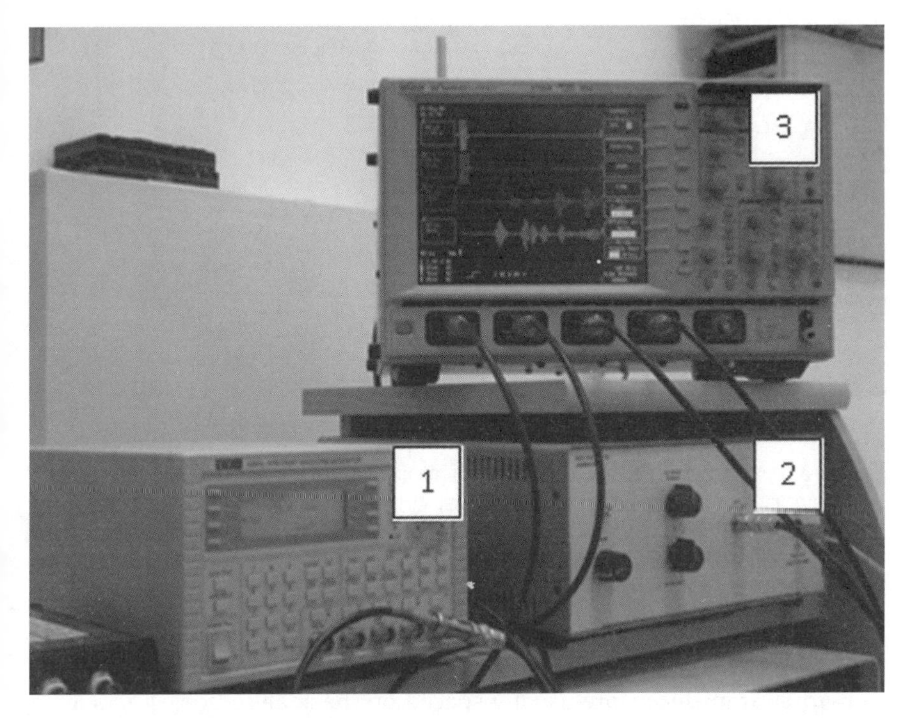

Figure 3.4 Laboratory station for generation and registration of elastic waves (Laboratory of Department of Mechanics of Intelligent Structures of IFFM): (1) signal generator, (2) signal amplifier, (3) oscilloscope

integrated signal generator, amplifiers, registering circuits and multiplexers are developed. Examples of such devices are presented in Figure 3.5. The presented devices allow for easy selection of both the parameters of a generated signal and channels to be used for generation of signals inducing elastic waves.

Figure 3.6 presents typical, easily available transducers used in our laboratory (Laboratory of Department of Mechanics of Intelligent Structures of IFFM). It is evident that individual transducers differ foremost in manufacturing and shape. From the angle of their application for elastic wave generation, the method of electrode manufacturing is very important. Transducers can have electrodes placed on the top and bottom surfaces, on the side surfaces or only on the top surface. Electrode placement predetermines how transducers need to be installed.

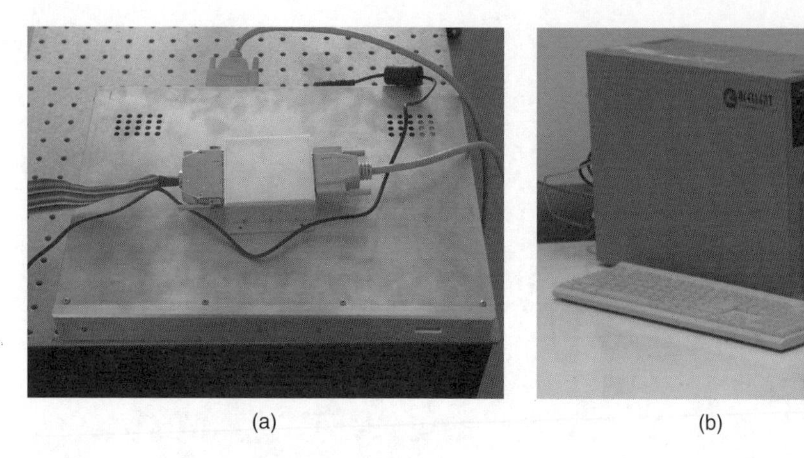

(a) (b)

Figure 3.5 Compact devices for generation and registration of elastic waves: (a) developed in the Laboratory of Department of Mechanics of Intelligent Structures of IFFM, (b) produced by Acellent Technologies, Inc. (photographed in the Laboratory of Department of Mechanics of Intelligent Structures of IFFM)

In problems of damage location using elastic wave propagation methods individual transducers are usually spaced on the analysed object in such a fashion that they form networks of different configurations [15]. In order to simplify installation and spacing on analysed structures, special matrices of piezoelectric transducers are manufactured. Figure 3.7 presents an example matrix designed and manufactured on the order of our laboratory (Laboratory of Department of Mechanics of Intelligent Structures of IFFM).

In some solutions piezoceramic transducers spaced in specific configurations are embedded in elastic material, usually in polyamide film. This approach produces ready-made matrices of relatively small thickness. Resulting matrices can be glued directly on to surface elements of the investigated structure or embedded between layers of composite material. Application of polyamide film protects transducers from humidity and aggressive agents. A commercial solution of this type is known as SMART layer (*Stanford multi-actuator–receiver transduction layer*) and has been developed at Stanford University [16–19]. A commercial version is produced by Acellent Technologies, Inc. Figure 3.8 presents a section of such a matrix used in the Laboratory of Department of Mechanics of Intelligent Structures of IFFM.

A competing solution is SAL (*smart active layer*) [20]. Piezoceramic transducers allow to generate Lamb waves, Rayleigh waves and SH (*shear horizontal*) waves.

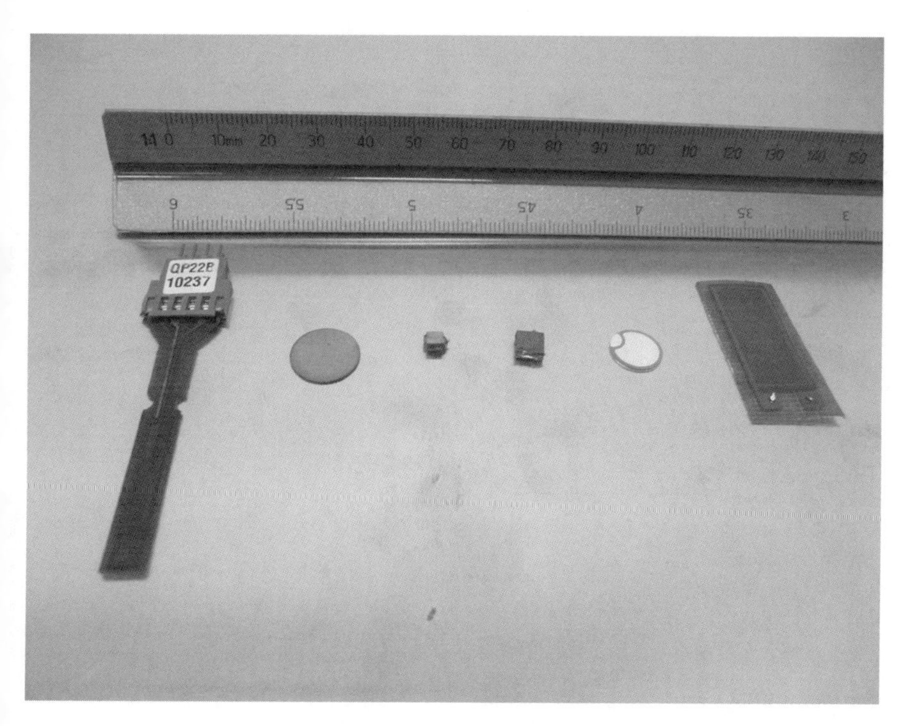

Figure 3.6 Easily available piezoelectric transducers used in laboratories (Laboratory of Department of Mechanics of Intelligent Structures of IFFM); from the left: Mide QP22B, T216–A4NO–273X Piezo Systems Inc. disc, Noliac CMAP06, Noliac CMAP11, Sonox P502 Ceramtec 8.7 mm disc, Macro Fibre Composite M2814P1MFC by Smart Material Corp

Another method of elastic wave generation employs interdigital transducers (IDTs), also known as comb-type transducers [21]. Piezoelectric comb-type transducers are constructed in a similar fashion to ultrasonic comb-type transducers (Figure 3.9). The wavelength of a generated wave depends on the spacing between electrodes [22]. IDTs are made of PVDF (polyvinylidene fluoride), which is very flexible [23]. Thanks to this, transducers can be mounted on curved elements like pipes or tanks. Transducers of this type can generate Lamb waves and Rayleigh waves [5].

A very similar solution involves using piezoelectric fibres embedded in polyamide. In practice two solutions are available: AFC (*active fibre composite*) [24, 25] and MFC (*macro fibre composite*) [26]. AFC transducers use piezoelectric fibres of circular cross-section, while MFC transducers use piezoelectric fibres of square cross-section. Such transducers were initially used only for

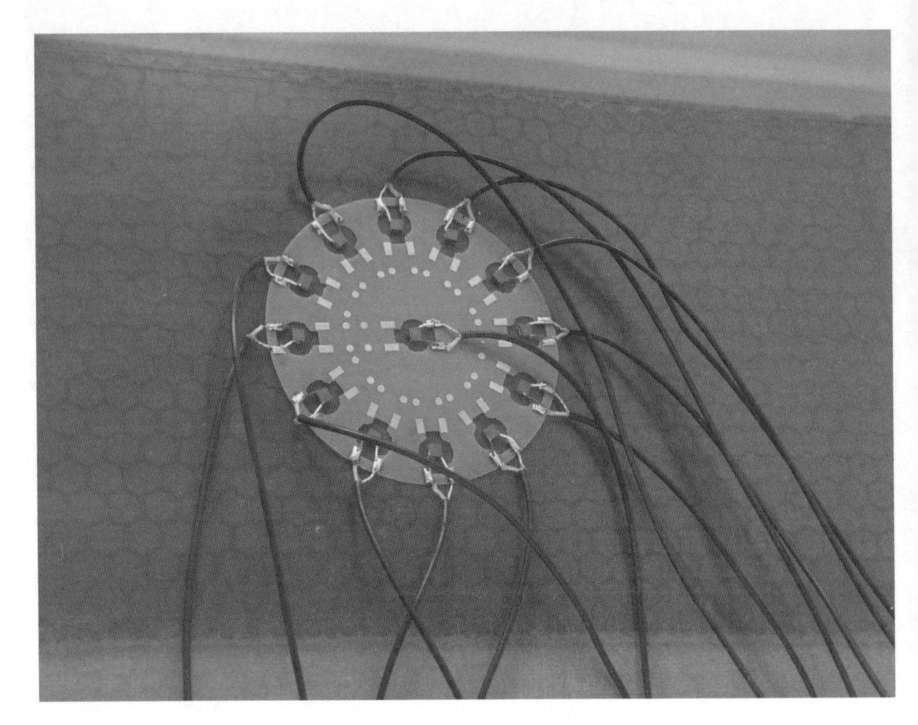

Figure 3.7 Matrix of piezoelectric transducers developed in the Laboratory of Department of Mechanics of Intelligent Structures of IFFM

controlling the shape of structural elements and for vibration attenuation. In time, the above transducers started being used for generating elastic waves [8]. The discussed transducers, like IDTs made of PVDF, are very flexible. This property allows them to be used for elastic wave generation in structures with curved surfaces.

3.1.5 Methods Based on the Magnetostrictive Effect

Besides the piezoelectric effect, magnetostriction is also employed for elastic wave generation [27]. Such transducers comprise fixed magnets, solenoids and nickel gratings. The most important advantages of the discussed transducers include the capability to generate high-power waves without an electrical connection between the magnetostrictive element and generator, low price and applicability to nonferromagnetic materials. The discussed transducers allow for generation of Lamb waves and SH (*shear horizontal*) waves [27].

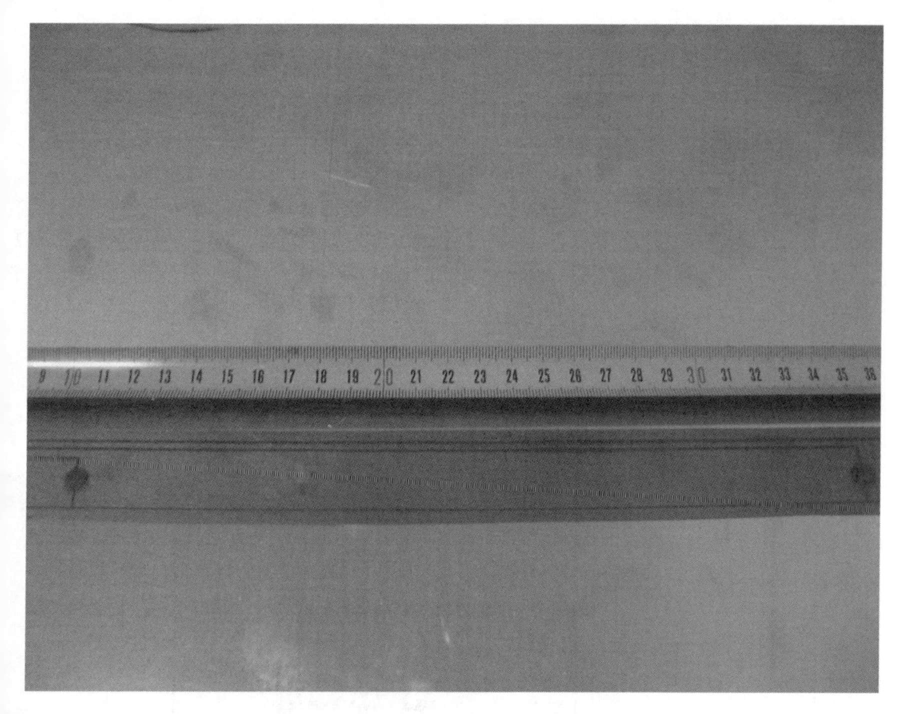

Figure 3.8 Section of the SMART layer matrix by Acellent Technologies Inc. (Laboratory of Department of Mechanics of Intelligent Structures of IFFM)

3.1.6 Photothermal Methods

Besides contact methods, contactless ones are also employed for elastic wave generation. One of the very interesting solutions is the photothermal method [9]. Its advantages include no mechanical contact, the ability to choose the shape of the elastic wave source and the ability to select the wave generation site [9]. The discussed method is very suitable for automatic diagnostic systems, because it is very easy to move the wave generation site. Elastic waves are generated using Nd:YAG laser sources [28–30].

The method of illuminating the surface with a laser beam is very important. What counts is the shape of the illuminated surface. The easiest approach involves point illumination, which is characterised by a widefrequency spectrum of generated elastic waves. Generated elastic waves are undirected in this case. Illumination along a line is also characterised by a wide frequency spectrum. In this way propagation of the generated elastic waves can be directed. Using a matrix of parallel sections or arcs allows both a narrow

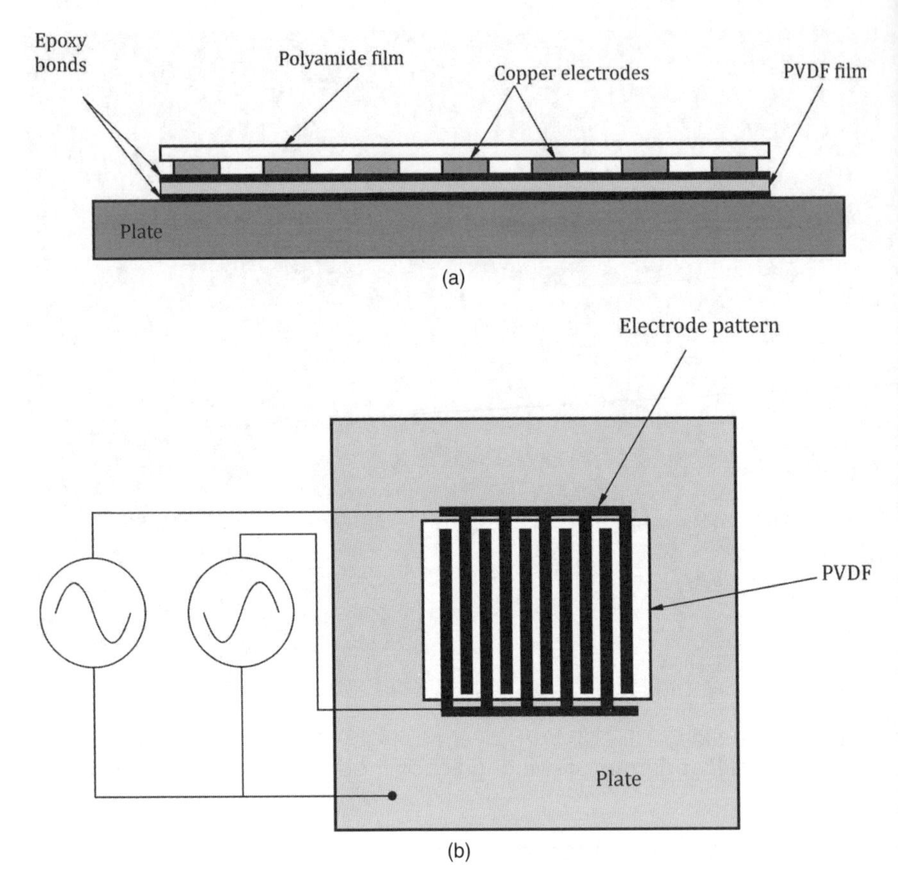

Figure 3.9 Scheme of the interdigital transducer (IDT) (5): (a) side view, (b) top view

frequency band and directionality of generation to be obtained [29]. In order to illuminate specific shapes on the surface, the laser beam must pass through a beam expander.

3.2 Review of Elastic Wave Registration Methods

Elastic wave registration and visualisation are important for scientific research. In particular, visualisation of propagating elastic waves is very valuable when analysing complex structures. Elastic wave registration methods listed in the present chapter can be divided into point methods and full-field methods. Point methods allow elastic waves to be registered at the point of sensor attachment (the sensor can be, for example, a piezoceramic transducer). Full-field methods, on the other hand, allow elastic wave propagation to be registered within a chosen area (e.g. *shearography* or laser scanning vibrometry). A point method requires the placement of sensors in many points on the structure. Conversely, full-field methods allow elastic wave propagation to be registered in the whole chosen area. However, when one considers employing full-field methods for elastic wave registration outside the laboratory, one should take into account the fact that their use is hampered by the need to access the analysed area of the structure. It may be particularly difficult to employ a full-field method for registration elastic waves in a structure during its regular operation. Thus, it is not possible to use only methods of one type. In the case of point methods, popularity of a given method depends on parameters such as the transducer weight and size, the need for surface preparation, the use of intermediary substances and the transducer price.

Some of the techniques described in Section 3.1 in association with elastic wave generation are used for elastic wave registration as well. Methods of elastic wave registration include ultrasonic ones, involving conventional plane bulk longitudinal wave transducers, or using wedge-coupled angle-adjustable ultrasonic transducers, as well as comb-type transducers (Figure 3.3). Also, methods using the electromagnetic effect are used for elastic wave registration. Such methods employ EMATs that were described in more detail in Section 3.1 [13]. Methods based on the piezoelectric effect are also used for elastic wave registration. A very common method of elastic wave registration utilises transducers made of piezoelectric materials. Most often these are piezoceramic transducers, described in Section 3.1. Elastic wave recording is achieved by means of the regular piezoelectric effect, that is the appearance of an electric charge in the transducer as a result of its mechanical deformations. Like for elastic wave generation, in this case elastic SMART layer matrices (Figure 3.8) and elastic transducers made of MFC and AFC piezoelectric fibres (Figure 3.6) are used. Also, interdigital transducers (IDTs) (Figure 3.9) are used for elastic wave registration. Another method of elastic wave registration involves piezoelectric paint. Paint is sprayed on to the chosen element [31]. Advantages of such an approach include the ability to cover structures having arbitrary shapes and the absence of an adhesive layer between the sensor and the structure. Disadvantages of piezoceramic paint include low sensitivity to deformation. Also, magnetostrictive transducers, covered in detail in Section 3.1 [27], are used for elastic wave registration. On the other hand, there are methods that are used only for registration elastic waves. These include optical methods.

3.2.1 Optical Methods

In 1864 August Toepler developed a method for visualising flows of variable density, which was called *Schlieren photography*. In this technique the base optical system includes a collimated light source illuminating a flowing fluid. Changes of the refraction index resulting from the fluid density gradient distort the collimated light beam. The above perturbations cause spatial changes of light intensity that in turn allow for visualising phenomena taking place in the fluid flow. This approach proved useful also for determining velocity of ultrasonic wave propagation in solids. Such experiments were performed by Barnes and Burton (1949), Chinnery, Humphrey and Beckett (1997) and Neubauer (1973) [9].

Theoretical research on visualisation of elastic wave propagation was also performed in 1922 by Brillouin. His research involved interactions between elastic waves in solids and electromagnetic waves (light). These interactions take the form of changes of medium permittivity resulting from elastic wave propagation. Permittivity changes affect the electric field of a light wave. Experimental research in this area was performed by Lucas and Biquard in France (1932) and by Debye and Sears in the US [32]. Later, this field was also investigated in India by Raman and Nath [32].

Another method for visualising elastic wave propagation is the photoelasticity technique. The photoelasticity technique allows observations of elastic wave propagation in glass to be made. This method uses polarised light divided into components propagating with different velocities. In effect, one can observe colourful striae associated with deformations in the given point of the material. The method is used in materials exhibiting birefringence resulting from material stress. Birefringence causes the material to exhibit two distinct refractive indexes. Values of refractive indexes in the given point depend on mechanical stress. It is thus possible to correlate the resulting striae with stress in the material. Ultrasonic wave visualisation using the described method was researched by Zhang, Shen and Ying (1988) and by Li and Negishi (1994) [9].

In recent years fibre optics sensors have also been used for elastic wave registration. Polarimetric sensors and fibre Bragg grating (FBG) sensors have been used for wave recording [33]. Both sensor types are well suited for elastic wave registration. Polarimetric sensors require only simple equipment and allow for registration waves on a large area. It should be underlined, however, that the results are difficult to analyse. Sensors with a Bragg grating require more complex equipment compared to the equipment used with polarimetric sensors. In this case measurement results are easy to process, but the area covered by sensors is much smaller than in the case of polarimetric sensors. For measurements performed with sensors of either type, measurement sensitivity depends on the angle of orientation of the sensor relative to the wavefront [33]. Registration of elastic waves requires meeting the condition that the sensor length should be at least seven times shorter than the wavelength of the elastic wave being registered [34]. When properties of FBG sensors were compared with those of sensors employing the Doppler effect (fibre optics Doppler (FOD) sensors), it was noticed that FOD sensors allow for registration the base mode of the shear wave SH_0 reflected from the damage site, while FBG sensors are unable to register it due to directionality issues [35]. Sensors of both types record the base symmetric mode S_0 and the base antisymmetric mode A_0 of Lamb waves.

Advantages of FBG sensors include their multiplexing capability. The process of multiplexing allows common equipment to be used for generation and registration the optical signal for all sensors. Fibre optics sensors have a low weight, very small diameter and are insensitive to electromagnetic interference. However, fibre optics sensors, just like piezoceramic transducers, are very brittle.

Despite the advances of multiple contact measurement methods (e.g. ultrasonic methods, methods based on the piezoelectric or electromagnetic effects), developments in measurement technologies caused research on elastic wave propagation to return to optical, contactless measurement methods. An example of such a method used for visualising elastic wave propagation is the *shearography* technique. This method allows contactless full-field measurement of element strains using a laser beam [36]. Among the advantages of this method contactless measurement is the foremost one. As measurement involves strain measurements this method is very well suited to detecting damage. As the shearography method is based on strain measurements, it is not sensitive to rigid body motions [37]. The discussed method utilises an image-shearing camera that creates a pair of laterally shifted images. As the structure being measured is being illuminated with a laser beam, the shifted images interfere with each other, creating a so-called speckle image. Comparing speckle images before and after the structure is loaded allows the strain distribution on its surface to be determined.

The discussed method is used for registration of elastic waves generated by piezoelectric transducers [39]. Elastic waves are registered using a combination of the spatial phase shifting technique and shearography. In practice this is obtained using a Mach–Zehnder interferometer. Another method being developed is electronic speckle pattern interferometry (ESPI), which also allows

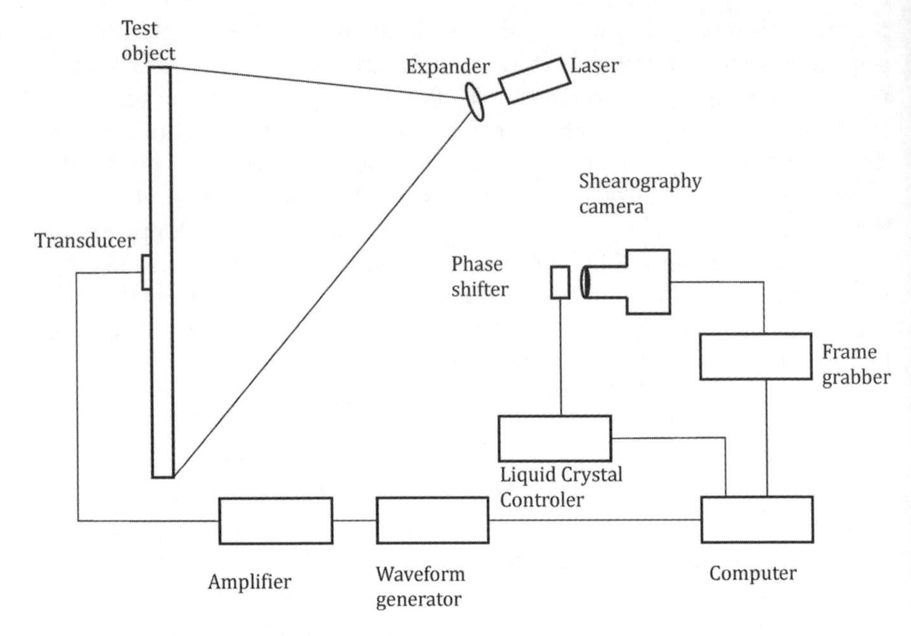

Figure 3.10 Scheme of equipment for the shearography method (38)

for visualising elastic wave propagation when combined with piezoelectric wave excitation [39] (Figure 3.10). Results of measurements visualising elastic wave propagation in an aluminium alloy plate are presented in Figure 3.11. Examples of visualisation of elastic wave propagation in a composite panel made of carbon fibre (CFRP) are presented in Figure 3.12. Compared with the previous method, ESPI allows displacement registration, which makes measurements sensitive to rigid body motions [37].

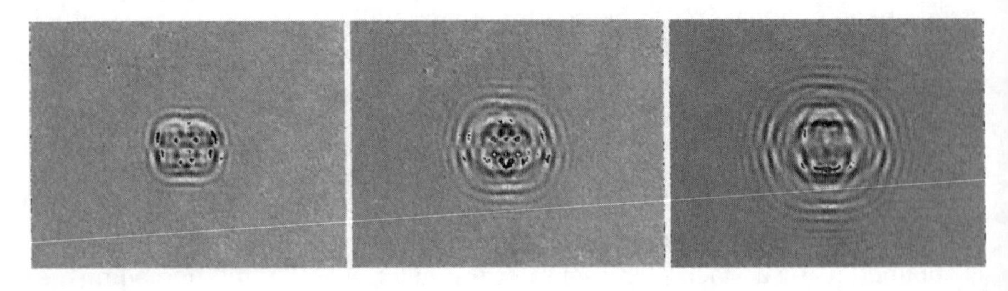

Figure 3.11 Visualisation of propagation of an antisymmetric mode of a Lamb wave in a plate – the shearography technique (39)

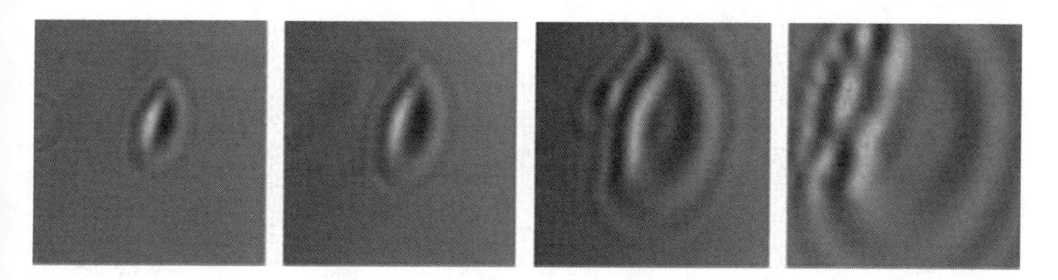

Figure 3.12 Visualisation of Lamb wave propagation in a composite panel made of carbon fibre (CFRP) – the ESPI technique (36)

The third contactless method of measuring elastic waves is based on laser vibrometry. This approach is becoming increasingly popular with researchers all around the world. Laser vibrometers come in multiple variants. Simple solutions allow for measuring only the velocity component along the laser beam incidence direction, at a single point (manual positioning of the laser head). More sophisticated versions, however, allow for automatic scanning of the investigated element. In the latter case the measurement covers either only the velocity component along the laser beam incidence direction [40, 41] or all three components simultaneously (three-dimensional (3D) measurement) [41, 42]. The principle of operation of laser vibrometers is presented in detail in Section 3.3.

3.3 Laser Vibrometry

Operation of a laser vibrometer is based on the Doppler effect. A laser vibrometer registers changes in the frequency of a light beam reflecting from a moving surface. If the beam reflects from a moving target, its frequency changes by Δf, which can be expressed as:

$$\Delta f = 2\cos(\alpha)\frac{V}{\lambda} \tag{3.2}$$

where V denotes the velocity of the object being measured relative to a reference point (observation point), α denotes the angle between the laser beam and the velocity vector of the surface being measured and λ denotes the wavelength of the emitted wave. In order to determine the velocity of the object being measured one measures the Doppler shift of electromagnetic wave

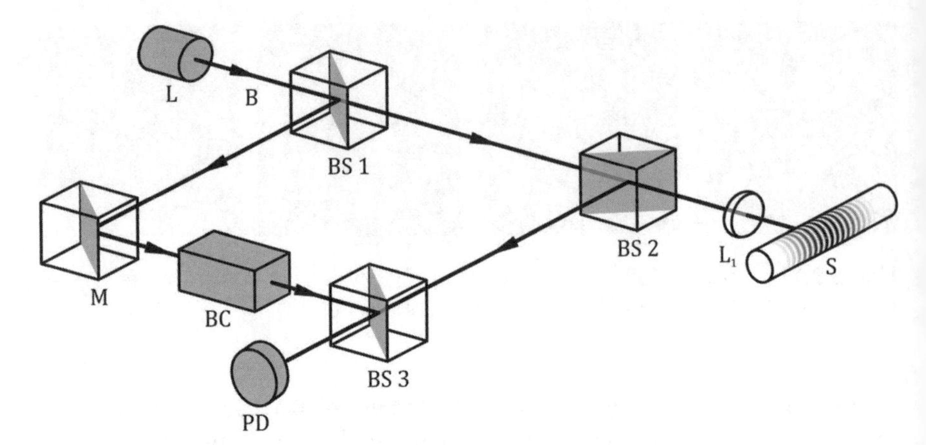

Figure 3.13 Principle of operation of the laser vibrometer

frequency for a known wavelength and fixed laser beam incidence angle:

$$V = \frac{\Delta f \lambda}{2\cos(\alpha)} \tag{3.3}$$

In the case of laser vibrometry the measurement of the light wave frequency change is performed by a laser interferometer. In that case two coherent light beams overlap. The resulting light intensity is not a simple sum of individual beam intensities, but rather the result of their interference. Intensity of such light I is expressed by the following formula:

$$I = I_1 + I_2 + 2\sqrt{I_1 I_2}\cos\left(2\pi\frac{r_1 - r_2}{\lambda}\right) \tag{3.4}$$

where I_1 and I_2 are intensities of individual beams and r_1 and r_2 are paths travelled by the respective beams. If the difference of travelled path lengths is a multiple of the laser light wavelength, the total light intensity rises four times compared to a single beam. If the difference of travelled path lengths is a multiple of half of the wavelength the light intensity becomes zero.

The scheme presented in Figure 3.13 illustrates how the Doppler effect is used for contactless measuring of vibration velocity using a laser vibrometer.

A laser beam (usually from an He–Ne (helium–neon) laser or a laser diode) of frequency f_L is separated by beam splitter (BS1) into a measurement beam (B1) and a reference beam (B2). The first one passes through another beam splitter (BS2) and is then focused by a lens system (L1) on to the investigated specimen (S). The laser beam is then reflected and its frequency is altered by Δf. The change in frequency results from the Doppler effect and is proportional to the velocity of the moving object being analysed. Part of this beam returns to the measuring circuit, directed by beam splitter BS2 towards the photodetector (PD). Reference beam B2 is directed by a mirror (M) through a system raising (or lowering) its frequency (a *Bragg cell* (BC)) by a known amount f_b. Both beams pass through another beam splitter (BS3) where they are combined, and finally reach the photodetector (PD). As interference striae in the photodetector would be identical whether the object moves towards or away from the vibrometer, the frequency of one of the beams is altered by a known amount – typically $f_b = 40$ MHz. This is usually achieved using an opto-acoustic modulator called a Bragg cell. This allows the velocity of the moving object to be determined on the basis of a modulation frequency of the signal in the photodetector. Thus, the photodetector generates a frequency modulated signal of carrier frequency corresponding to the frequency change in the Bragg cell f_b. The modulation frequency f_d depends on the length and sense of velocity vector of the moving specimen and results from the Doppler effect. The final measurement step involves demodulation of the recorded signal. A velocity component parallel to the beam is determined for the point being measured in time.

In order to allow for digital demodulation of the carrier signal it is transformed to the quadrature components I and Q. Quadrature signals should be sinusoidal, have the same amplitudes and be shifted in phase by 90°. Figure 3.14 presents a graphical representation of a revolving point plotting the I–Q signals. The point revolution angle is equal to the phase difference in the interferometric circuit $\Delta\varphi$, while theevolution direction depends on the sense of the vector of vibration of the investigated object.

The phase difference $\Delta\varphi$ is determined from the following trigonometric relationships:

$$I(\Delta\phi) = A\cos\Delta\phi \tag{3.5}$$

$$Q(\Delta\phi) = A\sin\Delta\phi \tag{3.6}$$

where A denotes amplitude. After dividing the equations by sides, one obtains:

$$\frac{Q(\Delta\phi)}{I(\Delta\phi)} = \frac{\sin\Delta\phi}{\cos\Delta\phi} = \tan\Delta\phi \tag{3.7}$$

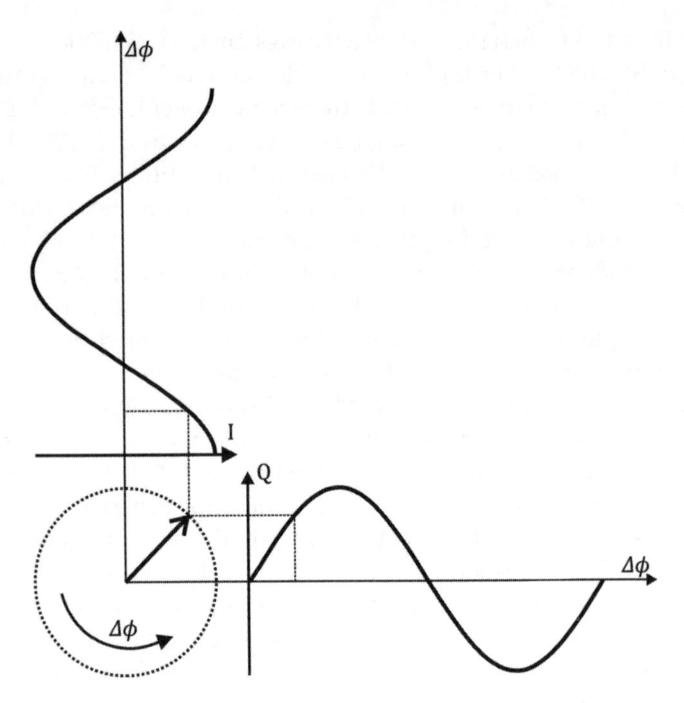

Figure 3.14 Quadrature signals *I* and *Q*

Introduction of the function inverse to tangent yields:

$$\Delta\phi = \operatorname{atan}\frac{Q\,(\Delta\phi)}{I\,(\Delta\phi)} \tag{3.8}$$

This phase difference $\Delta\varphi$ is proportional to the displacement Δx of the moving object and is described by the following relationship:

$$\Delta\phi = \frac{4\pi}{\lambda}\Delta x \tag{3.9}$$

Recently there is evidence that scanning laser vibrometers are being used increasingly often. These are equipped with a mirror system allowing for changes to be made to the measurement beam angle. Most frequently scanning vibrometers are equipped with a VGA camera or cooperate with one. This allows the measurement mesh to be set out directly on the chosen surface area on which vibrations or elastic wave propagation are being measured. The measurement mesh is also visualised on the screen. An advantage of such a system is the ability to perform measurements automatically in a large number of precisely defined points.

The principle of measuring vibrations by a laser vibrometer, described above, allows measurements to be taken of the velocity of a point along the axis of laser beam incidence on to the surface. In many cases information about vibrations of the investigated object in all three dimensions is important. For this aim, so-called 3D vibrometers, mentioned previously, are employed. A 3D vibrometer comprises three independent measurement heads of 1D scanning vibrometers, as well as circuits for data acquisition and steering. Velocity measurements are performed concurrently by all the heads oriented at different angles towards the investigated surface. Three independent laser beams crossing in set-out points on the surface of the investigated specimen allow all velocity components to be measured spatially. Knowing the geometry of the investigated object and the locations of individual heads in space, one can identify (separate) individual components of vibration velocity and compose the full vibration velocity vector for each of the measurement points. Figure 3.15 presents an example arrangement of measurement heads relative

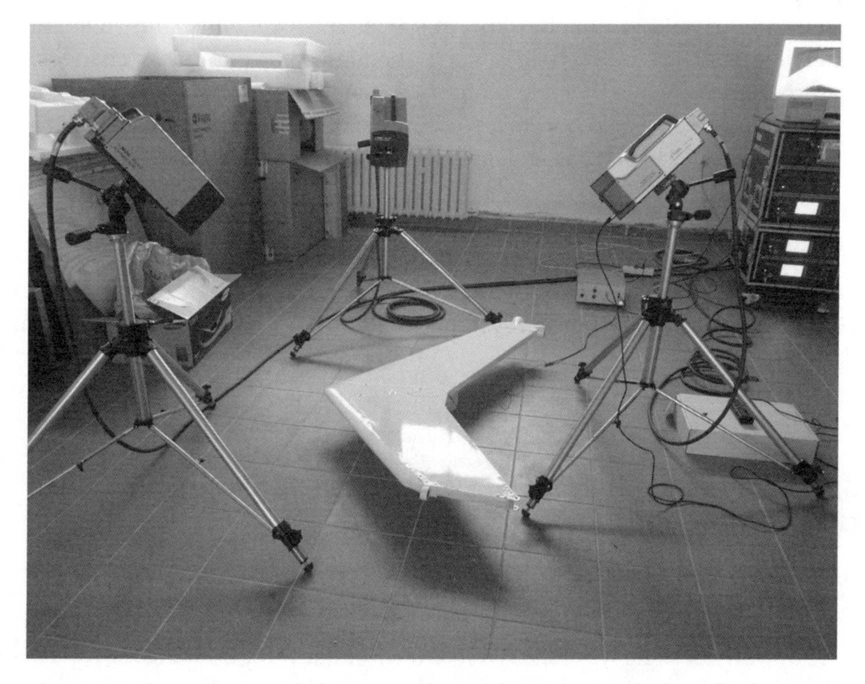

Figure 3.15 Sample arrangement of the measuring heads relative to the investigated specimen – Laboratory of Department of Mechanics of Intelligent Structures of IFFM

to the investigated specimen used for measuring the phenomenon of elastic wave propagation in all three directions.

A fundamental advantage of laser vibrometry is that measurement is contactless. This eliminates the detrimental effects of adding an additional mass of sensors in the point of contact of the transducer with the investigated structure or of locally altering structure rigidity. This fact is of key importance in the case of measuring elastic wave propagation, as each additional local change of investigated specimen properties can constitute a reflection site for the propagating wave or alter its velocity. Another unquestioned advantage of laser vibrometry is the 3D measurement capability. This measurement technique allows object vibration components to be recorded both in the plane perpendicular to the investigated surface and in the one parallel to it. This capability is important for researching the phenomenon of elastic wave propagation, for, depending on the mode of propagating wave, one of the wave components can dominate.

Scanning laser vibrometers allow automatic measurements to be performed for a very dense mesh of measurement points. In the case of measuring a propagating elastic wave this allows the phenomenon to be observed with precision, not only in time but also in space. This capability in turn enables a detailed analysis to be made of such phenomena as parameter changes of the propagating wave that encounters a damage site.

Laser vibrometry also boasts a wide range of measured frequencies, from close to 0 Hz to 24 MHz, as well as a wide range of vibration velocities from 20 nm/s to 20 m/s.

All these advantages make laser vibrometry one of the most effective measurement techniques for registration of the phenomenon of elastic wave propagation. The described measurement system allows numerical models to be verified and can also be used in systems for detecting, locating and identifying damage.

3.4 Analysis of Methods of Elastic Wave Generation and Registration

A review of the methods used for elastic wave generation and registration shows that there are methods that are only useful for either generation or registration of elastic waves, as well as ones suitable for both these processes. Table 3.1 presents the individual methods of elastic wave generation and registration.

Table 3.1 Methods of elastic wave generation and registration

Method/device/transducer	Elastic wave generation capability	Elastic wave registration capability
Ultrasonic transducers (conventional)	X	X
Comb-type ultrasonic transducers	X	X
Wedge-coupled angle-adjustable ultrasonic transducers	X	X
EMATs	X	X
Piezoceramic transducers	X	X
Piezoelectric paint	—	X
IDT piezoelectric transducers	X	X
MFC and AFC transducers	X	X
Magnetostrictive transducers	X	X
Force impulse (impact)	X	—
Photothermal	X	—
Shearography, ESPI	—	X
Laser vibrometry	—	X
Optic fibre sensors	—	X
Schlieren photography	—	X
Photoelasticity	—	X

A review of research papers shows that different methods of elastic wave generation and registration are used for individual types of elastic waves. Table 3.2 presents the individual types of elastic waves together with the methods for their generation and registration.

Table 3.2 Methods for generation and registration individual types of elastic waves

Elastic wave type	Methods of elastic wave generation/registration
Rayleigh waves	Comb-type ultrasonic transducers [5], EMAT transducers [5], IDT piezoelectric transducers [5]
Lamb waves	Conventional ultrasonic transducers [5], comb-type ultrasonic transducers [5], wedge-coupled angle-adjustable ultrasonic transducers [5], piezoceramic transducers [5, 15], EMATs [5, 6], magnetostrictive transducers [27], FOD sensors [35], fibre optics FBG sensors [33, 34], polarimetric fibre optics sensors [33], laser vibrometry [42], ESPI method [39]
Shear horizontal (SH) waves	piezoceramic transducers [15], EMATs [5], magnetostrictive transducers [27], FOD transducers [35]

3.5 Exemplary Results of Research on Elastic Wave Propagation Using 3D Laser Scanning Vibrometry

Experiments described below were performed in the Laboratory of Department of Mechanics of Intelligent Structures of IFFM. A 3D laser scanning vibrometer PSV 400 3D by Polytec® was used to perform the measurements. This device employs three scanning laser heads. Each of those houses a laser emitting a light beam of frequency $f_L = 4.74 \times 10^{14}$ Hz and power $P \leq 1$ mW/cw.

Elastic waves were induced using a ceramic piezoelectric disk of diameter 10 mm made of Sonox® by CeramTec®. The disc was glued to surfaces of investigated samples using wax used for attaching accelerometers. Inducing a signal of voltage $V_{pp} = 20$ V was fed from a TGA1241 generator by Thurlby Thandar Instruments through an EPA–104 amplifier by Piezo Systems® Inc. The voltage was gradually increased from $V_{pp} = 20$ V to the level of 200 V.

The inducing signal was generated from a sinusoidal signal multiplied by a Hanning window of width of three to five periods of the sinusoidal signal. For example, Figure 3.16 presents the plot of forcing for a frequency of 100 kHz and a window width corresponding to five periods of thebase signal.

Figure 3.17 presents a scheme of the measuring circuit. Three scanning measuring heads oriented towards the specimen are connected to a data acquisition system and steering circuit. A synchronisation cable connects the digital

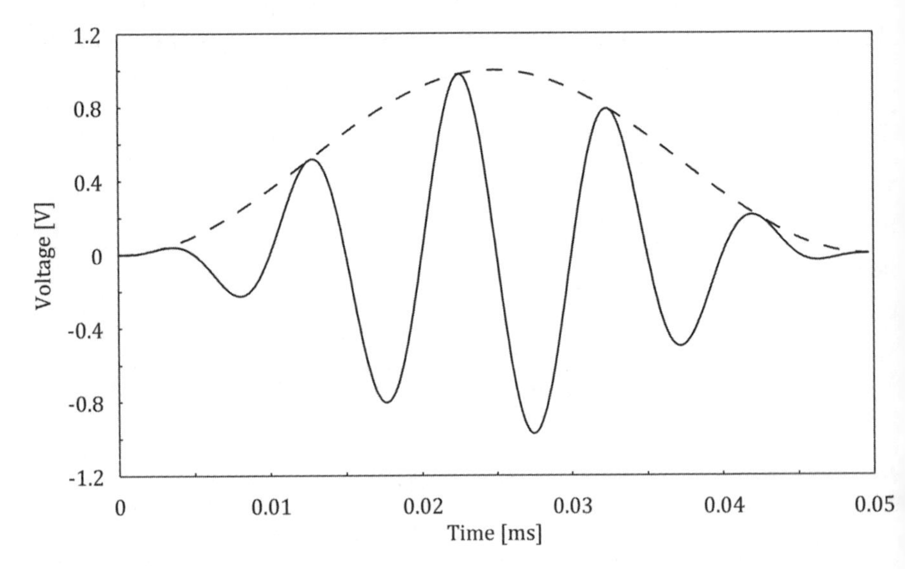

Figure 3.16 Typical plot of the forcing signal in time

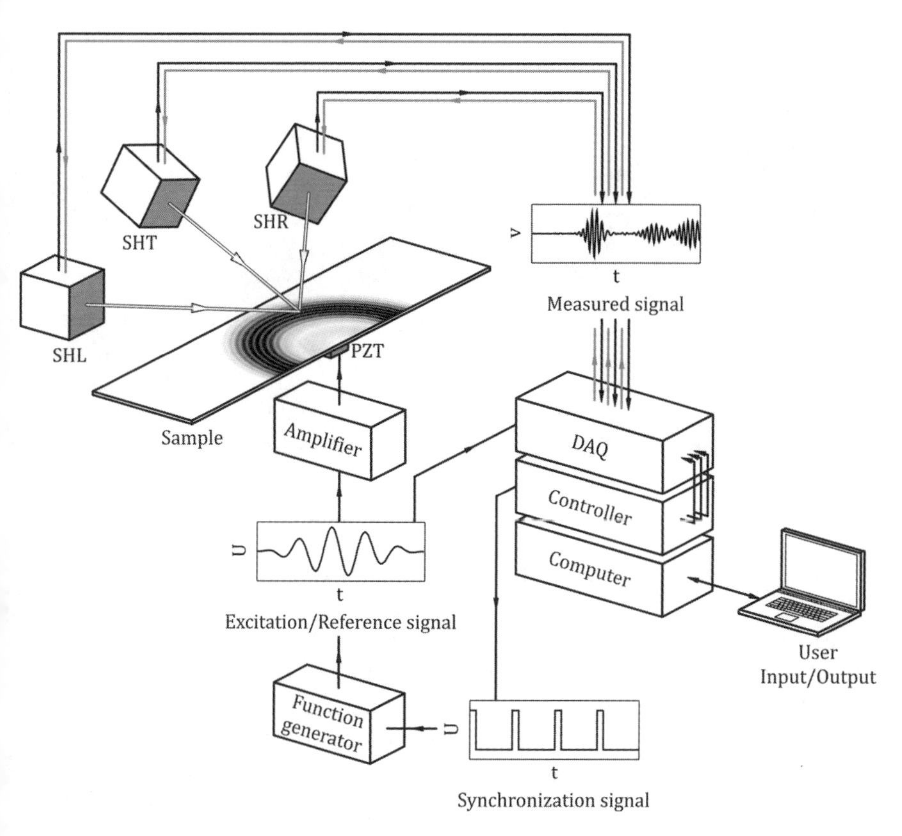

Figure 3.17 Scheme of the measurement circuit

generator with a steering circuit through the amplifier that feeds the inducing signal to the piezoelectric element. Additionally, a signal from the generator is fed into the data acquisition system. A computer system integrated with the data acquisition system and steering circuit provides communication with the user and allows measurement data to be processed.

An important parameter that directly affects measurement accuracy is the distance of the vibrometer heads from the investigated object. The best results are obtained when measurement heads are positioned at such a distance from the investigated object that maximum visibility VM can be reached. This parameter is determined for the given employed laser vibrometer from the following relationship:

$$VM_n = 99\,\text{mm} + n \times 204\,\text{mm}, \text{ for } n = 1, 2, 3 \ldots \tag{3.10}$$

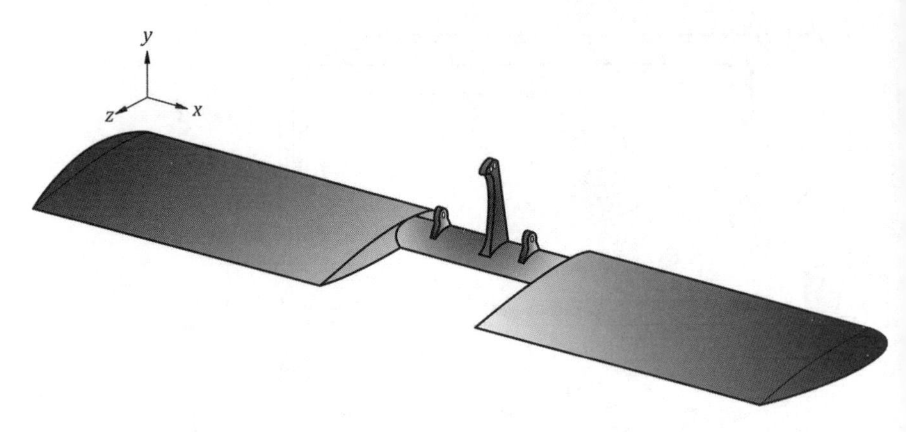

Figure 3.18 Tail plane of the PZL W–3A helicopter

For the measurement circuit used in our research this distance (average for all measurement points) was equal to 1731 mm.

Additionally, in order to improve measurement accuracy the investigated objects were covered with self-adhesive retroreflective film (*reflective film 5500*) by ORALITE®. This is aimed at improving the laser vibrometer signal level in each measurement point regardless of the angle of incidence of the measurement beam on to the surface being measured. The effect is improvement of the signal-to-noise ratio (SNR).

The investigated real-life specimen was a tail plane of a PZL W–3A helicopter (Figure 3.18) made of a glass-epoxy composite covered with epoxide enamel on the outside. The tail plane surfaces are stiffened on leading edge with evenly spaced sandwich ribs and with cell filler segments in the trailing edge part. The general scheme of the whole tail plane design is presented in Figure 3.19.

Dimensions of a single aerofoil of the tail plane together with the layout of the internal stiffening elements are shown in Figure 3.19. The measurement area, comprising a mesh of 291 × 475 evenly spaced measurement points, is marked in grey. Elastic waves were registered for excitation signals of carrier frequencies of 17.5 kHz, 35 kHz and 100 kHz, respectively.

Example results obtained from measurements of elastic wave propagation are presented in Figures 3.20, 3.21 and 3.22 in the form of time frames for the investigated tail plane.

The presented examples demonstrate that an elastic wave propagates mostly along the stiffening element on which the piezoelectric transducer

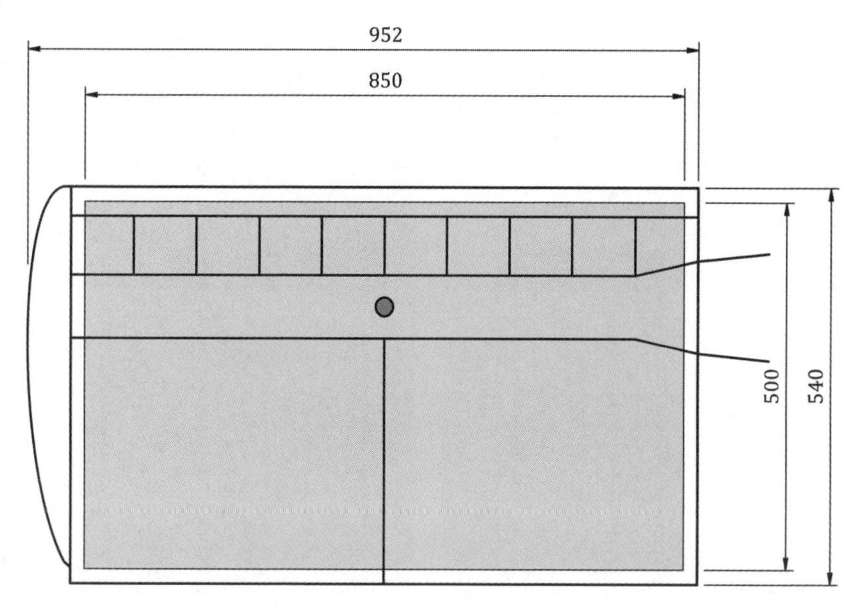

Figure 3.19 Design of an aerofoil of the tail plane of the PZL W–3A helicopter

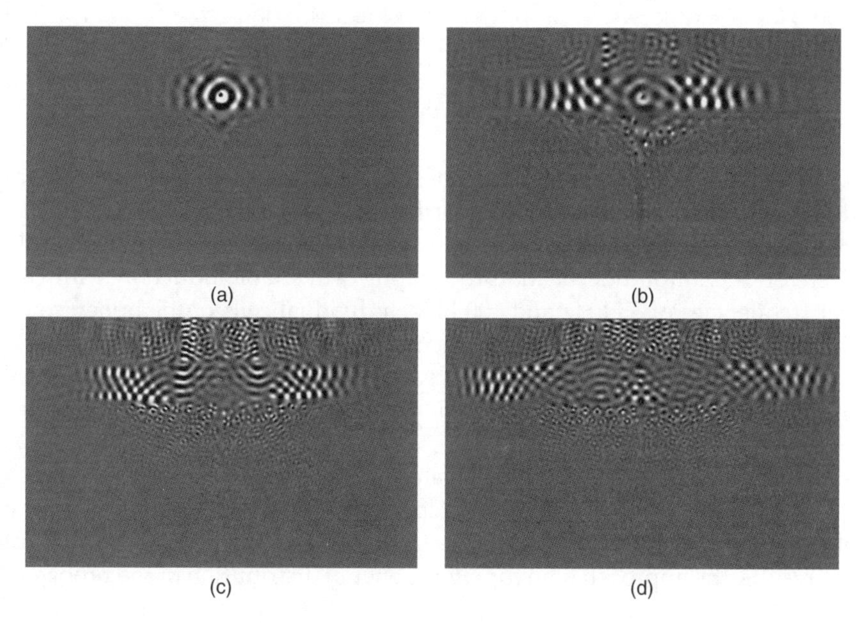

Figure 3.20 Elastic wave propagation in the tail plane for induction with a carrier frequency of 17.5 kHz at time instances: (a) 291 μs, (b) 486 μs, (c) 681 μs, (d) 876 μs

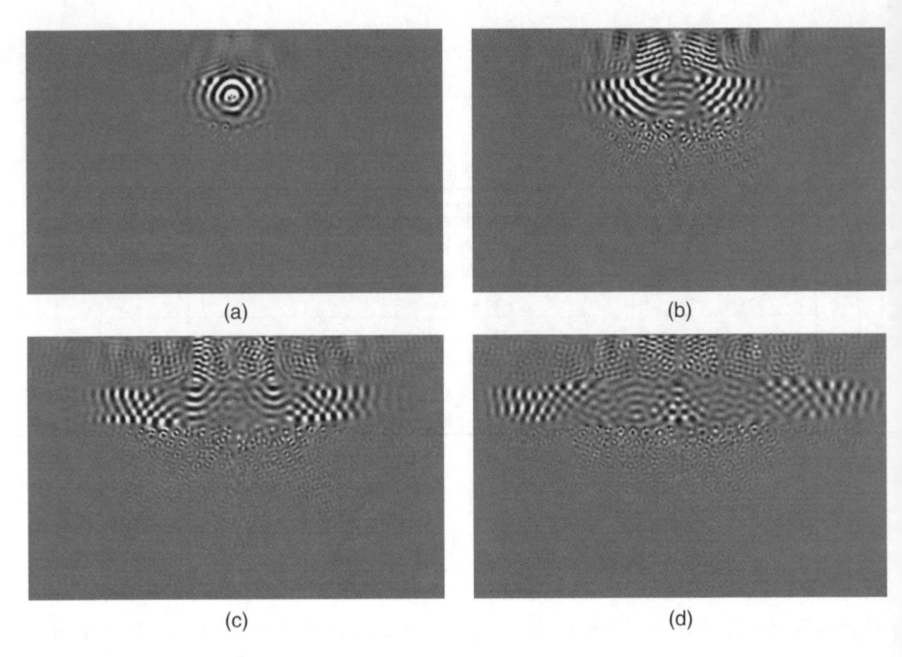

Figure 3.21 Elastic wave propagation in the tail plane for induction with a carrier frequency of 35 kHz at time instances: (a) 95 μs, (b) 193 μs, (c) 291 μs, (d) 388 μs

was installed. Both the wavelength of the generated wave and its propagation range are observed to decrease as the frequency increases. Stiffening ribs generate elastic wave reflections. Thus, ribs in the presented frames locally decrease the amplitude of the propagating wave. For induction with carrier frequencies of 35 kHz and 100 kHz, individual cells of the honey comb filler are also visible. In all the presented cases two base modes of elastic waves, symmetric and antisymmetric, are present simultaneously. However, this phenomenon is clearly visible only for induction with a carrier frequency of 100 kHz (Figure 3.22).

In the discussed case the investigated real-life object has fairly complex geometry. The propagating wave repeatedly reflects from the stiffening elements and the honey comb filler. In order to facilitate interpretation of the results, the approach of mapping the energy distribution of the propagating wave was used. Detailed information on the approach of mapping the energy distribution of the propagating wave including examples is presented in Chapter 7.

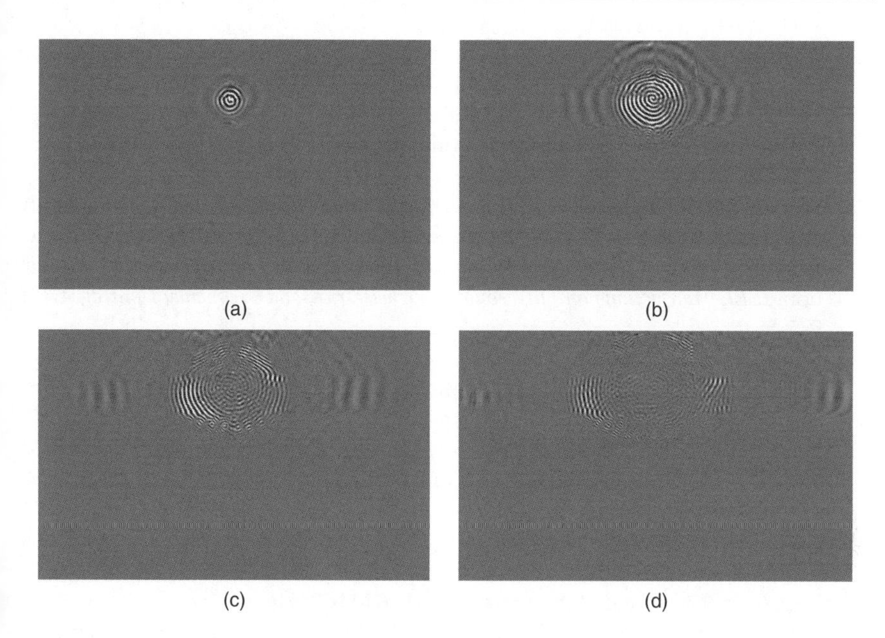

Figure 3.22 Elastic wave propagation in the tail plane for induction with a carrier frequency of 100 kHz at time instances: (a) 38 μs, (b) 77 μs, (c) 116 μs, (d) 155 μs

In the present book, piezoelectric transducers are used for elastic wave generation and laser vibrometry is used for registration and visualisation of the propagating waves. In Chapters 4 to 6 these methods are employed to verify the developed numerical models of elements of one-, two- and three-dimensional structures, respectively.

Both applications of piezoelectric transducers for inducing waves and the use of laser vibrometry for registration and visualisation of wave propagation are employed for detecting and locating damage in structural elements. Chapter 7 contains results of experiments employing piezoelectric transducers and involving laser vibrometry.

References

1. Kundu, T., Das, S., Martin, S.A. and Jata, K.V. (2008) Locating point of impact in anisotropic fiber reinforced composite plates. *Ultrasonics*, **48**, 193–201.
2. Jeong, H. (2001) Analysis of plate wave propagation in anisotropic laminates using a wavelet transform. *NDT&E International*, **34**, 185–190.

3. Acoustic Emission, Hsu-Nielsen source (2002) [online] Available at: http://www.ndt.net/ndtaz/content.php?id=474 (Accessed 30 June 2011).
4. Berthelot, J.-M., Souda, M.B. and Robert, J.L. (1992) Frequency response transducers used in acoustic emission testing of concrete. *NDT&E International*, **25**(6), 279–285.
5. Wilcox, P.D. (1998) Lamb wave inspection of large structures using permanently attached transducers. PhD thesis, Imperial College of Sciences, London.
6. Rose, J.L. (2000) Guided wave nuances for ultrasonic nondestructive evaluation. *IEEE Transactions on Ultrasonics, Ferroelectrics, and Frequency Control*, **47**(3), 575–583.
7. Pierce, S.G., Culshaw, B., Manson, G. *et al.* (2000) The application of ultrasonic Lamb wave techniques to the evaluation of advanced composite structures. *Proceedings of SPIE*, **3986**, 93–103.
8. Su, Z. and Ye, L. (2009) *Identification of Damage Using Lamb Waves. From Fundamentals to Applications*, Lecture notes in applied and computational mechanics, vol. 48, Springer-Verlag.
9. Rose, J.L. (1999) *Ultrasonic Waves in Solid Media*, Cambridge University Press, Cambridge.
10. http://apps.isiknowledge.com (accessed 16 May 2011).
11. Park, I.-K., Kim T.-H., Kim, H.-M. *et al.* 2006. Evaluation of hidden corrosion in a thin plate using a non-contact guided wave technique. *Key Engineering Materials*, **321–323**, 492–496.
12. Murayama, R. and Mizutani, K. (2002) Conventional electromagnetic acoustic transducer development for optimum Lamb wave modes. *Ultrasonic*, **40**, 491–495.
13. Dixon, S. and Palmer, S.B. (2004) Wideband low frequency generation and detection of Lamb and Rayleigh waves using electromagnetic acoustic transducers (EMATs). *Ultrasonics*, **42**, 1129–1136.
14. Achenbach, J.D. (2000) Quantitative nondestructive evaluation. *Internal Journal of Solids and Structures*, **37**, 13–27.
15. Wandowski, T., Malinowski, P. and Ostachowicz, W.M. (2011) Damage detection with concentrated configurations of piezoelectric transducers. *Smart Materials and Structures*, **20**, 025002 (14 pp).
16. Ihn, J.-B. and Chang, F.-K. (2008) Pitch catch active sensing methods in structural health monitoring for aircraft structures. *Structural Health Monitoring, An International Journal*, **7**(1), 5–19.
17. Lin, M. and Chang, F.-K. (2002) The manufacture of composite structures with a built-in network of piezoceramics. *Composites Science and Technology*, **62**, 919–939.
18. Qing, P.X, Beard, S., Shen, S.B. *et al.* (2009) Development of a real-time active pipeline integrity detection system. *Smart Materials and Structures*, **18**, 1–10.
19. Qing, X.P., Beard, S.J., Kumar, A. *et al.* (2006) Advances in the development of built-in diagnostic system for filament wound composite structures. *Composites Science and Technology*, **66**, 1694–1702.

20. Lee, Y.S., Yoon, D.J., Lee, S.I. and Kwon, J.H. (2005) An active piezo array sensor for elastic wave detection. *Key Engineering Materials*, **297–300**, 2004–2009.
21. Na, J.K., Blackshire, J.L. and Kuhr, S. (2008) Design, fabrication, and characterization of single-element interdigital transducers for NDT applications. *Sensors and Actuators A*, **148**, 359–365.
22. Mustapha, F., Manson, G., Worden, K. and Pierce, S.G. (2007) Damage location in an isotropic plate using a vector of novelty indices. *Mechanical Systems and Signal Processing*, **21**, 1885–1906.
23. Su, Z., Ye, L. and Lu, Y. (2006) Guided Lamb waves for identification of damage in composite structures: a review. *Journal of Sound and Vibration*, **295**, 753–780.
24. Melnykowycz, M. *et al.* (2006) Performance of integrated active fiber composites in fiber reinforced epoxy laminates. *Smart Materials and Structures*, **15**(1), 204–212.
25. Melnykowycz, M. *et al.* (2001) Packaging of active fiber composites for improved sensor performance, *Smart Materials and Structures*, **19**(1), 015001.
26. Wagg, D., Bond, I., Weaver, P. and Friswell, M. (2007) *Adaptive Structures. Engineering Applications*, John Wiley & Sons, Ltd.
27. Kim, I.K. and Kim, Y.Y. (2007) Shear horizontal wave transduction in plates by magnetostrictive gratings. *Journal of Mechanical Science and Technology*, **21**, 693–698.
28. Valle, C. and Littles Jr, J.W. (2002) Flaw localization using the reassigned spectrogram on laser-generated and detected Lamb modes. *Ultrasonics*, **39**, 535–542.
29. Hongjoon, K., Kyungyoung, J., Minjea, S. and Jaeyeol, K. (2006) A noncontact NDE method using a laser generated focused-Lamb wave with enhanced defect-detectionability and spatial resolution. *NDT&E International*, **39**, 312–319.
30. Hongjoon, K., Kyungyoung, J., Minjea, S. and Jaeyeol, K. (2006) Application of the laser generated focused-Lamb wave for non-contact imaging of defects in plate. *Ultrasonics*, **44**, 1265–1268.
31. Payoa, I. and Hale, J.M. (2010) Dynamic characterization of piezoelectric paint sensors under biaxial strain. *Sensors and Actuators A*, **163**, 150–158.
32. Royer, D. and Dieulesaint, E. (1999) *Elastic Waves in Solids II. Generation, Acousto-Optics Interaction, Applications*, Springer.
33. Thursby, G., Sorazu, B., Betz, D.C. *et al.* (2004) Comparison of point and integrated fiber optic sensing techniques for ultrasound detection and location of damage. *Proceedings of SPIE*, **5384**, 287–295.
34. Takeda, N., Okabe, Y., Kuwahara, J. *et al.* (2005) Development of smart composite structures with small-diameter fiber Bragg grating sensors for damage detection: quantitative evaluation of delamination length in CFRP laminates using Lamb wave sensing. *Composites Science and Technology*, **65**, 2575–2587.
35. Li, F., Murayama, H., Kageyama, K. and Shirai, T. (2009) Guided wave and damage detection in composite laminates using different fiber optic sensors. *Sensors*, **9**, 4005–4021.
36. Focke, O., Hildebrandt, A. and Von Kopylov, C. (2008) Inspection of laser generated Lamb waves using shearographic interferometry. 1st International

Symposium on Laser Ultrasonics: Science, Technology and Applications, 2008, Montreal, Canada.

37. Hung, Y.Y. (1996) Shearography for non-destructive evaluation of composite structures. *Optics and Lasers in Engineering*, **24**, 161–182.
38. Hung, Y.Y., Luo, W.D., Lin, L. and Shang, H.M. (2000) NDT of joined surfaces using digital time-integrated shearography with multiple-frequency sweep. *Optics and Lasers in Engineering*, **33**, 369–382.
39. Lammering, R. (2010) Observation of piezoelectrically induced wave propagation in thin plates by use of speckle interferometry. *Experimental Mechanics*, **50**(3), 377–387.
40. Staszewski, W.J., Lee, B.C., Mallet, L. and Scarpa, F. (2004) Structural health monitoring using scanning laser vibrometry: I. Lamb wave sensing. *Smart Materials and Structures*, **13**, 251–260.
41. Swenson, E., Sohn, H., Olson, S. and Desimio, M.A. (2010) Comparison of 1D and 3D laser vibrometry measurements of Lamb waves. *Proceedings of SPIE*, **7650**, 765003.
42. Staszewski, W.J., Lee, B.C. and Traynor, R. (2007) Fatigue crack detection in metallic structures with Lamb waves and 3D laser vibrometry. *Measurement Science and Technology*, **18**, 727–739.

4

One-Dimensional Structural Elements

4.1 Theories of Rods

The theories concerning propagation of longitudinal elastic waves in structural rod elements that are widely exploited in the literature can be generally classified as one-mode, two-mode, three-mode and higher-mode or higher-order theories. Thorough analysis of a general three-dimensional displacement field of structural rod elements can give rise to a number of theories based on different displacement fields. The relevant Maclaurin series expansion helps to reduce the number of unknown variables to a desired and necessary number. It should be emphasised that reducing the number of unknown variables not only lowers the complexity of displacement fields but also reduces the number of wave modes allowed by the theories, thus limiting their application range.

Using cylindrical coordinates (x, r, θ), the general form of the displacement field associated with a structural rod element can be described formally by vector $\boldsymbol{u} = [u_x, u_\theta, u_r]$ of the following form [1]:

$$\begin{aligned} u_x(x, r, \theta) &= U_x(x, r) \\ u_\theta(x, r, \theta) &= 0 \\ u_r(x, r, \theta) &= U_r(x, r) \end{aligned} \tag{4.1}$$

Guided Waves in Structures for SHM: The Time-Domain Spectral Element Method, First Edition.
Wieslaw Ostachowicz, Pawel Kudela, Marek Krawczuk and Arkadiusz Zak.

where $u_x(x, r, \theta)$, $u_\theta(x, r, \theta)$ and $u_r(x, r, \theta)$ are the longitudinal, angular and radial displacement components, respectively, while $U_x(x, r)$ and $U_r(x, r)$ are certain displacement functions dependent only on the spatial coordinates x and r.

For example, expansion of the function $U_x(x, r)$ around $r = 0$ into Maclaurin series leads to the following equation:

$$U_x(x, r) = U_x(x, 0) + \sum_{n=1}^{\infty} \frac{\partial^n U_x(x, 0)}{\partial r^n} \frac{r^n}{n!} \tag{4.2}$$

It should be mentioned at this point that in the case of function $U_x(x, r)$ the terms proportional to odd values of n are associated with antisymmetric behaviour and propagation of flexural waves, while the terms proportional to even values of n are associated with symmetric behaviour and propagation of longitudinal waves. Repeating the same expansion into Maclaurin series for the function $U_r(x, r)$ leads to the opposite conclusions [1–3].

The number of terms to be kept from the series given by (4.2) depends on the investigated phenomena and is directly related to the total number of degrees of freedom of any one-dimensional finite element approximation based on the series expansion. The expansion of function $U_x(x, r)$ at $n = 2$ yields a Maclaurin series of the following form:

$$U_x(x, r) = U_x(x, 0) + \frac{\partial U_x(x, 0)}{\partial r} r + \frac{1}{2} \frac{\partial^2 U_x(x, 0)}{\partial r^2} r^2 + E\left(r^3\right) \tag{4.3}$$

where $E\left(r^3\right)$ represents the truncation error of the expansion proportional to r^3. At this point a step towards finite element approximation can be made and then Equation (4.3) can be rewritten as:

$$U_x(x, r) = \varphi_0(x) + \varphi_1(x) r + \varphi_2(x) r^2 \tag{4.4}$$

where $\varphi_i(x)$ $(i = 0, 1, 2)$ may be interpreted as denoting degrees of freedom of a one-dimensional finite element associated with the Maclaurin expansion of the function $U_x(x, r)$ of the rod displacement field.

It is obvious that due to the truncation of the series (4.3) the obtained formula (4.4) is not exact and represents the three-dimensional displacement field of a one-dimensional structural element in an approximated sense only. However, it should be emphasised that solving most static or dynamic problems of engineering significance efficiently requires using finite elements that employ only the first two or three terms of the relevant Maclaurin series. In the case considered above one can immediately note that:

$$\begin{aligned} \varphi_0(x) &= U_x(x,0) \\ \varphi_1(x) &= \frac{\partial U_x(x,0)}{\partial r} \\ \varphi_2(x) &= \frac{1}{2}\frac{\partial^2 U_x(x,0)}{\partial r^2} \end{aligned} \tag{4.5}$$

Conversely, a great majority of problems involving propagation of elastic waves in one- or two-dimensional structural elements require a much more accurate representation of the three-dimensional behaviour of a solid element. This requirement is directly associated with modelling different modes of elastic waves propagating within such three-dimensional solid structures.

Wave propagation is associated with coupled interaction of shear and extensional waves propagating within a structure with structural lateral boundaries. As a result of this coupled interaction, the propagation of various modes of elastic waves can be observed. Appropriate representation of these modes in a broad range of wave propagation frequencies requires more terms of the Maclaurin series in order to capture the complexity of the interaction phenomena. For that reason special types of new finite elements have been developed that are known in the literature as *spectral finite elements*.

4.2 Displacement Fields of Structural Rod Elements

As mentioned before, the propagation of elastic longitudinal waves in one-dimensional structural rod elements is associated with symmetric rod behaviour. Therefore on the basis of Equation (4.1) and of the considerations specified below the general form of the displacement field of a one-dimensional rod spectral finite element for analysing propagation of elastic longitudinal waves can be written in the following way:

$$\begin{aligned} u_x(x,r) &= \varphi_0(x) + \varphi_2(x)r^2 + \varphi_4(x)r^4 \\ u_r(x,r) &= \psi_1(x)r + \psi_3(x)r^3 + \psi_5(x)r^5 \end{aligned} \tag{4.6}$$

where only six terms of Maclaurin series expansions of the displacement components $u_x(x,r)$ and $u_r(x,r)$ are used. One should remember that thanks to rotational symmetry of the rod with respect to its longitudinal x all displacement and strain components must be independent of angle θ.

The functions φ_i $(i = 0, 2, 4)$ and $\psi_i(x)$ $(i = 1, 3, 5)$ defined in Equations (4.6) represent the independent nodal variables or degrees of freedom of the rod spectral finite element. It can be seen that in the current formulation the rod element has as many as 6 degrees of freedom in a single node. This number of independent nodal variables may be reduced, however, by taking into account the zero traction condition, rewritten here:

$$\sigma_{rr}(x, r) = \tau_{xr}(x, r) = 0, \quad 0 \leq x \leq l, \quad r = a = \frac{d}{2} \tag{4.7}$$

where l now denotes length and d is the diameter of the rod spectral finite element (see Figure 1.16).

Starting from the form of the displacement field given by Equations (4.6), displacement fields for various one-mode, two-mode and other multi-mode rod theories can be constructed. Additionally, using Equations (4.7) representing the zero traction conditions on the lateral boundaries of the rod element allows one to enrich the displacement fields with some additional higher-order terms. However, in most cases the resulting system of two differential equations is very complicated and cannot be solved analytically. This problem can be avoided by simple mathematical manipulations (scaling and substitution), thanks to which the zeroth-order terms for both displacement components $u_x(x, r)$ and $u_r(x, r)$ can be represented as sums of other-order terms. In the present case this condition takes the following form:

$$\begin{aligned}
&\tilde{\varphi}_0(x) = \varphi_0(x) - \tilde{\varphi}_2(x) - \tilde{\varphi}_4(x), \quad \tilde{\varphi}_2(x) = -a^2\varphi_2(x), \quad \tilde{\varphi}_4(x) = -a^4\varphi_4(x) \\
&\tilde{\psi}_1(x) = \psi_1(x) - \tilde{\psi}_3(x) - \tilde{\psi}_5(x), \quad \tilde{\psi}_3(x) = -a^2\psi_3(x), \quad \tilde{\psi}_5(x) = -a^4\psi_5(x)
\end{aligned} \tag{4.8}$$

Taking into account Equations (4.8), rotational symmetry of the beam with respect to its longitudinal axis and antisymmetrical behaviour of bending elastic waves, a new form of the displacement field of a one-dimensional beam spectral finite element can be expressed as follows:

$$\begin{cases}
u_x(x, r) = \varphi_0(x) + \varphi_2(x)\left[1 - \left(\frac{r}{a}\right)^2\right] + \varphi_4(x)\left[1 - \left(\frac{r}{a}\right)^4\right] \\
u_r(x, r) = \psi_1(x)r + \psi_3(x)\left[1 - \left(\frac{r}{a}\right)^2\right]r + \psi_5(x)\left[1 - \left(\frac{r}{a}\right)^4\right]r
\end{cases} \tag{4.9}$$

Specific theories of rod symmetric behaviour known from the literature can be easily obtained starting from Equations (4.7) and (4.9). Different rod theories can be associated with different forms of the functions $\varphi_i(x)\,(i = 0, 2, 4)$ and $\psi_i(x)\,(i = 1, 3, 5)$. As an example, the dispersion curves for the elementary single-mode theory, the single-mode Love theory, the two-mode Mindlin–Herrmann theory, the higher order two-mode theory, as well as the three-mode theory, are presented in Figures 4.1 to 4.5. Their displacement fields can be presented in the following manner:

- *Elementary single-mode theory:*

$$\begin{gathered} u_x(x, r) = \varphi_0(x) \\ \varphi_2(x) = \varphi_4(x) = 0, \quad \psi_1(x) = \psi_3(x) = \psi_5(x) = 0 \end{gathered} \tag{4.10}$$

- *Single-mode Love theory* [4]:

$$\begin{gathered} u_x(x, r) = \varphi_0(x) \\ \ddot{u}_r(x, r) = -\nu r \frac{d}{dx} \dot{\varphi}_0(x) \\ \varphi_2(x) = \varphi_4(x) = 0, \quad \psi_1(x) = \psi_3(x) = \psi_5(x) = 0 \end{gathered} \tag{4.11}$$

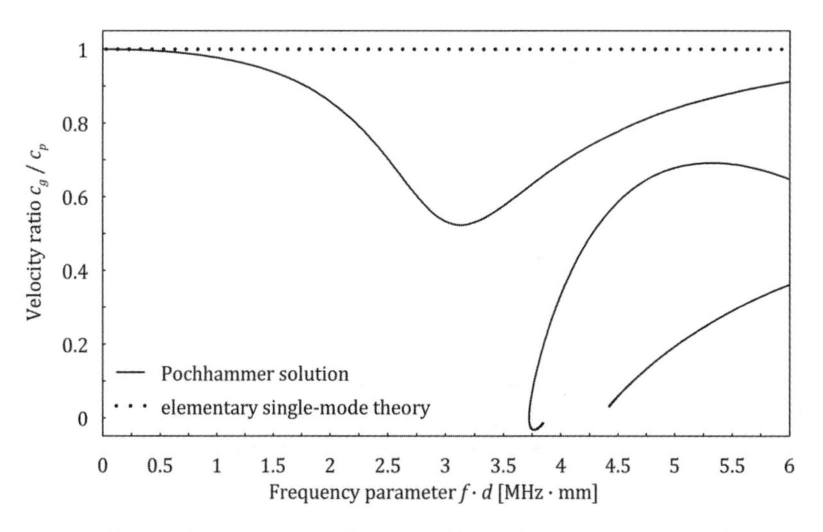

Figure 4.1 Dispersion curve for the velocity ratio c_g/c_p for the elementary single-mode theory of rods ($c_l = 6.3$ km/s, $c_s = 3.2$ km/s)

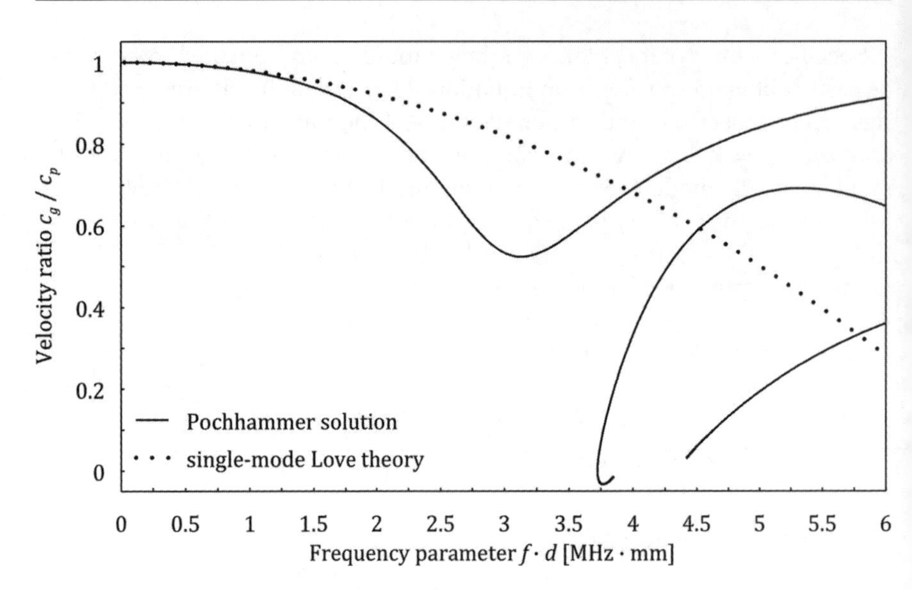

Figure 4.2 Dispersion curve for the velocity ratio c_g/c_p for the single-mode Love theory of rods ($c_l = 6.3$ km/s, $c_s = 3.2$ km/s)

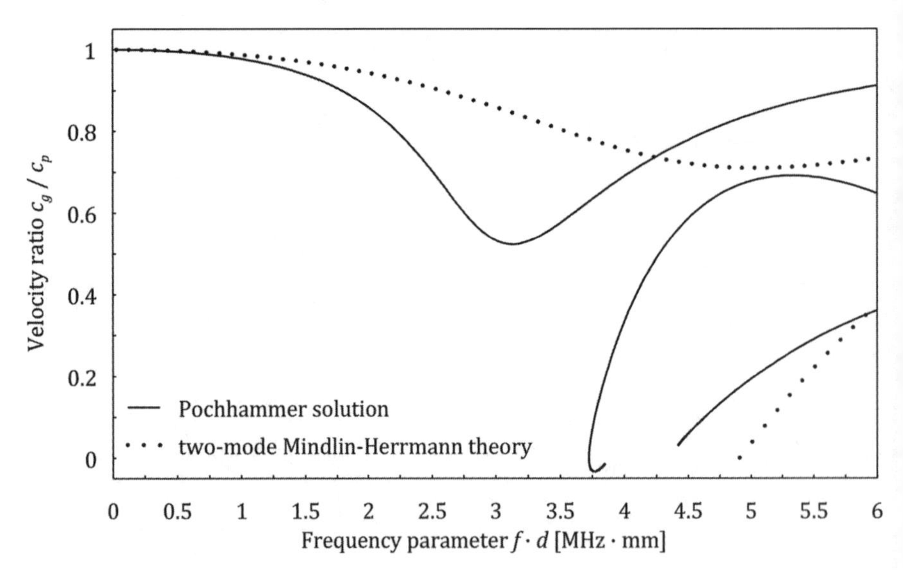

Figure 4.3 Dispersion curves for the velocity ratio c_g/c_p for the two-mode Mindlin–Herrmann theory of rods ($c_l = 6.3$ km/s, $c_s = 3.2$ km/s)

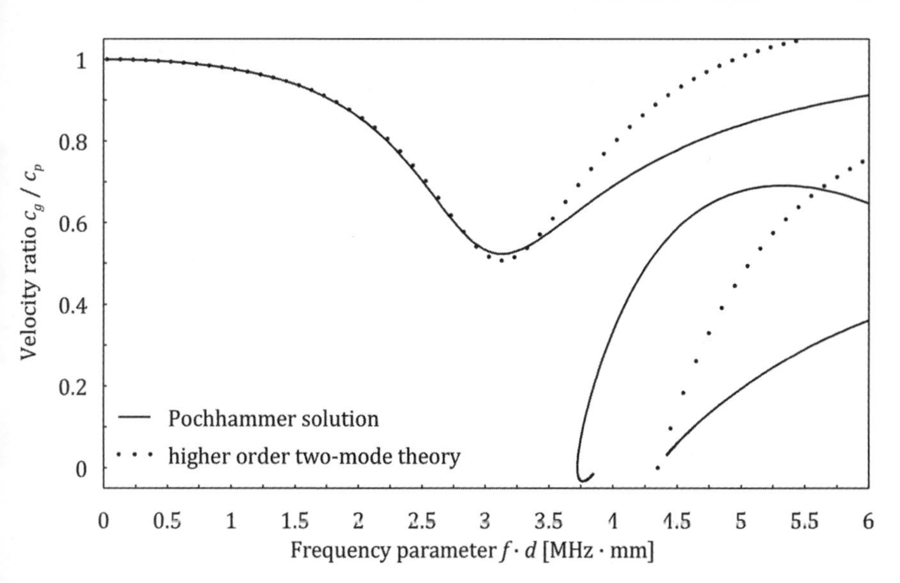

Figure 4.4 Dispersion curves for the velocity ratio c_g/c_p for the higher order two-mode theory of rods ($c_l = 6.3$ km/s, $c_s = 3.2$ km/s)

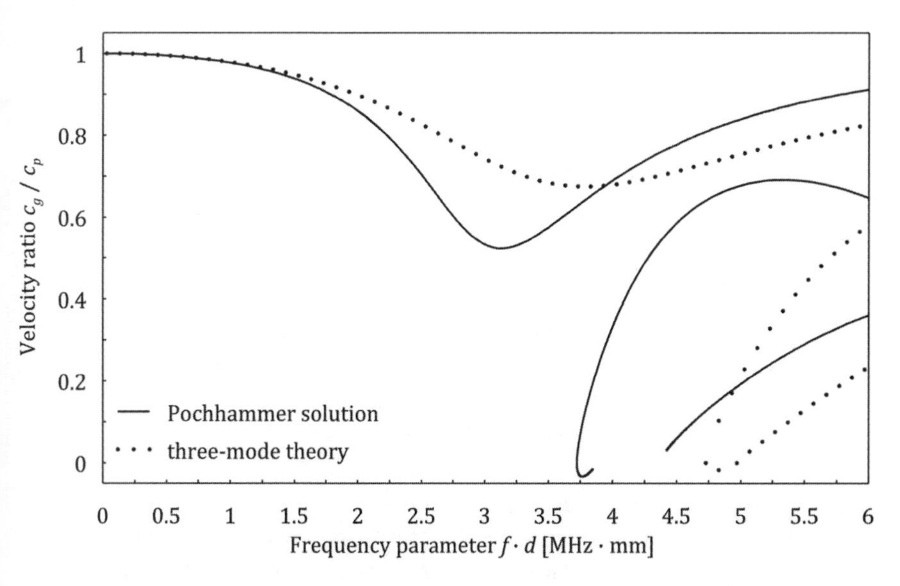

Figure 4.5 Dispersion curves for the velocity ratio c_g/c_p for the three-mode theory of rods ($c_l = 6.3$ km/s, $c_s = 3.2$ km/s)

obtained after adding an additional equation representing the assumption regarding coupling between the longitudinal velocity $\dot{u}_x$ and the transverse velocity $\dot{u}_r$ through the Poisson ratio effect $\dot{\varepsilon}_{rr} = -\nu\,\dot{\varepsilon}_{xx}$ influencing the rod kinetic energy:

$$\dot{u}_r(x,r) = -\nu r \frac{d}{dx}\varphi_0(x)$$

- *Two-mode Mindlin–Herrmann theory* [5]:

$$\begin{gathered} u_x(x,r) = \varphi_0(x) \\ u_r(x,r) = \psi_1(x)r \\ \varphi_2(x) = \varphi_4(x) = 0, \quad \psi_3(x) = \psi_5(x) = 0 \end{gathered} \tag{4.12}$$

- *Higher order two-mode theory* (Authors):

$$\begin{gathered} u_x(x,r) = \varphi_0(x) + \varphi_2(x)\left[1 - \left(\tfrac{r}{a}\right)^2\right] \\ u_r(x,r) = \psi_1(x)r + \psi_3(x)\left[1 - \left(\tfrac{r}{a}\right)^2\right]r \\ \varphi_2(x) = \frac{a^2}{2}\frac{d}{dx}\psi_1(x), \quad \psi_3(x) = \frac{\mu+\lambda}{2\mu+\lambda}\psi_1(x) + \frac{\lambda}{2\,(2\mu+\lambda)}\frac{d}{dx}\varphi_0(x), \\ \varphi_4(x) = 0, \quad \psi_5(x) = 0 \end{gathered} \tag{4.13}$$

- *Three-mode theory* [1]:

$$\begin{gathered} u_x(x,r) = \varphi_0(x) + \varphi_2(x)\left[1 - \left(\frac{r}{a}\right)^2\right] \\ u_r(x,r) = \psi_1(x)r \\ \varphi_4(x) = 0, \quad \psi_3(x) = \psi_5(x) = 0 \end{gathered} \tag{4.14}$$

It should be emphasised that the physical meaning of the higher order terms $\tilde{\varphi}_i(x)\,(i = 2, 4)$ and $\tilde{\psi}_i(x)\,(i = 1, 3, 5)$ must always be associated with the form of the particular displacement field under consideration and results from certain mathematical manipulations that influence it. In the current approach these terms express higher order corrections to the initially assumed distributions of longitudinal and transverse displacement components.

4.3 Theories of Beams

As with rod theories, the theories of propagation of bending (flexural) elastic waves in structural beam elements that are discussed in the literature can generally be classified as one-mode, two-mode, three-mode and higher-mode or higher order theories. A thorough analysis of a general three-dimensional displacement field of structural beam elements can give rise to a number of theories based on different displacement fields. The relevant Maclaurin series expansion helps to reduce the number of unknown variables to a desired and necessary number. It should be emphasised that reducing the number of unknown variables not only lowers the complexity of displacement fields but also reduces the number of wave modes allowed by the theories, thus limiting their application range.

Using cylindrical coordinates (x, r, θ), the general form of the displacement field associated with a structural beam element can be described formally by the vector $\boldsymbol{u} = [u_x, u_\theta, u_r]$ of the following form [1]:

$$\begin{aligned} u_x\,(x, r, \theta) &= U_x(x, r)\sin\theta \\ u_\theta\,(x, r, \theta) &= U_\theta(x, r)\cos\theta \\ u_r\,(x, r, \theta) &= U_r(x, r)\sin\theta \end{aligned} \tag{4.15}$$

where $u_r(x, r)$, $u_\theta(x, r)$ and $u_x(x, r)$ denote the radial, angular and longitudinal displacement components, respectively, while $U_x(x, r)$, $U_\theta(x, r)$ and $U_r(x, r)$ are certain displacement functions dependent only on the spatial coordinates x and r.

Expansion into the Maclaurin series, for example, of the displacement function $U_r(x, r)$ about $r = 0$ leads to the following equation:

$$U_r(x, r) = U_r\,(x, 0) + \sum_{n=1}^{\infty} \frac{\partial^n U_r\,(x, 0)}{\partial r^n}\frac{r^n}{n!} \tag{4.16}$$

It should be mentioned at this point that in the case of the displacement function $U_r(x, r)$ the terms proportional to even values of n are associated with antisymmetric behaviour and the propagation of flexural waves, while the terms proportional to odd values of n are associated with symmetric behaviour and the propagation of longitudinal waves. Repeating the same expansion into the Maclaurin series for displacement functions $U_x(x, r)$ and $U_\theta(x, r)$ leads to analogous conclusions [1–3].

The number of terms to be kept in the series given by (4.16) depends on the investigated phenomena and is directly related to the total number of degrees of freedom of any one-dimensional finite element approximation based on the series expansion. The expansion of the displacement function $U_r(x, r)$ at $n = 2$ yields the Maclaurin series of the following form:

$$U_r(x, r) = U_r(x, 0) + \frac{\partial U_r(x, 0)}{\partial r} r + \frac{1}{2}\frac{\partial^2 U_r(x, 0)}{\partial r^2} r^2 + E\left(r^3\right) \tag{4.17}$$

where $E\left(r^3\right)$ represents the truncation error of the expansion proportional to r^3. At this point a step towards a finite element approximation can be made and then Equation (4.3) can be rewritten as:

$$U_r(x, r) = \psi_0(x) + \psi_1(x)r + \psi_2(x)r^2 \tag{4.18}$$

where now $\psi_i(x)\,(i = 0, 1, 2)$ may be interpreted as denoting degrees of freedom of a one-dimensional finite element associated with the Maclaurin expansion of the displacement function $U_r(x, r)$ of the beam displacement field.

It is obvious that the obtained formula (4.18) is not exact due to the truncation of the series (4.17) and represents the three-dimensional displacement field of a one-dimensional structural element in an approximated sense only. However, it should be emphasised that solving most static or dynamic problems of engineering significance efficiently requires the use of finite elements that employ only the first two or three terms of the relevant Maclaurin series. In the case considered above one can immediately note that:

$$\begin{aligned} \psi_0(x) &= U_r(x, 0) \\ \psi_1(x) &= \frac{\partial U_r(x, 0)}{\partial r} \\ \psi_2(x) &= \frac{1}{2}\frac{\partial^2 U_r(x, 0)}{\partial r^2} \end{aligned} \tag{4.19}$$

Conversely, a great majority of problems involving propagation of elastic waves in one- or two-dimensional structural elements require a much more accurate representation of the three-dimensional behaviour of a solid element. This requirement is directly associated with modelling different modes of elastic waves propagating within such three-dimensional solid structures.

Wave propagation is associated with coupled interaction of shear and extensional waves propagating within a structure with structural lateral boundaries. As a result of this coupled interaction, the propagation of various modes of elastic waves can be observed. Appropriate representation of these modes in a broad range of wave propagation frequencies requires more terms of the Maclaurin series in order to capture the complexity of the interaction phenomena. For that reason special types of new finite elements have been developed that are known in the literature as *spectral finite elements*.

4.4 Displacement Fields of Structural Beam Elements

As mentioned before, the propagation of elastic flexural waves in one-dimensional structural beam elements is associated with antisymmetric beam behaviour. Therefore, on the basis of Equation (4.15) and the considerations specified below, the general form of the displacement field of a one-dimensional beam spectral finite element for the needs of analysing propagation of elastic bending waves can be written in the following way:

$$\begin{aligned} U_x(x,r) &= \varphi_1(x)r + \varphi_3(x)r^3 + \varphi_5(x)r^5 \\ U_\theta(x,r) &= \vartheta_0(x) + \vartheta_2(x)r^2 + \vartheta_4(x)r^4 \\ U_r(x,r) &= \psi_0(x) + \psi_2(x)r^2 + \psi_4(x)r^4 \end{aligned} \tag{4.20}$$

where only nine terms of Maclaurin series expansions of displacement functions $U_x(x,r)$, $U_\theta(x,r)$ and $U_r(x,r)$ are taken.

The displacement functions $\varphi_i(x)\,(i=1,3,5)$, $\vartheta_i(x)\,(i=0,2,4)$ and $\psi_i(x)\,(i=0,2,4)$ defined in Equations (4.20) represent the independent nodal variables or degrees of freedom of the beam spectral finite element. It can be seen that in the current formulation the beam element has as many as 9 degrees of freedom in a single node. This number of independent nodal variables may be reduced, however, by taking into account the same zero traction condition, rewritten here:

$$\sigma_{rr}(x,r) = \tau_{xr}(x,r) = 0, \quad 0 \le x \le l, \quad r = a = \frac{d}{2} \tag{4.21}$$

where l now denotes length and d is the diameter of the beam spectral finite element (see Figure 1.16).

Starting from the form of the displacement field given by Equations (4.20), displacement fields for various one-mode, two-mode and other multi-mode rod theories can be constructed. Additionally, using Equation (4.21) representing the zero traction conditions on the lateral boundaries of the rod element allows one to enrich the displacement fields with some additional higher-order terms. However, in most cases the resulting system of two differential equations is very complicated and cannot be solved analytically, but this can be avoided by simple substitutions and rearrangements of terms, thanks to which the zeroth-order terms for the displacement functions $U_x(x,r)$, $U_\theta(x,r,\theta)$ and $U_r(x,r)$ can be represented as sums of all order terms in a similar manner, as presented in Equations (4.8).

Taking into account Equations (4.8), the rotational symmetry of the beam with respect to its longitudinal axis and antisymmetric behaviour of bending elastic waves, a new form of the displacement field of a one-dimensional beam spectral finite element for analysing the propagation of elastic bending (flexural) waves can be expressed as follows:

$$\begin{aligned}
U_x(x,r) &= \varphi_1(x)r + \varphi_3(x)\left[1-\left(\frac{r}{a}\right)^2\right]r + \varphi_5(x)\left[1-\left(\frac{r}{a}\right)^4\right]r \\
U_\theta(x,r) &= (\vartheta_0(x)-\vartheta_2(x)-\vartheta_4(x)) + \vartheta_2(x)\left[1-\left(\frac{r}{a}\right)^2\right] + \vartheta_4(x)\left[1-\left(\frac{r}{a}\right)^4\right] \\
U_r(x,r) &= (\psi_0(x)-\psi_2(x)-\psi_4(x)) + \psi_2(x)\left[1-\left(\frac{r}{a}\right)^2\right] + \psi_4(x)\left[1-\left(\frac{r}{a}\right)^4\right]
\end{aligned} \tag{4.22}$$

Specific theories of beam flexural behaviour known from the literature can be easily obtained starting from Equations (4.21) and (4.22). Different beam theories can be associated with different forms of the functions $\varphi_i(x)\,(i=1,3,5)$, $\vartheta_i(x)\,(i=0,2,4)$ and $\psi_i(x)\,(i=0,2,4)$. As an example, the dispersion curves for the classical single-mode theory, the single-mode Bernoulli–Rayleigh theory, the two-mode Timoshenko theory, the higher order two-mode Reddy theory, as well as the higher-mode theory, are presented in Figures 4.6 to 4.10. Their displacaement fields can be presented in the following manner:

- *Classical single-mode theory:*

$$\begin{gathered}
u_x(x,r,\theta) = \varphi_1(x)r\sin\theta \\
u_\theta(x,r,\theta) = \vartheta_0(x)\cos\theta \\
u_r(x,r,\theta) = \psi_0(x)\sin\theta \\
\varphi_1(x) = -\frac{d}{dx}\psi_0(x), \quad \vartheta_0(x) = \psi_0(x) \\
\varphi_3(x) = \varphi_5(x) = 0, \quad \vartheta_2(x) = \vartheta_4(x) = 0, \quad \psi_2(x) = \psi_4(x) = 0
\end{gathered} \tag{4.23}$$

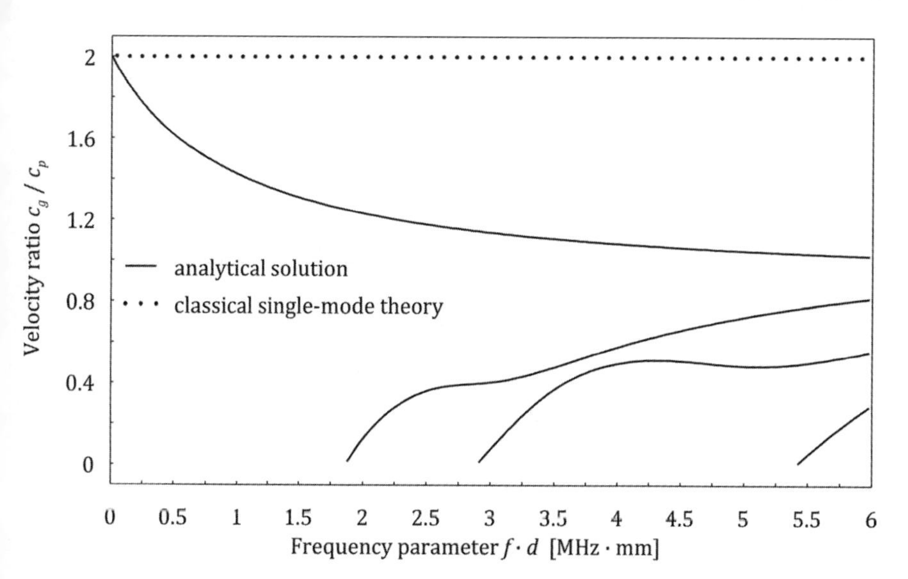

Figure 4.6 Dispersion curve for the velocity ratio c_g/c_p for the classical single-mode theory of beams ($c_l = 6.3$ km/s, $c_s = 3.2$ km/s)

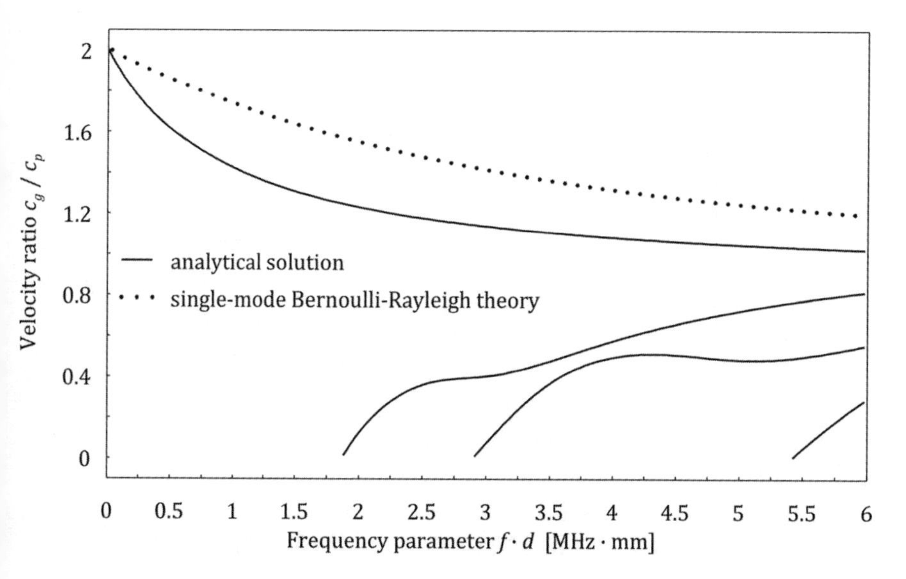

Figure 4.7 Dispersion curve for the velocity ratio c_g/c_p for the single-mode Bernoulli–Rayleigh theory of beams ($c_l = 6.3$ km/s, $c_s = 3.2$ km/s)

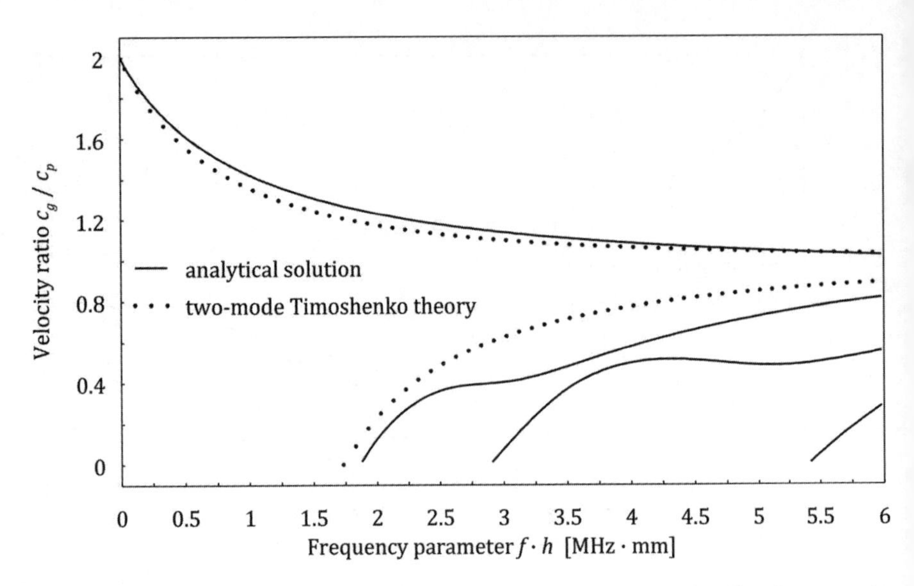

Figure 4.8 Dispersion curves for the velocity ratio c_g/c_p for the two-mode Timoshenko theory of beams ($c_l = 6.3$ km/s, $c_s = 3.2$ km/s)

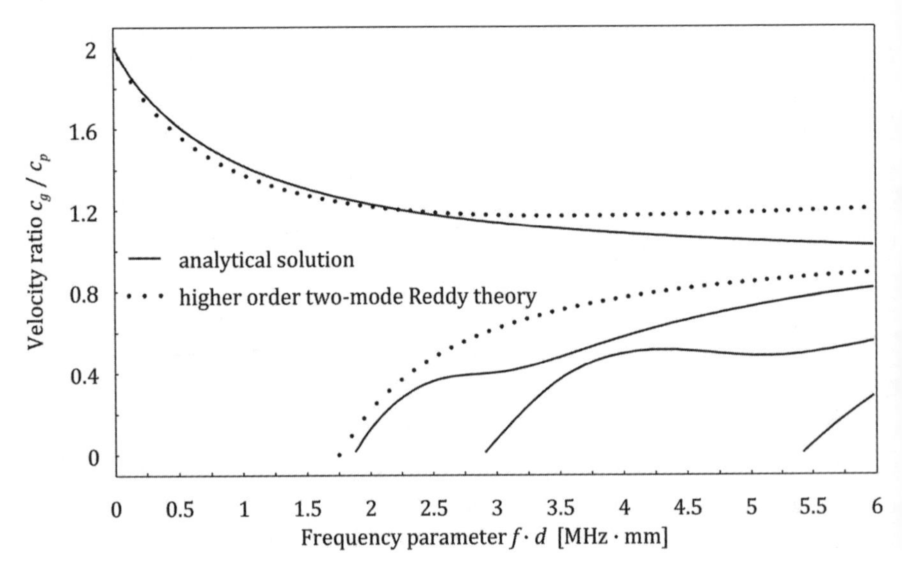

Figure 4.9 Dispersion curves for the velocity ratio c_g/c_p for the higher order two-mode Reddy theory of beams ($c_l = 6.3$ km/s, $c_s = 3.2$ km/s)

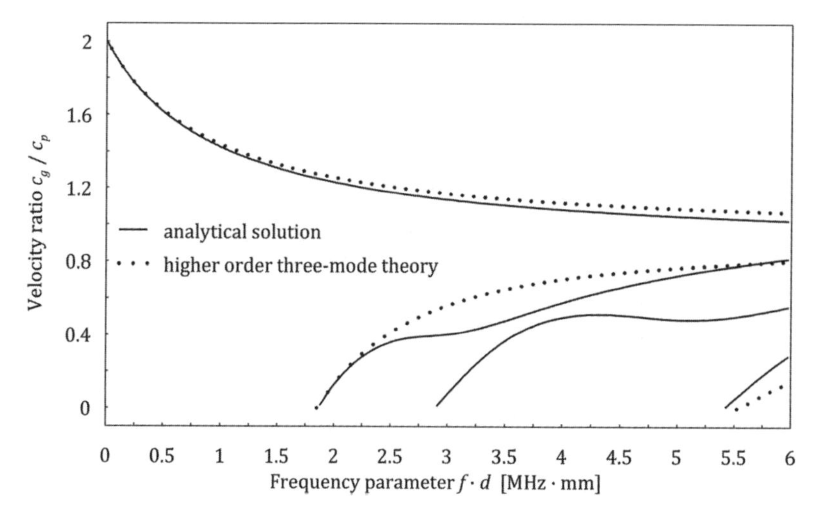

Figure 4.10 Dispersion curves for the velocity ratio c_g/c_p for the higher order three-mode theory of beams ($c_l = 6.3$ km/s, $c_s = 3.2$ km/s)

keeping in mind that the contribution to the beam kinetic energy from longitudinal velocity $\dot{u}_x$ is neglected:

$$\tilde{T} = \left(\frac{\partial \vartheta_0}{\partial t} \cos\theta\right)^2 + \left(\frac{\partial \psi_0}{\partial t} \sin\theta\right)^2 \tag{4.24}$$

- *Single-mode Bernoulli–Rayleigh theory* [6]:

$$\begin{gathered} u_x(x, r, \theta) = \varphi_1(x) r \sin\theta \\ u_\theta(x, r, \theta) = \vartheta_0(x) \cos\theta \\ u_r(x, r, \theta) = \psi_0(x) \sin\theta \\ \varphi_1(x) = -\frac{d}{dx}\psi_0(x), \quad \vartheta_0(x) = \psi_0(x) \\ \varphi_3(x) = \varphi_5(x) = 0, \quad \vartheta_2(x) = \vartheta_4(x) = 0, \quad \psi_2(x) = \psi_4(x) = 0 \end{gathered} \tag{4.25}$$

in which the contribution to the beam kinetic energy from the longitudinal velocity $\dot{u}_x$ is taken into account:

$$\tilde{T} = \left(\frac{\partial \varphi_1}{\partial t} r \sin\theta\right)^2 + \left(\frac{\partial \vartheta_0}{\partial t} \cos\theta\right)^2 + \left(\frac{\partial \psi_0}{\partial t} \sin\theta\right)^2 \tag{4.26}$$

- *Two-mode Timoshenko theory* [7]:

$$
\begin{aligned}
u_x(x,r,\theta) &= \varphi_1(x) r \sin\theta \\
u_\theta(x,r,\theta) &= \vartheta_0(x)\cos\theta \\
u_r(x,r,\theta) &= \psi_0(x)\sin\theta \\
\vartheta_0(x) &= \psi_0(x) \\
\varphi_3(x) = \varphi_5(x) = 0, \quad \vartheta_2(x) &= \vartheta_4(x) = 0, \quad \psi_2(x) = \psi_4(x) = 0
\end{aligned}
\tag{4.27}
$$

- *Higher order two-mode Reddy theory* [8]:

$$
\begin{aligned}
u_x(x,r,\theta) &= \left\{\varphi_1(x) r + \varphi_3(x)\left[1-\left(\frac{r}{a}\right)^2\right] r\right\}\sin\theta \\
u_\theta(x,r,\theta) &= \vartheta_0(x)\cos\theta \\
u_r(x,r,\theta) &= \psi_0(x)\sin\theta \\
\vartheta_0(x) = \psi_0(x), \quad \varphi_3(x) &= \frac{1}{2}\left[\frac{d}{dx}\psi_0(x) + \varphi_1(x)\right] \\
\varphi_5(x) = 0, \quad \vartheta_2(x) &= \vartheta_4(x) = 0, \quad \psi_2(x) = \psi_4(x) = 0
\end{aligned}
\tag{4.28}
$$

- *Higher order three-mode theory* (Authors):

$$
\begin{aligned}
u_x(x,r,\theta) &= \left\{\varphi_1(x) r + \varphi_3(x)\left[1-\left(\frac{r}{a}\right)^2\right] r + \varphi_5(x)\left[1-\left(\frac{r}{a}\right)^4\right] r\right\}\sin\theta \\
u_\theta(x,r,\theta) &= \left\{(\vartheta_0(x) - \vartheta_2(x)) + \vartheta_2(x)\left[1-\left(\frac{r}{a}\right)^2\right]\right\}\cos\theta \\
u_r(x,r,\theta) &= \left\{(\psi_0(x) - \psi_2(x)) + \psi_2(x)\left[1-\left(\frac{r}{a}\right)^2\right]\right\}\sin\theta \\
\vartheta_0(x) &= \psi_0(x), \quad -\vartheta_2(x) = \psi_2(x) = \frac{a^2}{4}\frac{\lambda}{\lambda+\mu}\frac{d}{dx}\varphi_1(x) \\
\varphi_3(x) &= \frac{1}{4}\left[\varphi_1(x) - 2\varphi_3(x) + \frac{d}{dx}\psi_0(x) - \frac{d}{dx}\psi_2(x)\right] \\
\varphi_5(x) &= 0, \quad \vartheta_4(x) = 0, \quad \psi_4(x) = 0
\end{aligned}
\tag{4.29}
$$

It should be emphasised that the physical meaning of the higher order terms associated with the displacement functions $\varphi_i(x)\,(i=3,5)$, $\vartheta_i(x)\,(i=2,4)$ and $\psi_i(x)\,(i=2,4)$ must always be associated with the form of a particular displacement field under consideration and results from certain mathematical manipulations that influence it. In the current approach these terms express higher order corrections to the initially assumed distributions of longitudinal and transverse displacement components.

4.5 Dispersion Curves

Dispersion curves for a particular theory carry very important information about certain frequency characteristics of the theory, but the most important pieces are the range of its application and consistency with known analytical solutions. In the case of displacement fields presented in this book and associated with the different theories discussed, dispersion curves can be evaluated following a very simple procedure.

In the first step one needs to formulate equations of motion associated with the investigated theory and this can be easily achieved by applying Hamilton's principle. Based on the given displacement field, the virtual work W related to the deformation and motion of a structural element may be expressed in terms of its strain energy U, kinetic energy T as well as the work of some external forces F. Application of Hamilton's principle at this point leads to a set of equations of motion that are derived for each component of the displacement field, as presented by Doyle [2].

In the next step, the propagation of harmonic waves within the structural element is assumed. This allows the equations of motion to be transformed from a set of partial differential equations, defined in the time domain for each displacement component, to a set of linear homogeneous equations defined in the frequency domain but for amplitudes of each displacement component. This system can be solved only when its determinant vanishes, which leads directly to a characteristic polynomial equation. The roots of the characteristic polynomial equation define the dispersion relations between particular modes of the harmonic waves that can propagate within the structural element, the wave number k and the angular frequency ω of these waves.

The dispersion curves for each theory discussed in this book were obtained using the Mathematica® package [9] that was employed to perform all required analytical manipulations, while for necessary numerical calculations associated with evaluation of the dispersion curves the authors employed the MATLAB® package [10].

The procedure mentioned above is discussed here in more detail for the two-mode Mindlin–Herrmann theory of rod behaviour from the previous section 4.2. In the case of the two-mode Mindlin–Herrmann theory, taking into account Equations (4.9) as well as the relations given by Equations (4.12) leads to the strain energy U and the kinetic energy T, expressed as:

$$T = \frac{1}{2}\iiint_V \rho\tilde{T}V, \quad U = \frac{1}{2}\iiint_V \tilde{U}dV, \tag{4.30}$$

where ρ is the rod material density and V denotes the rod volume – for clarity and simplicity of presentation the argument x has been omitted hereinafter:

$$\tilde{T} = \left(\frac{\partial \varphi_0}{\partial t}\right)^2 + r^2\left(\frac{\partial \psi_1}{\partial t}\right)^2$$

$$\tilde{U} = (2\mu + \lambda)\left(\frac{\partial \varphi_0}{\partial x}\right)^2 + r^2\left(\frac{\partial \psi_1}{\partial x}\right)^2 + 4\lambda\left(\frac{\partial \varphi_0}{\partial x}\right)\psi_1 + 4(\mu + \lambda)\psi_1^2 \tag{4.31}$$

Application of Hamilton's principle and integration of Equations (4.30) by parts leads to equations of motion associated with the two-mode Mindlin–Herrmann theory of rod behaviour. These equations can be written as the following set of two partial differential equations:

$$\begin{aligned}
\rho\frac{\partial^2 \varphi_0}{\partial t^2} &= (2\mu + \lambda)\frac{\partial^2 \varphi_0}{\partial x^2} + 2\lambda\frac{\partial \psi_1}{\partial x} \\
a^2\rho\frac{\partial^2 \psi_1}{\partial t^2} &= a^2\mu\frac{\partial^2 \psi_1}{\partial x^2} - 4\lambda\frac{\partial \varphi_0}{\partial x} - 8(\mu + \lambda)\psi_1
\end{aligned} \tag{4.32}$$

The equations in (4.32) describe the motion of rod structural elements according to the two-mode Mindlin–Herrmann theory [5] and couple spatial changes in the displacement components φ_0 and ψ_1 with changes in time t. However, in order to obtain the dispersion curves, which express changes in the phase c_p and group c_g velocities as a function of the angular frequency ω or frequency $f = \omega/2\pi$ for the two modes of elastic longitudinal waves associated with the Mindlin–Herrmann theory of rods, the equations of motion (4.32) must be transformed from the time domain into the frequency domain. For that purpose it is convenient to assume that the displacement components φ_0 and ψ_1 can be expressed as solutions of the equations of motion:

$$\begin{aligned}
\varphi_0 &= \langle\varphi_0\rangle \exp[-i(kx - \omega t)] \\
\psi_1 &= \langle\psi_1\rangle \exp[-i(kx - \omega t)]
\end{aligned} \tag{4.33}$$

where $i = \sqrt{-1}$ is the imaginary unit and ω and k denote the angular frequency and wave number, respectively.

A system of two linear homogeneous equations can be obtained for each harmonic amplitude component $\langle\varphi_0\rangle$ and $\langle\psi_1\rangle$ by simple substitution of Equations (4.33) into Equations (4.32) and some simplifications:

$$\begin{aligned} \left[\rho\omega^2 - k^2(\lambda + 2\mu)\right]\langle\varphi_0\rangle - (2\lambda ik)\langle\psi_1\rangle = 0 \\ (4\lambda ik)\langle\varphi_0\rangle - \left[8\lambda + \left(8 + a^2k^2\right)\mu - a^2\rho\omega^2\right]\langle\psi_1\rangle = 0 \end{aligned} \tag{4.34}$$

This system has a nontrivial solution only when its determinant vanishes, which leads to the characteristic polynomial equation associated with the current problem:

$$\begin{aligned} \mu(\lambda + 2\mu)a^2k^4 + \left[a^2\rho\omega^2 - 8(\lambda + \mu)\right]\rho\omega^2 \\ + \left[8\mu(3\lambda + 2\mu) - (\lambda + 3\mu)a^2\rho\omega^2\right]k^2 = 0 \end{aligned} \tag{4.35}$$

which is a fourth-order polynomial equation with respect to the wave number k and a function of the angular frequency ω. This characteristic polynomial equation has two real and positive roots that are associated with the two modes of elastic longitudinal waves that can propagate within a rod structural element and are allowed by the two-mode Mindlin–Herrmann theory. These roots can be calculated numerically for any chosen value of angular frequency ω and thanks to the obtained relation $k = k(\omega)$, the phase velocity $c_p = \omega/k$ as well as group velocity $c_p = d\omega/dk$ can be easily calculated and plotted, as presented in Figure 4.3.

Exactly the same procedure was employed to calculate the remaining dispersion curves associated with the other rod theories presented and discussed in the previous section.

4.6 Certain Numerical Considerations

The spectral finite element method has been employed by many authors for analysing problems involving wave propagation in various structural elements [11–14]. However, problems associated with the accuracy of wave propagation modelling using the spectral finite element method have not been discussed or investigated. Although the results of numerical investigations presented below have been obtained for rod structural elements, in the opinion of the authors the general conclusions that can be made based on these results are also valid for other types of one-, two- or three-dimensional structural elements. This results from the inherent properties of the element shape functions used by the spectral finite element method as well as the finite element method. In a general case the elemental shape functions of two- or three-dimensional structural elements are constructed by simple multiplication of appropriate one-dimensional shape functions; therefore relevant numerical imprints of these functions will also manifest in the case of two or three dimensions. However, due to element geometry complexities (shell or solid elements) and the multi-mode nature of elements such an analysis is always very complicated and cumbersome.

For this very reason the results of numerical simulations associated with the application of simple single-mode, two-mode and higher order rod behaviour theories performed by the authors and presented below are discussed and studied in detail. The influence of such parameters as the node distribution within elements, the order of approximation polynomials, inertia matrix diagonalisation, and more, all of which apply to numerical problems concerning wave propagation, have been carefully analysed in terms of their impact on appropriate modelling of the wave propagation phenomena.

4.6.1 Natural Frequencies

Any dynamic response of a linear structure to a time-dependent mechanical load may be thought of and represented by a sum of its harmonic components of different frequencies and amplitudes. The amplitudes of these components are directly correlated with the frequency content of the load, which on the other hand may be represented by its Fourier spectrum. The frequency spectrum of the load excites those natural frequencies and modes of vibrations of the structure that are covered by this range. Therefore appropriate modelling of dynamic structural responses is directly linked with accurate modelling of structural natural frequency spectra and this is also why the natural frequency analysis has such great importance in the field of computational structural dynamics.

As the first example, we will discuss the effect of the form of elemental inertia matrix $\mathbf{M}^e$ on changes of natural frequencies of a simple aluminium rod, as presented in Figure 1.16. It has been assumed for that purpose that a two-sided free rod structural element has been divided into 20 eight-node rod finite elements, resulting in 141 degrees of freedom of the numerical model. The elementary single-mode theory of rod behaviour has been employed for computations [2, 14] and the Chebyshev node distribution has been selected for the purpose of the study. A consistent form of the inertia matrix $\mathbf{M}^e$ has been considered [15, 16], together with diagonal forms obtained via two

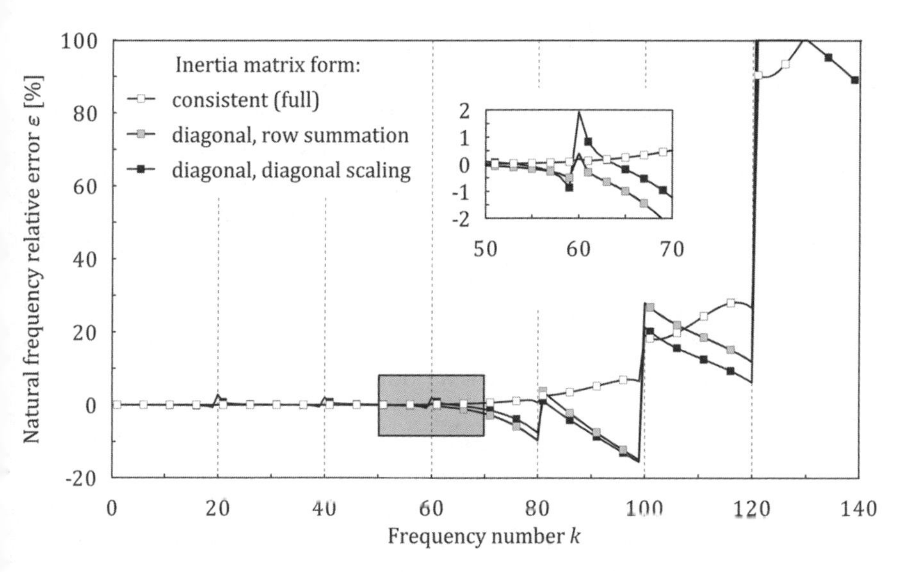

Figure 4.11 Natural frequency relative error as a function of the rod natural frequency number for various inertia matrix diagonalisation methods in the case of the Chebyshev node distribution and the elementary single-mode theory of rods

different diagonalisation techniques that are often used and known in the literature, that is row summation [17–20] and diagonal scaling [15, 21–23].

The results presented in Figure 4.11, limited to the first hundred natural frequencies of the rod, show that the smallest relative errors ε between the kth natural frequency of the rod calculated analytically, according to the theory applied, and numerically, using the spectral finite element method, are obtained for the consistent form of the inertia matrix $\boldsymbol{M}^e$. This error slowly increases with the natural frequency number k and exhibits certain small and periodic discontinuities when the natural frequency number k is a multiple of the total number of spectral elements used in the numerical model. This effect is directly linked with dimensions of the elemental characteristic matrices (of stiffness $\boldsymbol{K}^e$ and of inertia $\boldsymbol{M}^e$) as well as with the structures of the global characteristic matrices (of stiffness $\boldsymbol{K}$ and of inertia $\boldsymbol{M}$) and their bandwidths. This correlates closely with the order n of approximation polynomials. It can also be clearly seen from Figure 4.11 that in the cases of both diagonalisation techniques the relative error ε oscillates around the exact analytical values and the amplitudes of these oscillations are smaller for the row summation technique than for the diagonal scaling. Periodic discontinuities associated with multiples of the finite element number of the model manifest more strongly in those two cases.

The two next numerical examples show the influence of the approximation polynomial order n on changes in the natural frequency relative error ε. Both Lobatto and Chebyshev node distributions are considered for the rod of the same boundary conditions, while the total number of degrees of freedom of the numerical models is kept constant as 316, that is 105 four-node elements, 63 six-node elements, 45 eight-node elements and 35 ten-node elements. In the case of a Lobatto node distribution Lobatto quadrature has been applied for calculating the rod element inertia matrix $\mathbf{M}^e$, while in the case of a Chebyshev node distribution Gaussian quadrature has been used for that purpose.

The results of numerical calculations presented in Figures 4.12 and 4.13 show that in both cases the natural frequency relative error ε is smaller for higher orders n of approximation polynomials up to half of the frequency spectrum, which in the current case is equivalent to the 158th natural frequency. Above this value the natural frequency relative error ε starts to oscillate, rapidly increasing its amplitude. It is noteworthy that the amplitudes of these oscillations are smaller for lower orders n of approximation

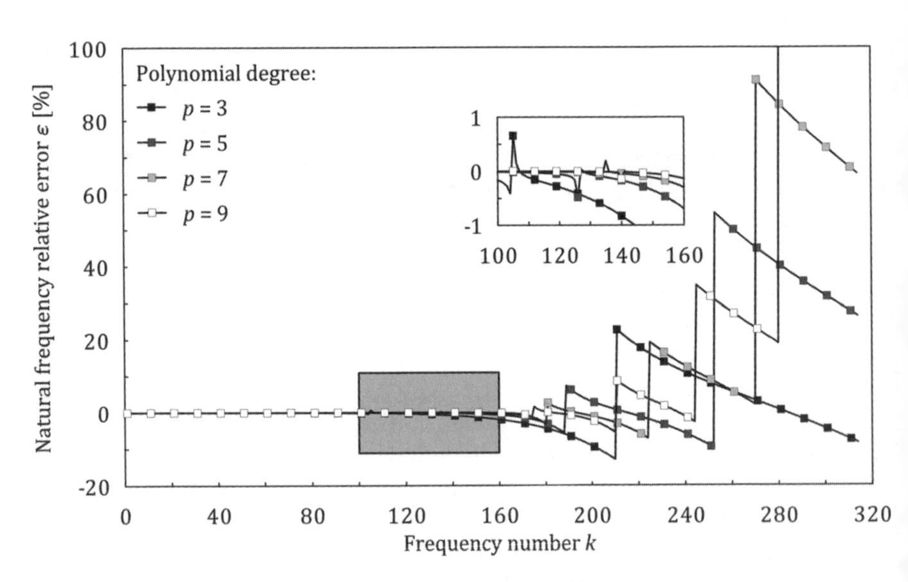

Figure 4.12 Natural frequency relative error as a function of the rod natural frequency number for various approximation polynomial orders in the case of the Lobatto node distribution and the elementary single-mode theory of rods

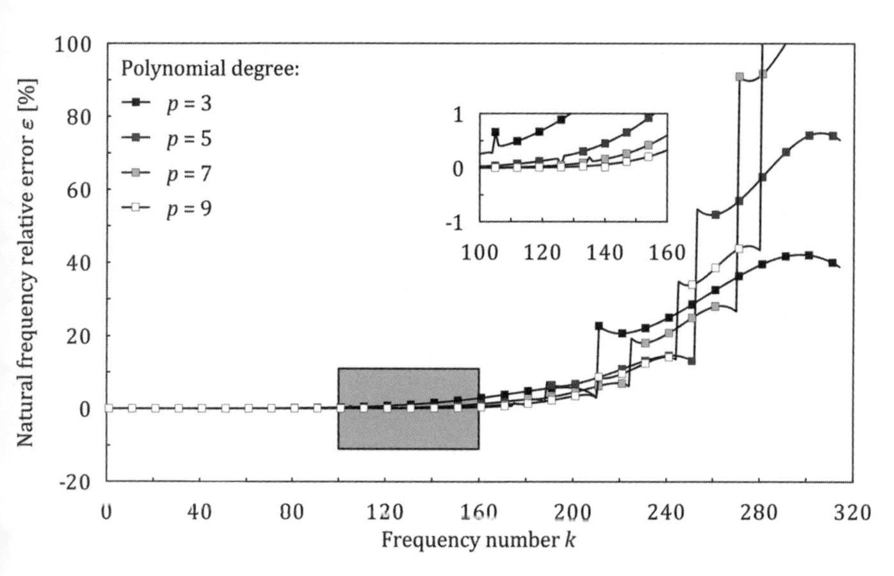

Figure 4.13 Natural frequency relative error as a function of the rod natural frequency number for various approximation polynomial orders in the case of the Chebyshev node distribution and the elementary single-mode theory of rods

polynomials. Moreover, in the case of the Lobatto node distribution the natural frequency relative error ε has negative values in the first half of the frequency spectrum, while it is always positive for the Chebyshev node distribution. This is because Lobatto quadrature is exact for approximation polynomials up to the order $2n - 1$ [21, 24], while at the same time numerical evaluation of the rod element inertia matrix $\boldsymbol{M}^e$ requires a quadrature exact up to the order $2n$. In this way the inertia distribution within the inertia matrix $\boldsymbol{M}^e$ is underestimated.

4.6.2 Wave Propagation

It should be understood that meaningful modelling of propagation of elastic waves in structural elements requires very accurate numerical models. These models must be capable of properly capturing and mimicking the high frequency content of propagating signals without any signal loss or distortion. This is why an analysis of elastic wave propagation in structural elements is still a challenge and draws the attention of many researchers all around the world.

To illustrate certain problems that may be encountered during such an analysis, the propagation of longitudinal elastic waves in the same rod structural element is investigated as the fourth example, based on the numerical model employed for the first example. In the first step, an analysis of the natural frequencies of the rod has been performed, taking into account the results presented in Figure 4.11. It can be seen that the modes ξ_k and natural frequencies f_k calculated numerically using the spectral finite element method correspond well with the analytical solutions: $\xi_k = (x/l)_k = \cos(k\pi\xi)$ and $f_k = k/(2l\sqrt{E/\rho})$, where $k = 1, 2, \ldots$, for which natural frequencies of the rod vary linearly with the frequency number [25].

It can also be clearly seen from Figure 4.11 that approximately the first half of the natural frequency spectrum closely follows the analytical solution for each form of the inertia matrix $\boldsymbol{M}^e$. In the second half of the spectrum and above the 70th natural frequency of the rod the numerical solution starts to diverge significantly from the analytical solution up to a point of a sudden jump at the 80th natural frequency. At this point the numerical solution returns back to the vicinity of the analytical solution. All subsequent deviations of the numerical solution from the analytical solution repeat at the multiples of the element number of the numerical model and are much more significant for the cases when diagonalisation (row summing or diagonal scaling) of the original consistent inertia matrix of the rod $\boldsymbol{M}^e$ has been performed.

The observed behaviour of natural frequencies of the rod is in fact a consequence of the well-known Nyquist theorem [26]. It can be said that the correct reconstruction process of rod dynamics in terms of its natural frequencies and modes of vibration requires spatial sampling that is at least equal to half of the mode characteristic wavelength. In other words, the shortest wavelength (i.e. the highest mode) that can be accurately reconstructed requires at least two spatial sampling points. This means that the first half of the natural frequency spectrum fulfils the Nyquist condition, while the natural frequencies in the second half of the spectrum are affected by aliasing.

The mode aliasing problem is well illustrated by Figures 4.14 and 4.15, which compare numerically calculated modes of rod vibrations and their spectra for the 50th, 70th and 90th natural frequencies. These natural frequencies correspond to three selected points of the frequency spectrum well below, around and above the Nyquist frequency $f_{Nyq} = 182.56$ kHz. It should be emphasised here that the mode aliasing problem has a direct and profound impact on the results of numerical investigations involving wave propagation. In the case when the excitation frequency spectrum covers the natural frequency spectrum affected by aliasing for a given numerical model, the results of numerical simulations may be significantly affected or distorted. In the

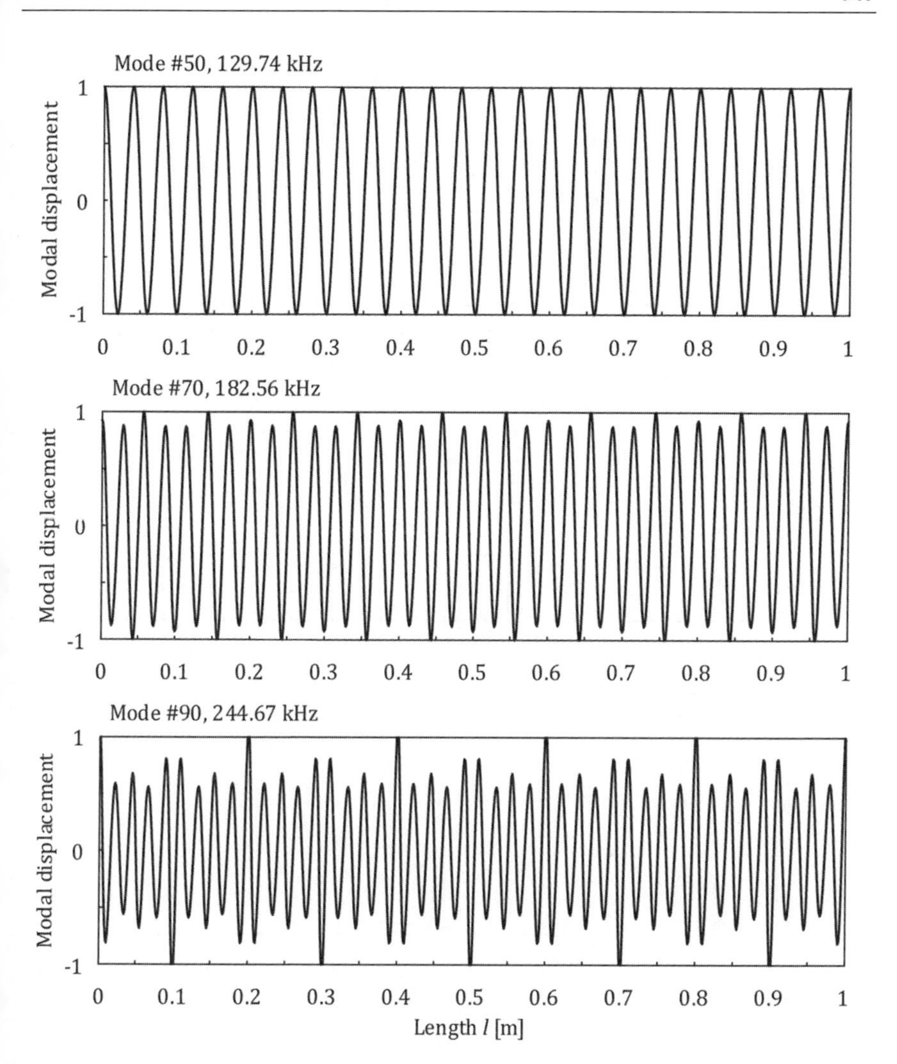

Figure 4.14 Selected modes of rod natural vibrations calculated numerically in the case of Chebyshev node distribution and the elementary single-mode theory of rods

worst scenario the numerical model may not be able to generate any results due to serious instability and/or convergence problems.

In order to demonstrate the importance of this problem, in the following numerical example the analysis of propagation of longitudinal elastic waves within the rod has been investigated for three different excitation cases, as

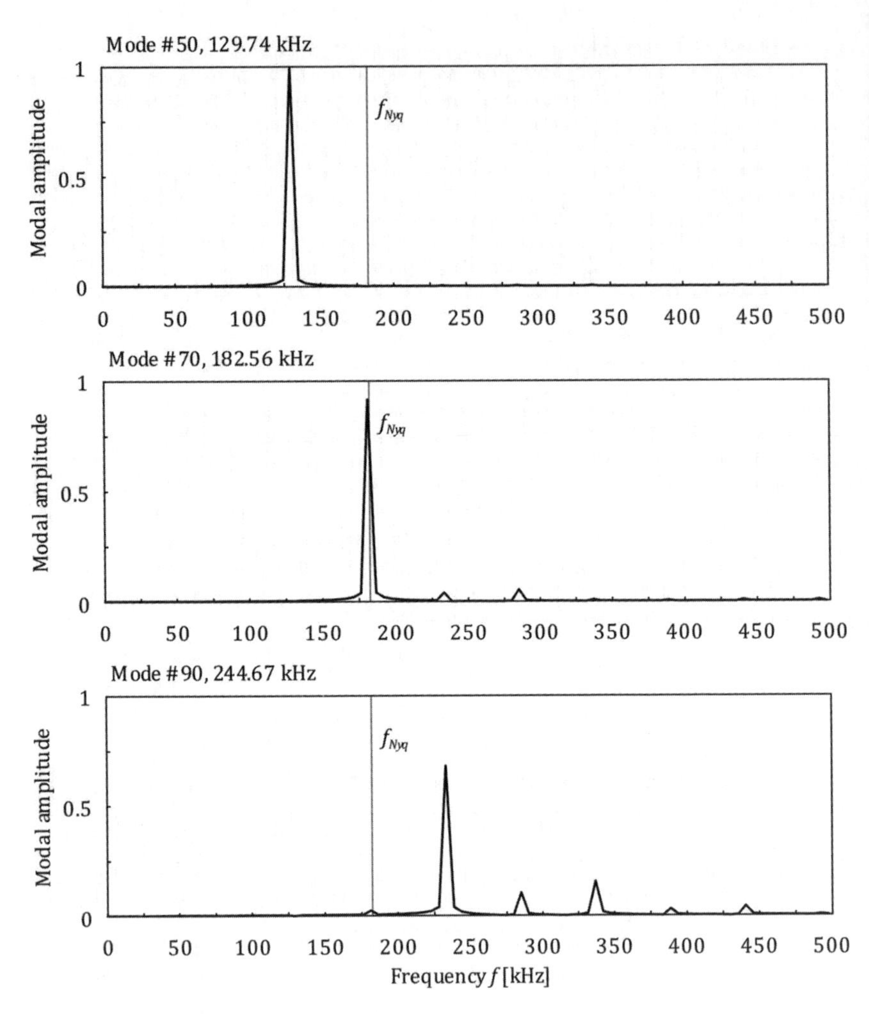

Figure 4.15 Spectra of selected modes of rod natural vibrations calculated numerically in the case of the Chebyshev node distribution and the elementary single-mode theory of rods

shown in Figure 4.16. In all these cases it has been assumed that the excitation $F(t)$, acting over time $t_1 = 1/f_m$ at the left free end of the rod, has the form of an eight-cycle sine pulse modulated by a Hann window. The carrier frequency f_c of the excitation signal $F(t)$ has been chosen as 100 kHz for the first case and then increased to 150 kHz and 200 kHz, respectively. As a consequence of the

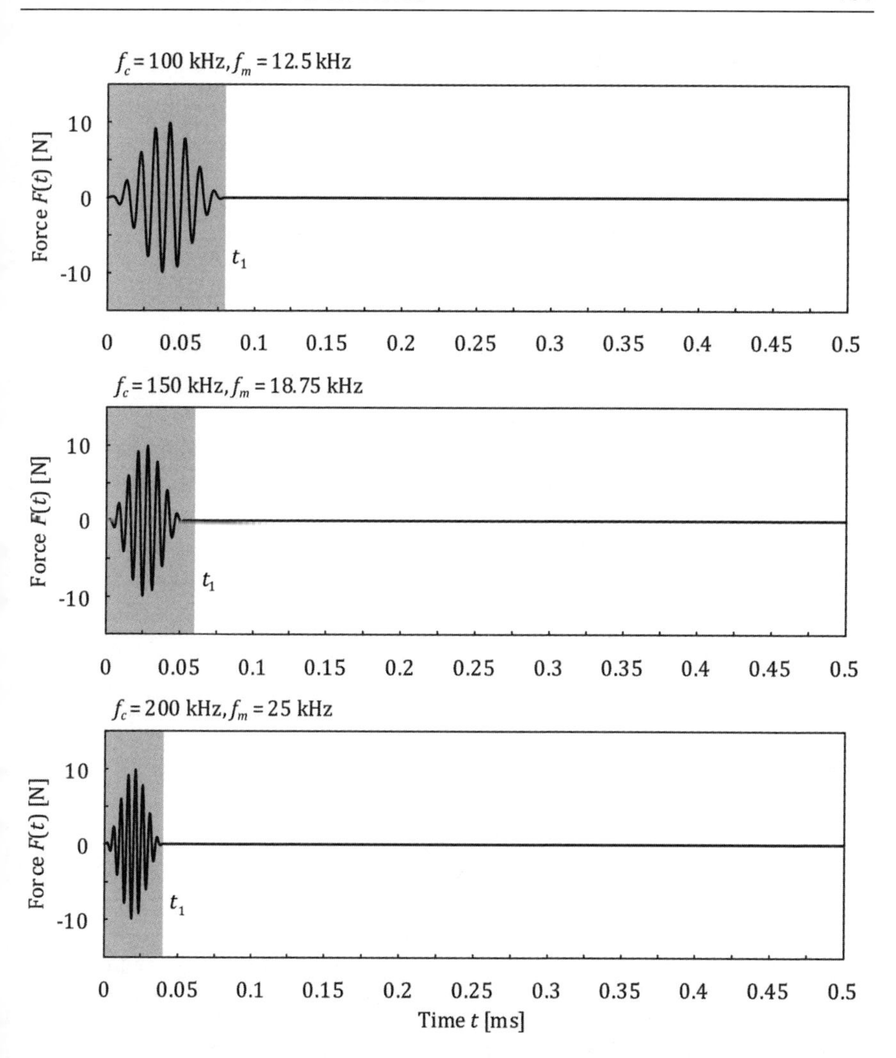

Figure 4.16 Excitation signals in the time domain used for numerical calculations of rod responses in the form of eight-cycle sine pulses modulated by a Hann window for carrier frequencies: (a) 100 kHz, (b) 150 kHz, (c) 200 kHz

fixed eight-cycle length of the excitation, the resulting modulation frequencies f_m are 12.5 kHz, 18.75 kHz and 25 kHz. Increasing the carrier frequency f_c results in the position and width of the excitation in the frequency domain, as shown in Figure 4.17, where $f_1 = f_c - 2f_m$ and $f_2 = f_c + 2f_m$. This, on the other hand, affects the range of the natural frequencies of the rod that are

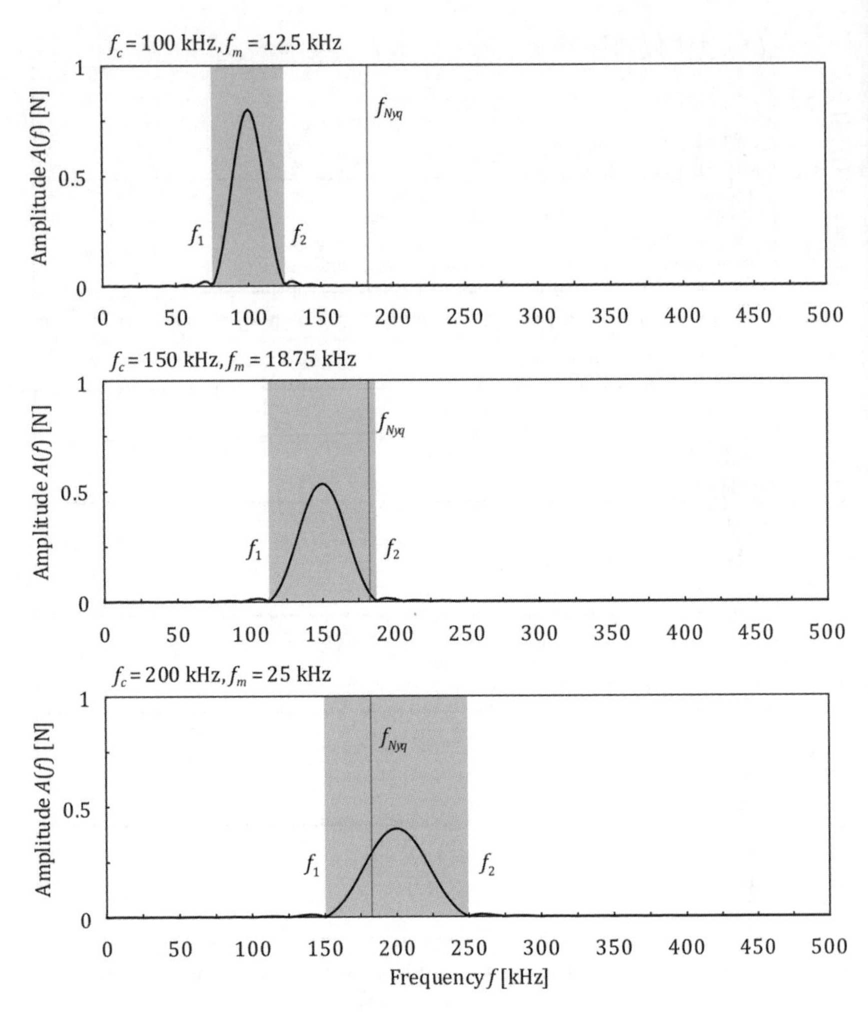

Figure 4.17 Spectra of the excitation signals in the frequency domain used for numerical calculations of rod responses in the form of eight-cycle sine pulses modulated by a Hann window for carrier frequencies: (a) 100 kHz, (b) 150 kHz, (c) 200 kHz

excited, influencing the rod response. This response $q(t)$, also calculated at the left free end of the rod as longitudinal displacements, is presented in the time and space domains in Figures 4.18 and 4.19, respectively. The equation of motion associated with the wave propagation problem has been solved numerically using the central difference method [27].

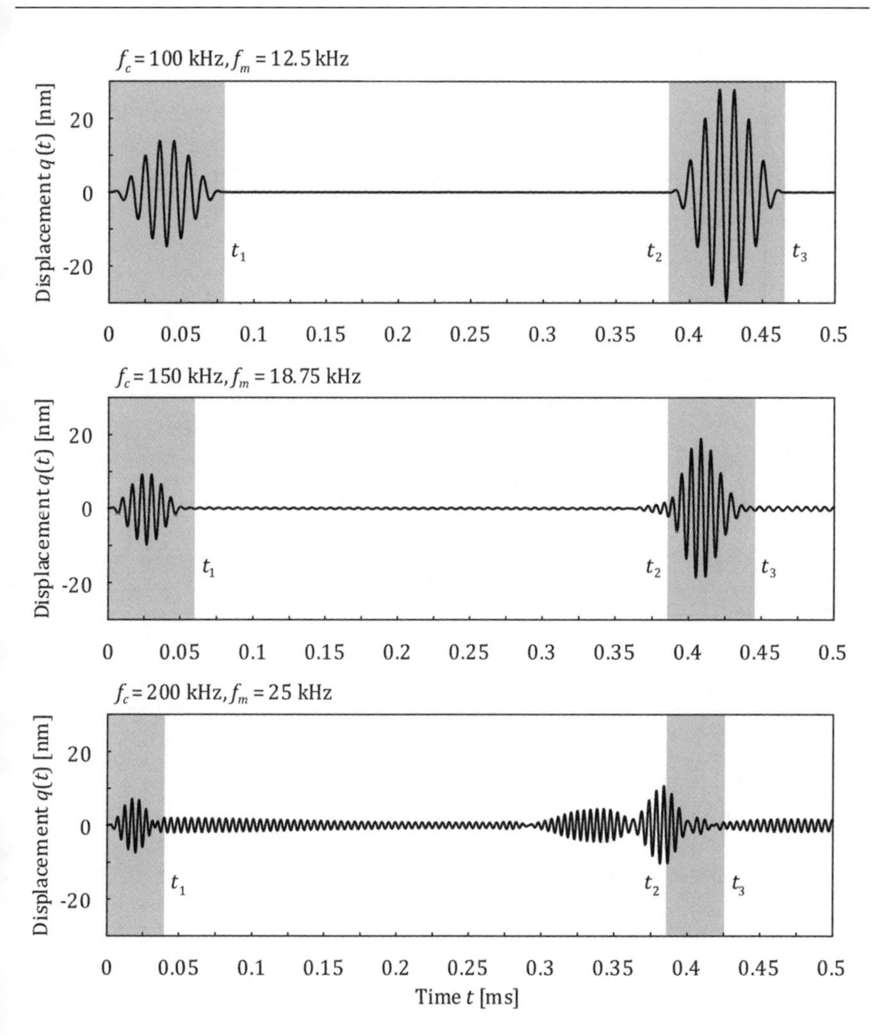

Figure 4.18 Rod responses in the time domain calculated numerically for carrier frequencies: (a) 100 kHz, (b) 150 kHz, (c) 200 kHz, in the case of the Chebyshev node distribution and the elementary single-mode theory of rods

The results of numerical calculations presented in Figures 4.18 and 4.19 show that the frequency band of excitation should always be closely correlated with the natural frequency spectrum of a given numerical model in order to avoid problems resulting from aliasing, as shown in both cases. When the carrier frequency of the excitation signal $f_c = 100$ kHz, it can be seen that

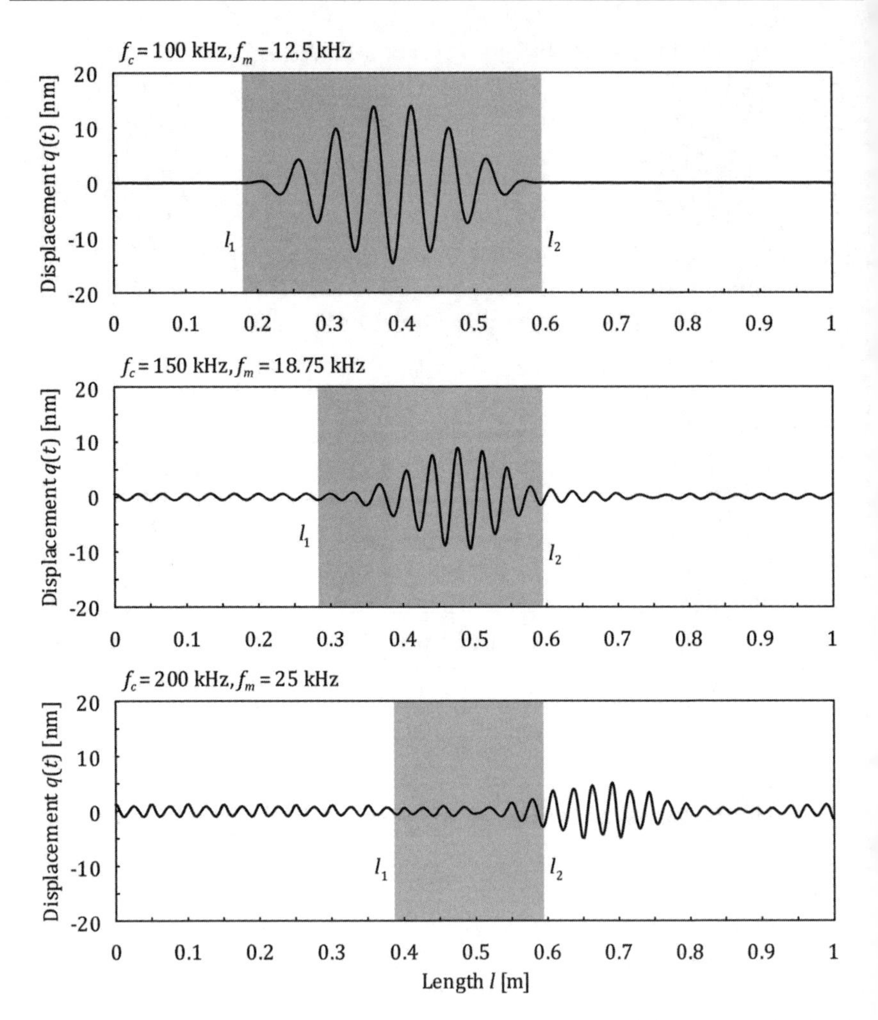

Figure 4.19 Rod responses in the space domain calculated numerically for carrier frequencies: (a) 100 kHz, (b) 150 kHz, (c) 200 kHz, in the case of the Chebyshev node distribution and the elementary single-mode theory of rods

the highest frequencies within the excitation frequency band shown in Figure 4.17 stay well below the Nyquist frequency $f_{Nyq} = 182.56$ kHz, that is the aliasing point, and within the first half of the natural frequency spectrum. As a consequence, the calculated response of the rod $q(t)$ is free of any distortions and agrees well with the analytical prediction concerning the signal time of flight or signal propagation distance. The analytically calculated time of flight t_2 required for the signal to travel from the excitation point (i.e. the left end) to the right end of the rod and back again is $t_2 = 2l/c_p$ =0.385 ms, whereas time t_3 is defined as $t_3 = 2l/c_p - t_1$ and depends on the excitation duration t_1. The signal propagation distance corresponding to the analysed time window Δt of 0.5 ms can be calculated as $l_3 = c_p \Delta t - 2l = 0.5945$ m, where the distance l_2 is shorter by the total spatial width of the excitation signal, that is $l_2 = c_p (\Delta t - t_1) - 2l$. However, when the carrier frequency of the excitation signal $f_c = 150$ kHz, the highest frequencies within the excitation frequency band are only slightly above the aliasing point that splits the natural frequency band into two halves at the Nyquist frequency f_{Nyq}, that is $f_2 > 182.56$ kHz. Small distortions of the calculated responses of the rod $q(t)$ are clearly visible in this case, but their highest amplitudes are still relatively small in comparison with the highest amplitudes of the calculated propagating signal. The numerically calculated times of flight t_1 and t_2 or distances l_1 and l_2 of signal arrivals are only slightly affected in comparison with their values calculated analytically. Finally, when the carrier frequency of the excitation signal $f_c = 200$ kHz, the excitation frequency band significantly exceeds the Nyquist frequency f_{Nyq}, which results in high distortions of the calculated signal response of the rod $q(t)$, that is $f_2 \gg 182.56$ kHz. In this case also the times of flight t_1 and t_2 resulting from the numerical simulation are significantly and visibly shortened.

4.7 Examples of Numerical Calculations

In this section the effectiveness of the spectral finite element method is demonstrated by a number of selected numerical examples that illustrate wave propagation phenomena in one-dimensional structural elements like rods and beams. The results of numerical simulations have also been compared with experimental measurements employing laser scanning vibrometry.

In a general case an aluminium structural rod element is considered, as shown in Figure 4.20, composed of material with the following properties: Young's modulus $E = 68.0$ GPa, Poisson's ratio $\nu = 0.33$, density $\rho = 2800$ kg/m^3. Both ends of the rod are assumed to be free, which corresponds to the free–free type of boundary conditions. The total length of the rod is l and its diameter is d.

The excitation signal $F(t)$ acting at a free end of the rod results in the propagation of elastic waves within the rod. As found previously, it has been assumed that the excitation $F(t)$, acting over time $t_1 = 1/f_m$ at the left free end of the rod, has the form of a sine pulse modulated by the Hann window

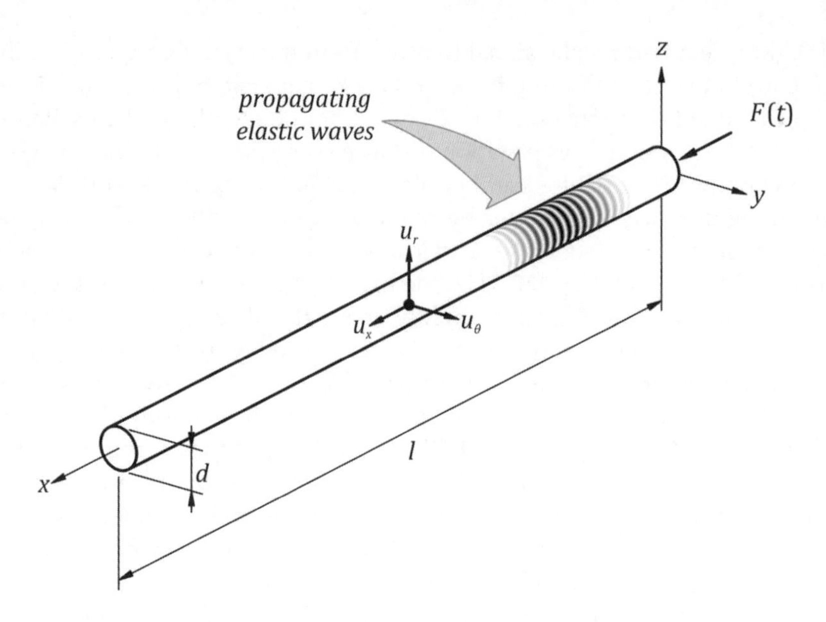

Figure 4.20 Geometry of an aluminium structural rod element

of the carrier frequency f_c. The carrier f_c and modulation f_m frequencies, as well as the total calculation time t_T could vary in each simulation following the total number and type of spectral finite elements employed.

4.7.1 Propagation of Longitudinal Elastic Waves in a Cracked Rod

It has been assumed for this simulation that the total length of the rod l is 2000 mm, while the rod diameter d is 10 mm. The carrier and modulation frequencies f_c and f_m of the excitation signal $F(t)$ of amplitude 1 N have been assumed as $f_c = 50$ kHz and $f_m = 10$ kHz, respectively, which corresponds to five cycles of the carrier signal. This is equivalent to the range of the frequency parameter fd from $(f_c - 2f_m)d$ to $(f_c + 2f_m)d$, that is from 0.3 to 0.7 MHz·mm, in the dispersion curve for velocity ratio c_g/c_p presented in Figure 4.2.

The calculation time t_T divided into 8000 time steps covers 2 ms in order to allow for multiple reflections of propagating waves from both ends of the rod. The result of numerical simulation for the case of a cracked aluminium rod, in the form of a two-dimensional map of the longitudinal displacement component u_x, is presented in Figure 4.21.

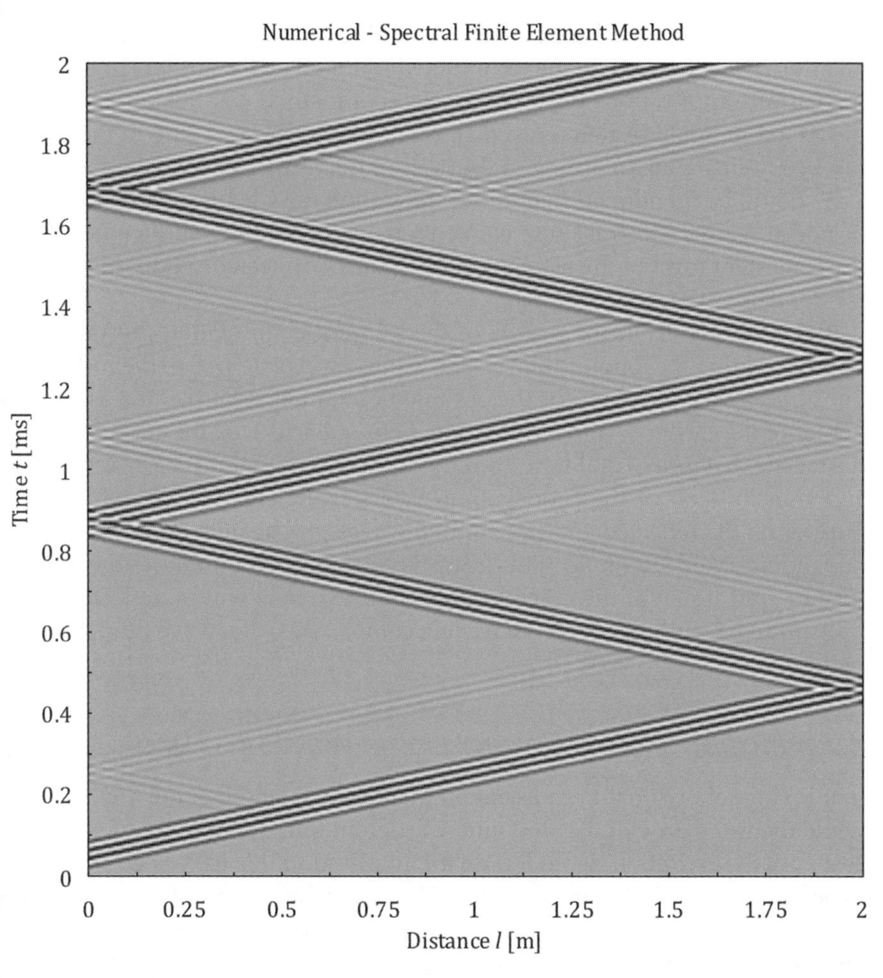

Figure 4.21 Two-dimensional map of the longitudinal displacement component u_x for the propagation of longitudinal elastic waves in a cracked aluminium rod according to the single-mode Love theory of rods – results of numerical simulation using the spectral finite element method

In total, 82 spectral finite elements following the single-mode Love rod theory have been used in this simulation (refer to Equation (4.11) for more details). A transverse and open crack located at a distance of 500 mm from the right free end of the rod has been considered; its depth has been assumed to be 2.5 mm, which constitutes 25 % of the rod diameter. The cracked section of the rod has been modelled using a special rod spectral finite element with an internal crack based on the technique developed by the authors in the past for various applications of the finite element method [28–31], an approach described in detail in Reference [32]. The rod stiffness loss due to the crack has been calculated according to the laws of fracture mechanics.

It should be mentioned that in the simulation the Chebyshev distribution of nodes within spectral finite elements has been employed, together with the consistent form of the element inertia matrix $\boldsymbol{M}^e$ (refer to Equation (4.11) for details).

It can be seen from the results of the numerical simulation presented in Figure 4.21 that the longitudinal elastic waves propagate within the rod without noticeable dispersion for the assumed excitation form (i.e. the assumed values of the carrier f_c and modulation f_m frequencies). At the crack location (i.e. at the distance of 500 mm from the right end of the rod) clear reflection and transmission of the propagating waves at the crack can be observed. Further on the reflected and transmitted waves propagate within the rod independently and after reflections from either of the rod boundaries the waves further split down at the crack location into two new waves, reflected and transmitted, resulting in ever-increasing complexity of the wave propagation pattern.

4.7.2 Propagation of Flexural Elastic Waves in a Rod

In this simulation it has been assumed that the total length of the rod l is 2 m, while the rod diameter d is 100 mm. Carrier and modulation frequencies f_c and f_m of the excitation signal $F(t)$ of the amplitude of 1 N have been assumed as before as $f_c = 50$ kHz and $f_m = 10$ kHz, respectively, which corresponds to five cycles of the carrier signal. This is equivalent to the range of the frequency parameter fd from $(f_c - 2f_m)d$ to $(f_c + 2f_m)d$, that is from 3 to 7 MHz·mm, in the dispersion curve for velocity ratio c_g/c_p presented in Figure 4.8.

The calculation time t_T divided into 12 000 time steps covers 2 ms in order to allow for multiple reflections of the propagating waves from both ends of the rod. The result of the numerical simulation, in the form of a two-dimensional map of the transverse displacement component, is presented in Figure 4.22

In total 80 spectral finite elements following the two-mode Timoshenko beam theory have been used in this simulation and the Lobatto distribution of nodes within the spectral finite elements has been employed (refer to Equation (4.27) for details).

It can be clearly seen from the results of the numerical simulation presented in Figure 4.22 that flexural elastic waves propagate within the rod

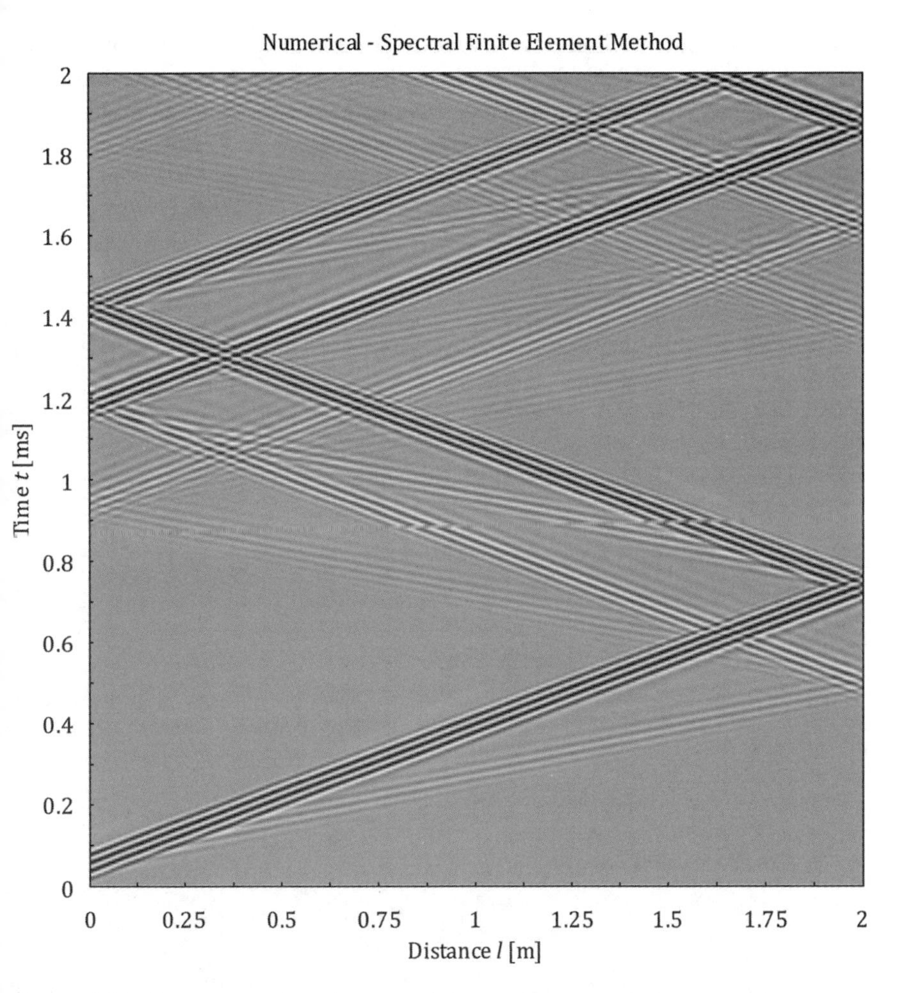

Figure 4.22 Two-dimensional map of the transverse displacement component u_r for the propagation of flexural elastic waves in an aluminium rod according to the two-mode Timoshenko beam theory – results of numerical simulation using the spectral finite element method

without noticeable dispersion for the assumed excitation form (i.e. the assumed values of the carrier frequency f_c and modulation frequency f_m) as two independent wave modes: faster and slower. This is a direct consequence of the dimensional and excitation characteristics (refer to Figure 4.8 for more details). Each time, either of the modes is reflected from the boundaries as two distinct modes of precisely determined amplitudes. This phenomenon is known in the subject literature as mode conversion. It is noteworthy that when the faster mode reflects from the boundaries, it converts into two modes (slower and faster), and the slower mode amplitude is higher than that of the incident mode, so energy of the incident and reflected waves is conserved.

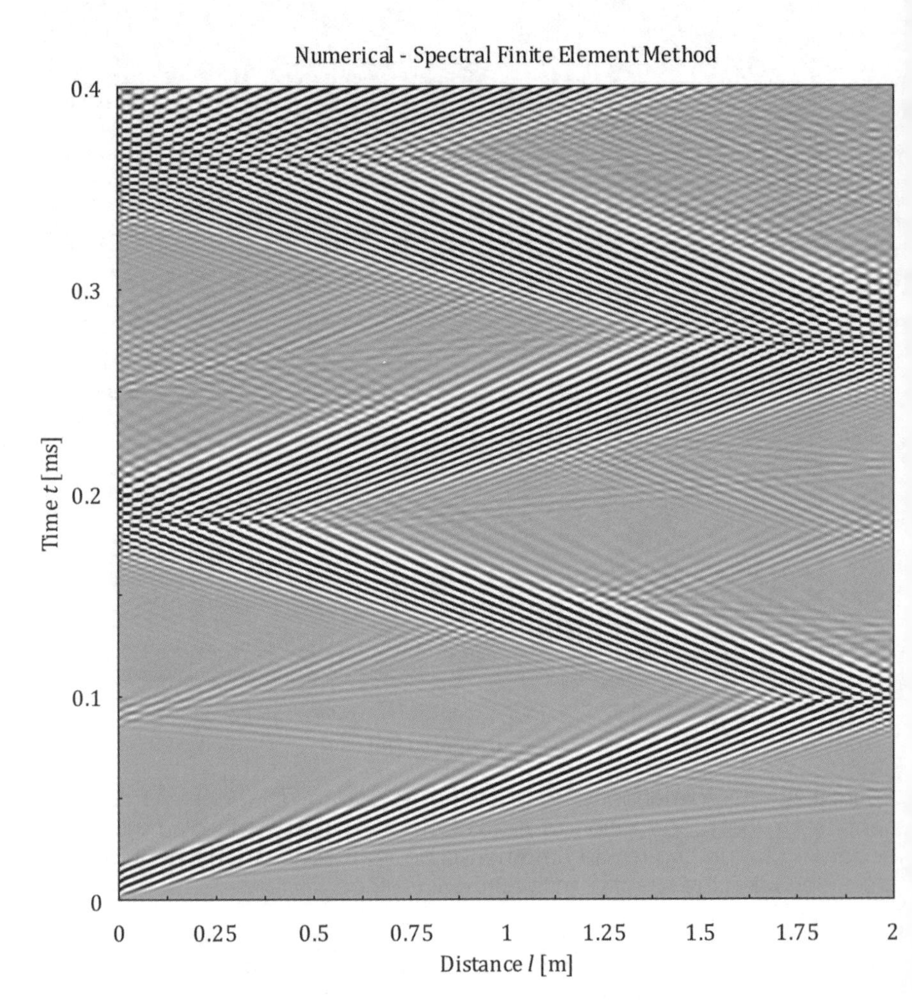

Figure 4.23 Two-dimensional map of the transverse velocity component $\dot{u}_r$ for the propagation of longitudinal and flexural elastic waves in an aluminium rod according to the two-mode Mindlin–Herrmann rod theory and the two-mode Timoshenko beam theory – results of numerical simulation using the spectral finite element method

It is worth remembering that in the current case of the propagation of flexural elastic waves within a rod each reflection of waves from the boundaries doubles the total number of propagating signals. After several such reflections and additionally in the presence of any discontinuities within the rod (i.e. crack, corrosion, change of cross-section, etc.) a rapid increase in complexity of the wave propagation pattern can be observed.

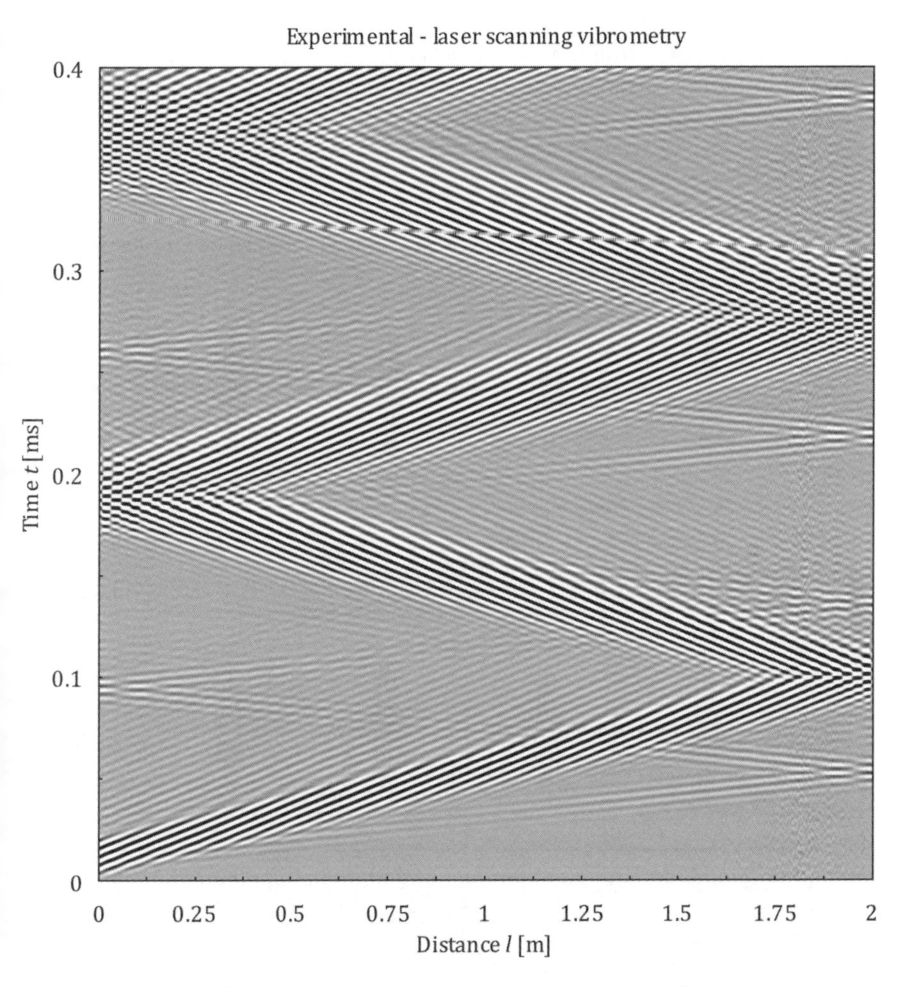

Figure 4.24 Two-dimensional map of the transverse velocity component $\dot{u}_r$ for the propagation of longitudinal and flexural elastic waves in an aluminium rod – results of experimental measurements by laser scanning vibrometry

4.7.3 Propagation of Coupled Longitudinal and Flexural Elastic Waves in a Rod

In this case the results of numerical simulation employing the spectral finite element method have been compared with results of experimental measurements using laser scanning vibrometry. The comparison concerned wave propagation patterns for the transverse velocity component $\dot{u}_r$, as well as the longitudinal velocity component $\dot{u}_x$ sampled at the free end of the rod at two points – the neutral axis of the rod and close to the rod circumference. Additionally an experimental wave propagation map has been recorded by laser scanning measurements in 665 points located along the rod length and at 4096 time instances.

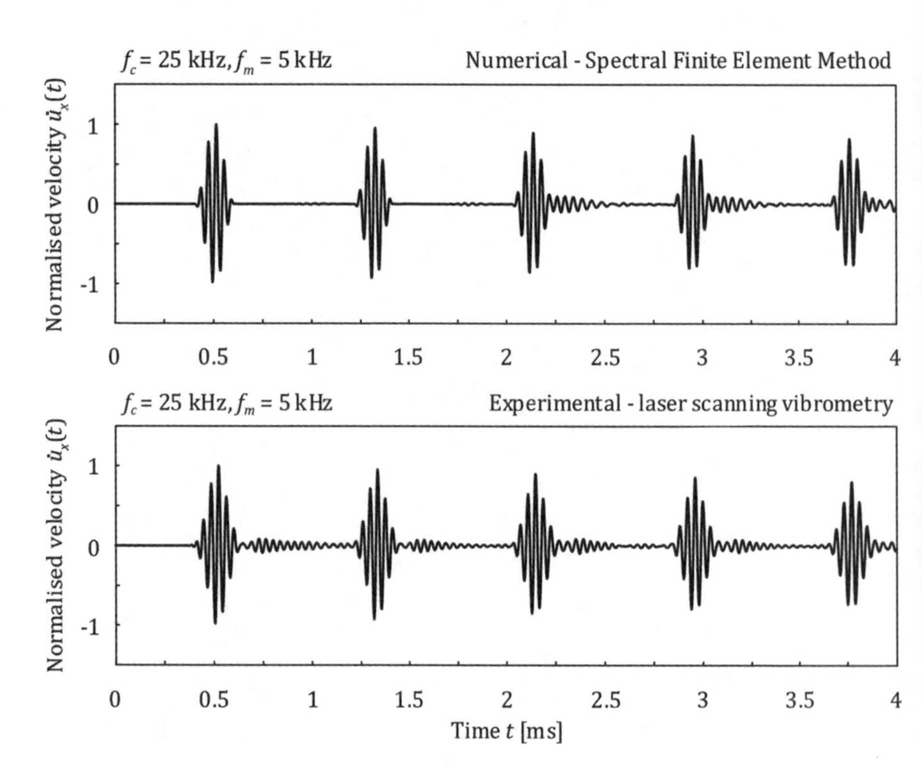

Figure 4.25 Time signals of the longitudinal velocity components $\dot{u}_x$ at the free end of the rod at its neutral axis for propagation of longitudinal elastic waves – results of numerical simulation using the spectral finite element method and experimental measurements by laser scanning vibrometry

In this simulation it has been assumed that the total length of the rod l is 2 m, while the rod diameter d is 10 mm. The carrier and modulation frequencies f_c and f_m of the excitation signal $F(t)$ remain unchanged as $f_c = 25$ kHz and $f_m = 5$ kHz, respectively, which corresponds to five cycles of the carrier signal. This corresponds to a range of the frequency parameter fd from $(f_c - 2f_m)\,d$ to $(f_c + 2f_m)\,d$, that is from 0.15 to 0.35 MHz·mm, in the dispersion curves for the velocity ratio c_g/c_p presented in Figures 4.3 and 4.8.

The calculation time t_T divided into 12 000 time steps covers 4 ms in order to allow for multiple reflections of the propagating waves from both ends of the rod.

In total, 80 spectral finite elements following two-mode theories of rod and beam behaviour by Mindlin–Herrmann and Timoshenko have been used in

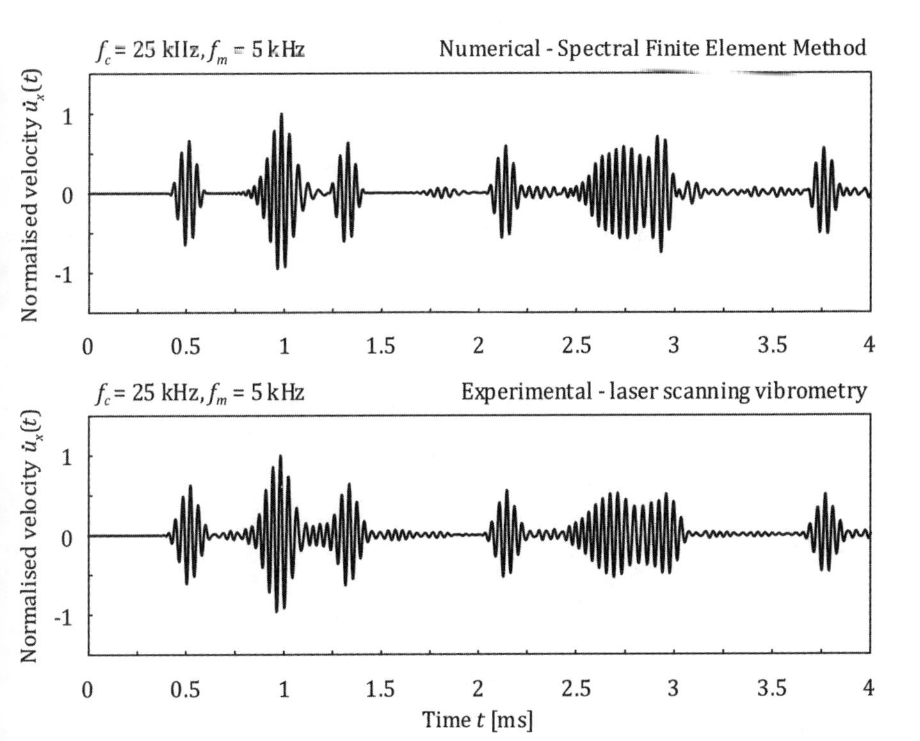

Figure 4.26 Time signals of the longitudinal velocity components $\dot{u}_x$ at the free end of the rod close to the rod circumference for the propagation of longitudinal and flexural elastic waves – results of numerical simulation using the spectral finite element method and experimental measurements by laser scanning vibrometry

this simulation (refer to Equations (4.12) and (4.27) for more details). The Lobatto distribution of nodes within spectral finite elements has been employed. Values of unknown parameters related to geometrical coupling due to the presence of a piezoelectric transducer at one end of the rod, longitudinal and flexural components of excitation, together with material damping for the frequency range under investigation, have been estimated experimentally. In the numerical model damping proportional only to the inertia matrix $\boldsymbol{M}^e$ has been considered with the value of the parameter $\alpha = 100$.

A two-dimensional map of the transverse displacement component u_r calculated numerically is presented in Figure 4.23 and the corresponding map obtained from laser scanning measurements is presented in Figure 4.24. Both maps only present a 0.4 ms long part of the total time response. The reason is strong dispersion of the propagating waves that become unreadable after only several reflections from the rod boundaries.

It can be clearly seen from Figures 4.23 and 4.24, as well as in Figures 4.25 and 4.26, that very good agreement has been achieved between the results of numerical simulations employing the spectral finite element method and laser scanning vibrometry. This is true for maps of the transverse velocity component $\dot{u}_r$ as well as time signals of the longitudinal velocity component $\dot{u}_x$. The dispersion of propagating waves observed in the experimental signals is slightly higher than in the case of numerical ones since the numerical model used by the spectral finite element method takes into account the influence of the bond between the piezoelectric transducer and the rod [33] in a simplified manner. However, in all cases the additional signal features observed, such as coupling between longitudinal and flexural wave modes through the piezoelectric transducers, mode conversion or the distribution of flexural waves across the rod thickness, have been captured very well.

References

1. Rose, J.L. (1999) *Ultrasonic Waves in Solid Media*, Cambridge University Press, Cambridge.
2. Doyle, J.F. (1997) *Wave Propagation in Structures*, Springer-Verlag New York Inc., New York.
3. Achenbach, J.D (1973) *Wave Propagation in Elastic Solids*, North-Holland Publishing Company, Amsterdam.
4. Love, A.E. (1927) *A Treatise on the Mathematical Theory of Elasticity*, Dover Publications.
5. Mindlin, R.D. and Herrmann, G. (1952) A one-dimensional theory of compressional waves in an elastic rod. *Proceedings of the First U.S. National Congress of Applied Mechanics – 1951*, pp. 187–191.

6. Rayleigh, J.W.S. (1945) *The Theory of Sound*, Dover Publications, Inc.
7. Timoshenko, S.P. (1921) On the correction factor for shear of the differential equation for transverse vibrations of bars of uniform cross-section. *Philosophical Magazine*, **41**, 744–746.
8. Ochoa, O.O. and Reddy, J.N. (1992) *Finite Element Analysis of Composite Laminates*, Kluwer Academic Publishers, Dordrecht.
9. Wolfram Research: Mathematica, Technical and Scientific Software (2011) Available at: http://www.wolfram.com (accessed 30.06.2011).
10. MathWorks – MATLAB and Simulink for Technical Computing (2011) Available at: http://www.mathworks.com (accessed 30.06.2011).
11. Kudela, P., Krawczuk, M. and Ostachowicz W. (2007) Wave propagation modelling in 1D structures using spectral finite elements. *Journal of Sound and Vibration*, **300**, 88–100.
12. Żak, A., Krawczuk, M. and Ostachowicz, W. (2006) Propagation of in-plane waves in an isotropic panel with a crack. *Finite Elements in Analysis and Design*, **42**, 929–941.
13. Żak, A., Krawczuk, M. and Ostachowicz, W. (2006) Propagation of in-plane waves in a composite panel. *Finite Elements in Analysis and Design*, **43**, 154–154.
14. Żak, A. and Krawczuk, M. (2010) Assessment of rod behaviour theories used in spectral finite element modelling. *Journal of Sound and Vibration*, **329**, 2099–2113.
15. Rao, S.S. (1981) *The Finite Element Method in Engineering*, Pergamon Press.
16. Zienkiewicz, O.C. (1989) *The Finite Element Method*, McGraw-Hill, Inc.
17. Dauksher, W. and Emery, A.F. (1997) Accuracy in modelling the acoustic wave equation with Chebyshev spectral finite elements. *Finite Elements in Analysis and Design*, **26**, 115–128.
18. Christon, M.A. (1999) The influence of the mass matrix on the dispersive nature of the semi-discrete, second-order wave equation. *Computer Methods in Applied Mechanics and Engineering*, **173**, 147–166.
19. Wu, S.R. (2006) Lumped mass matrix in explicit finite element method for transient dynamics of elasticity. *Computer Methods in Applied Mechanical Engineering*, **195**, 5983–5994.
20. Żak, A. (2009) A novel formulation of a spectral plate element for wave propagation in isotropic structures. *Finite Elements in Analysis and Design*, **45**, 650–658.
21. Pozrikidis, C. (2005) *Introduction to Finite and Spectral Element Methods Using MATLAB*, Chapman & Hall/CRC.
22. Saruna, K.S. (1978) Lumped mass matrices with non-zero inertia for general shell and axisymmetric shell elements. *International Journal for Numerical Methods in Engineering*, **12**, 1635–1650.
23. Zboiński, G. and Kubiak J.A. (1989) Application of the thick shell and transition elements for analysis of the long turbine blades: Part I – an algorithm. *Latest Advances in Steam Turbine Design, Blading, Repairs, Condition Assessment, and Condenser Interaction*, **7**, 23–30.
24. Ralston A. (1965) *A First Course of Numerical Analysis*, McGraw-Hill, Inc.

25. Blevins, R.D. (1995) *Formulas for Natural Frequency and Mode Shape*, Krieger Publishing Company.
26. Shin, K. and Hammond, J.K. (2008) *Fundamentals of Signal Processing for Sound and Vibration Engineers*, John Wiley & Sons, Ltd.
27. Kleiber, M. (1989) *Incremental Finite Element Modelling in Non-linear Solid Mechanics*, E. Horwood.
28. Krawczuk, M. (1992) Modeling and identification of cracks in truss constructions. *Finite Elements in Analysis and Design*, **12**, 41–50.
29. Krawczuk, M. and Ostachowicz, W. (1995) Modeling and vibration analysis of a cantilever composite beam with a transverse open crack. *Journal of Sound and Vibration*, **183**, 69–89.
30. Krawczuk, M., Ostachowicz, W. and Żak, A. (1997) Dynamics of cracked composite material structures. *Computational Mechanics*, **20**, 79–83.
31. Krawczuk, M. and Ostachowicz, W. (1997) Natural vibrations of a clamped–clamped arch with an open transverse crack. *Trans. ASME, Journal of Vibration and Acoustics*, **119**, 145–151.
32. Ostachowicz, W. and Krawczuk, M. (2009) Modeling for detection of degraded zones in metallic and composite structures, in *Encyclopaedia of Structural Health Monitoring* (eds C. Boller, F.-K. Chang and Y. Fujino), John Wiley & Sons, Ltd, pp. 851–866.
33. Lee, B.C., Palacz, M., Krawczuk, M. *et al.* (2004) Wave propagation in a sensor/actuator diffusion bond model. *Journal of Sound and Vibration*, **276**, 671–687.

5

Two-Dimensional Structural Elements

5.1 Theories of Membranes, Plates and Shells

As discussed earlier for the case of structural rod elements, the theories concerning the propagation of elastic waves in two- or three-dimensional structural membrane, plate and shell elements that are widely employed in the literature can be generally classified as one-mode, two-mode, three-mode and higher-mode or higher order theories. A thorough analysis of a general three-dimensional displacement field of structural three-dimensional elements can give rise to a number of theories based on different displacement fields. Using the Cartesian coordinates (x, y, z), the general form of the displacement field associated with a three-dimensional structural element can be described formally by the vector $\boldsymbol{u} = [u_x, u_y, u_z]$ of the following form [1]:

$$\begin{cases} u_x(x, y, z) = F_1(x, y, z) \\ u_y(x, y, z) = F_2(x, y, z) \\ u_z(x, y, z) = F_3(x, y, z) \end{cases} \tag{5.1}$$

where $u_x(x, y, z)$, $u_y(x, y, z)$ and $u_z(x, y, z)$ can be thought of as representing in this case two in-plane and one transverse displacement components, respectively, while $F_1(x, y, z)$, $F_2(x, y, z)$ and $F_3(x, y, z)$ are certain displacement functions dependent only on the spatial coordinates x, y and z.

Guided Waves in Structures for SHM: The Time-Domain Spectral Element Method, First Edition.
Wieslaw Ostachowicz, Pawel Kudela, Marek Krawczuk and Arkadiusz Zak.

For example, expansion of the function $F_1(x, y, z)$ around $z = 0$ into a Maclaurin series leads to the following equation:

$$F_1(x, y, z) = F_1(x, y, 0) + \sum_{n=1}^{\infty} \frac{\partial^n F_1(x, y, 0)}{\partial z^n} \frac{z^n}{n!} \tag{5.2}$$

As before, it can be noticed at this point that in the case of the two in-plane displacement component functions $F_1(x, y, z)$ and $F_2(x, y, z)$ the terms proportional to odd values of n are associated with antisymmetric behaviour and the propagation of flexural waves and antisymmetric shear waves, while the terms proportional to even values of n are associated with symmetric behaviour and the propagation of longitudinal waves and symmetric shear waves. Repeating the same expansion into a Maclaurin series for the function $F_3(x, y, z)$ leads to the opposite conclusions [1–3].

The number of terms to be kept from the series given by (5.2) depends on the investigated phenomena and is directly related to the total number of degrees of freedom of any finite element approximation based on the series expansion. The expansion of the function $F_1(x, y, z)$ at $n = 2$ yields a Maclaurin series of the following form:

$$F_1(x, y, z) = F_1(x, y, 0) + \frac{\partial F_1(x, y, 0)}{\partial z} r + \frac{1}{2} \frac{\partial^2 F_1(x, y, 0)}{\partial z^2} z^2 + E(z^3) \tag{5.3}$$

where $E(z^3)$ represents the truncation error of the expansion proportional to z^3. Again at this point a step towards a spectral finite element approximation can be made and then Equation (5.3) can be rewritten as:

$$F_1(x, y, z) = \varphi_0(x, y) + \varphi_1(x, y)z + \varphi_2(x, y)z^2 \tag{5.4}$$

where $\varphi_i(x, y)(i = 0, 1, 2)$ may be interpreted as denoting the degrees of freedom of a spectral finite element associated with a Maclaurin expansion of the in-plane displacement component function $F_1(x, y, z)$ of the three-dimensional displacement field.

It is obvious that due to the truncation of the series (5.3) the obtained formula (5.4) is not exact and represents the three-dimensional displacement field of a structural element in an approximate sense only. However, it should be emphasised that solving most static or dynamic problems of engineering significance efficiently requires using spectral finite elements that employ only the first two or three terms of the relevant Maclaurin series. In the case considered above one can immediately note that:

$$\begin{aligned}\varphi_0(x,y) &= F_1(x,y,0)\\ \varphi_1(x,y) &= \frac{\partial F_1(x,y,0)}{\partial z}\\ \varphi_2(x,y) &= \frac{1}{2}\frac{\partial^2 F_1(x,y,0)}{\partial z^2}\end{aligned} \tag{5.5}$$

Wave propagation is associated with coupled interaction of shear and extensional waves propagating within a structure with lateral boundaries. As a result of this coupled interaction, the propagation of various modes of elastic waves can be observed. Appropriate representation of these modes in a broad range of wave propagation frequencies requires more terms of the Maclaurin series in order to capture the complexity of the interaction phenomena. For that reason special new types of spectral finite elements have been developed [4–7].

5.2 Displacement Fields of Structural Membrane Elements

As mentioned earlier, the propagation of elastic longitudinal and shear waves in two-dimensional structural membrane elements is associated with their symmetric behaviour. Therefore, on the basis of Equations (5.1) and of the considerations specified below, the general form of the displacement field of a two-dimensional membrane spectral finite element for analysing the propagation of elastic waves can be written down in the following form:

$$\begin{cases} u_x(x,y,z) = \varphi_0(x,y) + \varphi_2(x,y)z^2 + \varphi_4(x,y)z^4\\ u_y(x,y,z) = \psi_0(x,y) + \psi_2(x,y)z^2 + \psi_4(x,y)z^4\\ u_z(x,y,z) = \theta_1(x,y)z + \theta_3(x,y)z^3 + \theta_5(x,y)z^5\end{cases} \tag{5.6}$$

where only nine terms of Maclaurin series expansions of the displacement components $u_x(x,y,z)$, $u_y(x,y,z)$ and $u_z(x,y,z)$ are used.

The functions $\varphi_i(x,y)(i=0,2,4)$, $\psi_i(x,y)(i=0,2,4)$ and $\theta_i(x,y)(i=1,3,5)$ defined in Equations (5.6) represent the independent nodal variables or degrees of freedom of the membrane spectral finite element. It can be seen that in the current formulation the membrane element has as many as nine degrees of freedom in a single node. This number of independent nodal variables may be reduced, however, by taking into account the zero traction condition, rewritten here:

$$\begin{aligned}&\sigma_{zz}(x, y, z) = \tau_{yz}(x, y, z) = \tau_{zx}(x, y, z) = 0\\&0 \le x \le l, \quad 0 \le y \le b, \quad z = \pm a = \pm\frac{h}{2}\end{aligned} \tag{5.7}$$

where l and b now denote the length and width, respectively, while h is the thickness of the membrane spectral finite element.

Starting from the form of the displacement field given by Equations (5.6), displacement fields for various one-mode, two-mode and other multi-mode membrane theories can be constructed. Additionally, using Equation (5.7) representing the zero traction conditions on the lateral boundaries of the membrane element allows one to enrich the displacement fields with some additional higher order terms. However, in most cases the resulting system of three differential equations is very complicated and cannot be solved analytically. This problem can be avoided by simple mathematical manipulations (scaling and substitution), thanks to which the zeroth-order terms of all displacement components $u_x(x, y, z)$, $u_y(x, y, z)$ and $u_z(x, y, z)$ can be represented as sums of other-order terms. In the present case this condition takes the following form:

$$\begin{aligned}&\tilde{\varphi}_0(x, y) = \varphi_0(x, y) - \tilde{\varphi}_2(x, y) - \tilde{\varphi}_4(x, y)\\&\tilde{\psi}_0(x, y) = \tilde{\psi}_0(x, y) - \tilde{\psi}_2(x, y) - \tilde{\psi}_4(x, y)\\&\tilde{\theta}_1(x, y) = \tilde{\theta}_1(x, y) - \tilde{\theta}_3(x, y) - \tilde{\theta}_5(x, y)\\&\tilde{\varphi}_2(x, y) = -a^2\varphi_2(x, y), \quad \tilde{\varphi}_4(x, y) = -a^4\varphi_4(x, y)\\&\tilde{\psi}_2(x, y) = -a^2\psi_2(x, y), \quad \tilde{\psi}_4(x, y) = -a^4\psi_4(x, y)\\&\tilde{\theta}_3(x, y) = -a^2\theta_3(x, y), \quad \tilde{\theta}_5(x, y) = -a^4\theta_5(x, y)\end{aligned} \tag{5.8}$$

Taking into account Equations (5.8), a new form of the displacement field of a two-dimensional membrane spectral finite element needed to analyse the propagation of elastic waves can be expressed as follows (for clarity and simplicity of presentation the tilde signs have been omitted hereinafter):

$$\begin{cases}u_x(x, y, z) = \varphi_0(x, y) + \varphi_2(x, y)\left[1 - \left(\frac{z}{a}\right)^2\right] + \varphi_4(x, y)\left[1 - \left(\frac{z}{a}\right)^4\right]\\u_y(x, y, z) = \psi_0(x, y) + \psi_2(x, y)\left[1 - \left(\frac{z}{a}\right)^2\right] + \psi_4(x, y)\left[1 - \left(\frac{z}{a}\right)^4\right]\\u_z(x, y, z) = \theta_1(x, y)z + \theta_3(x, y)\left[1 - \left(\frac{z}{a}\right)^2\right]z + \theta_5(x, y)\left[1 - \left(\frac{z}{a}\right)^4\right]z\end{cases} \tag{5.9}$$

Specific theories of membrane symmetric behaviour can be easily obtained starting from Equations (5.7) and (5.9). Different membrane theories can be associated with different forms of the functions $\varphi_i(x, y)(i = 0, 2, 4)$, $\psi_i(x, y)(i = 0, 2, 4)$ and $\theta_i(x, y)(i = 1, 3, 5)$. As an example, the dispersion curves for the elementary two-mode theory, the higher order two-mode theory, the three-mode theory, the higher order three-mode theory, as well as the six-mode theory, are presented in Figures 5.1 to 5.5. Their displacement fields can be presented conveniently in the following manner:

- *Elementary two-mode theory:*

$$\begin{cases} u_x(x, y, z) = \varphi_0(x, y) \\ u_y(x, y, z) = \psi_0(x, y) \end{cases}$$
$$\varphi_i(x, y) = \psi_i(x, y) = \theta_j(x, y) = 0, \quad i = 2, 4, \quad j = 1, 3, 5 \tag{5.10}$$

- *Higher order two-mode theory:*

$$\begin{cases} u_x(x, y, z) = \varphi_0(x, y) \\ u_y(x, y, z) = \psi_0(x, y) \end{cases}$$
$$\dot{u}_z(x, y, z) = -\nu \frac{\mathrm{d}}{\mathrm{d}x}\left[\dot{\varphi}_0(x) + \dot{\psi}_0(x)\right] z \tag{5.11}$$
$$\varphi_i(x, y) = \psi_i(x, y) = \theta_j(x, y) = 0, \quad i = 2, 4, \quad j = 1, 3, 5$$

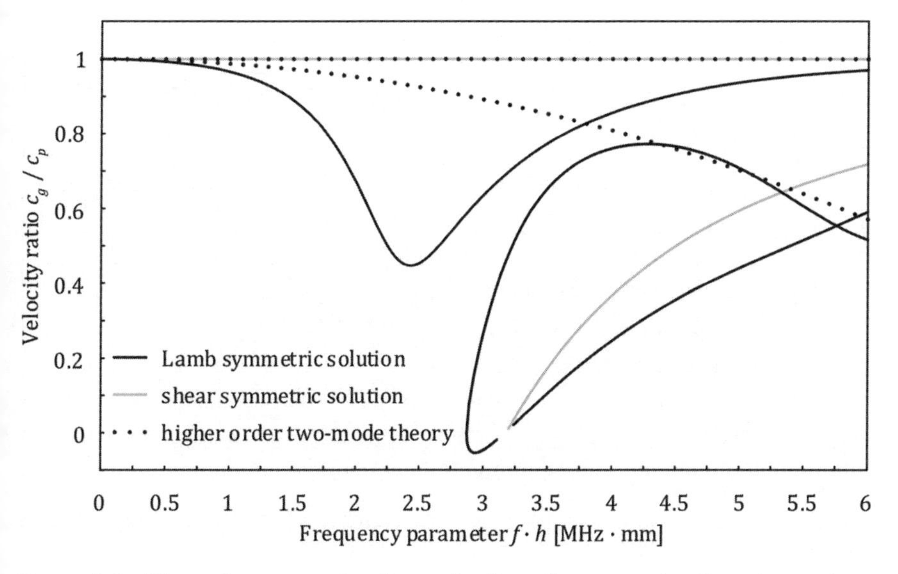

Figure 5.1 Dispersion curve for the velocity ratio c_g/c_p for the elementary two-mode theory of membranes ($c_l = 6.3$ km/s, $c_s = 3.2$ km/s)

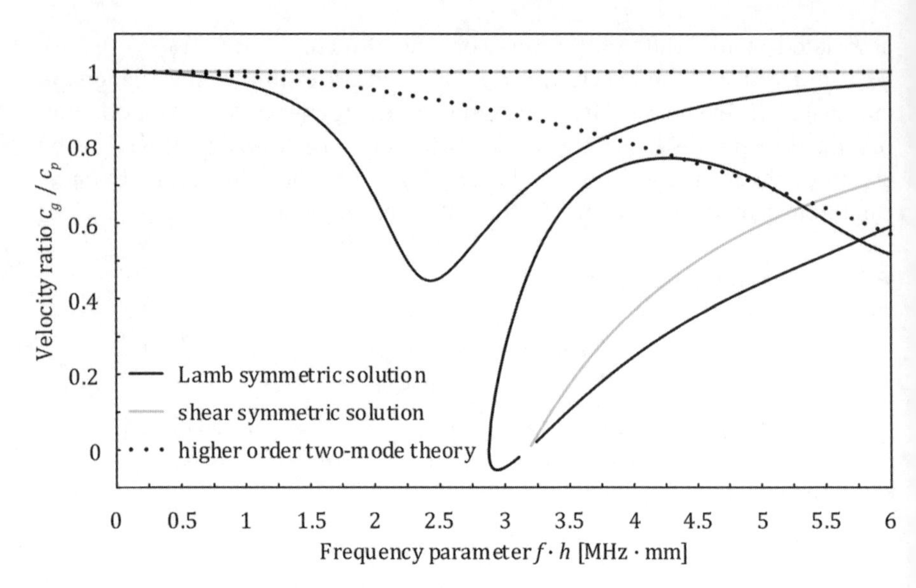

Figure 5.2 Dispersion curve for the velocity ratio c_g/c_p for the higher order two-mode theory of membranes ($c_l = 6.3$ km/s, $c_s = 3.2$ km/s)

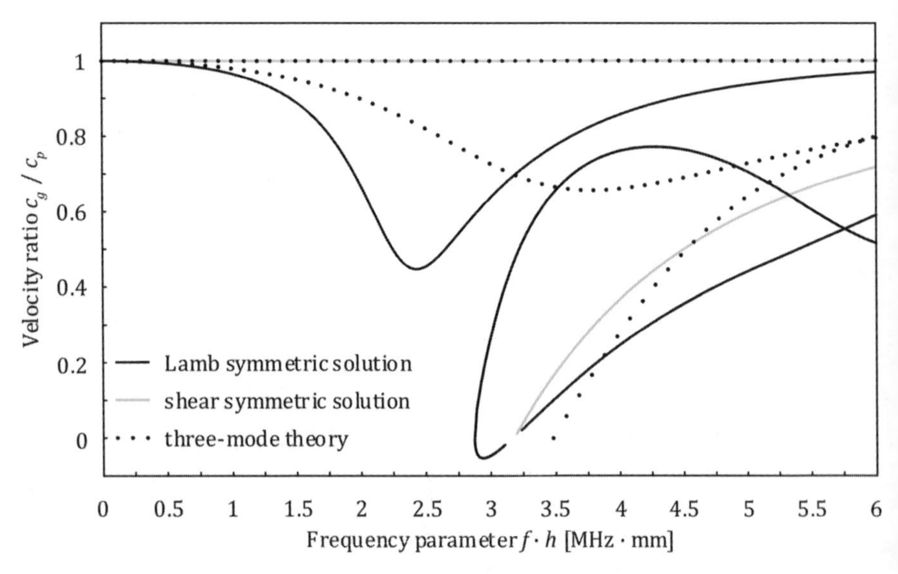

Figure 5.3 Dispersion curve for the velocity ratio c_g/c_p for the three-mode theory of membranes ($c_l = 6.3$ km/s, $c_s = 3.2$ km/s)

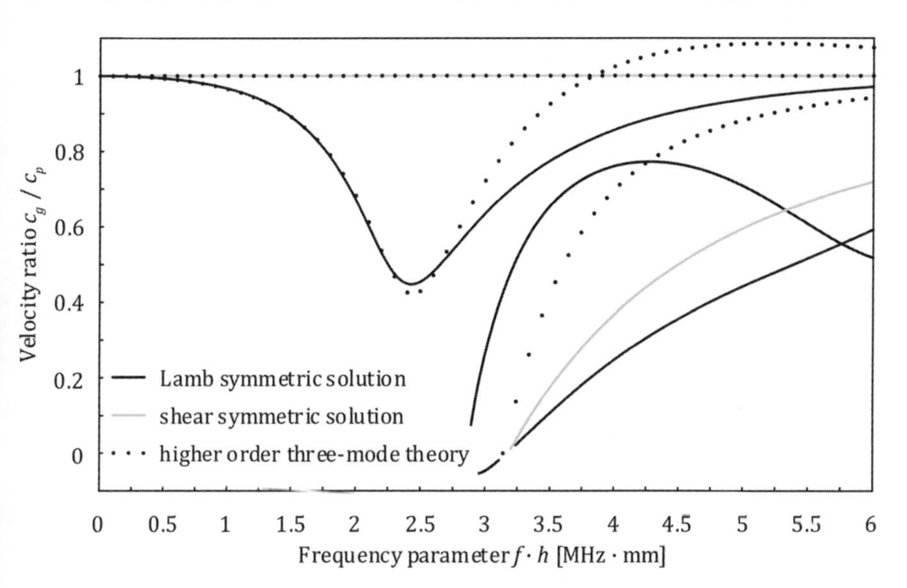

Figure 5.4 Dispersion curve for the velocity ratio c_g/c_p for the higher order three-mode theory of membranes ($c_l = 6.3$ km/s, $c_s = 3.2$ km/s)

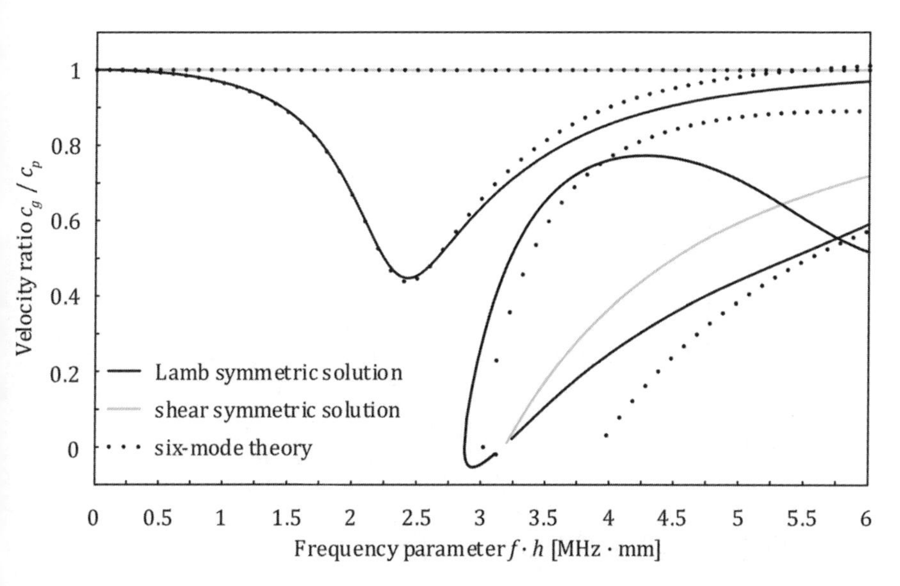

Figure 5.5 Dispersion curve for the velocity ratio c_g/c_p for the six-mode theory of membranes ($c_l = 6.3$ km/s, $c_s = 3.2$ km/s)

obtained after adding an additional equation representing the assumption regarding coupling between the in-plane velocities $\dot{u}_x$ and $\dot{u}_y$ and the transverse velocity $\dot{u}_z$ through the Poisson's ratio effect $\dot{\varepsilon}_{zz} = -\nu(\dot{\varepsilon}_{xx} + \dot{\varepsilon}_{yy})$ influencing rod kinetic energy (refer to the single-mode Love theory of rods [8]):

$$\dot{u}_z(x, y, z) = -\nu \frac{\mathrm{d}}{\mathrm{d}x}\left[\dot{\varphi}_0(x) + \dot{\psi}_0(x)\right] z$$

- *Three-mode theory:*

$$\begin{cases} u_x(x, y, z) = \varphi_0(x, y) \\ u_y(x, y, z) = \psi_0(x, y) \\ u_z(x, y, z) = \theta_1(x, y) z \end{cases}$$
$$\varphi_i(x, y) = \psi_i(x, y) = \theta_j(x, y) = 0, \quad i = 2, 4, \quad j = 3, 5 \tag{5.12}$$

- *Higher order three-mode theory:*

$$\begin{cases} u_x(x, y, z) = \varphi_0(x, y) + \varphi_2(x, y)\left[1 - \left(\frac{z}{a}\right)^2\right] \\ u_y(x, y, z) = \psi_0(x, y) + \psi_2(x, y)\left[1 - \left(\frac{z}{a}\right)^2\right] \\ u_z(x, y, z) = \theta_1(x, y) z + \theta_3(x, y)\left[1 - \left(\frac{z}{a}\right)^2\right] z \end{cases}$$
$$\varphi_2(x, y) = \frac{1}{2} a^2 \frac{\partial}{\partial x}\theta_1(x, y), \quad \psi_2(x, y) = \frac{1}{2} a^2 \frac{\partial}{\partial y}\theta_1(x, y)$$
$$\theta_3(x, y) = \frac{1}{2}\theta_1(x, y) + \frac{\lambda}{2(\lambda + 2\mu)}\left[\frac{\partial}{\partial x}\varphi_0(x, y) + \frac{\partial}{\partial y}\psi_0(x, y)\right]$$
$$\varphi_4(x, y) = \psi_4(x, y) = \theta_5(x, y) = 0 \tag{5.13}$$

- *Six-mode theory:*

$$\begin{cases} u_x(x, y, z) = \varphi_0(x, y) + \varphi_2(x, y)\left[1 - \left(\frac{z}{a}\right)^2\right] \\ u_y(x, y, z) = \psi_0(x, y) + \psi_2(x, y)\left[1 - \left(\frac{z}{a}\right)^2\right] \\ u_z(x, y, z) = \theta_1(x, y)z + \theta_3(x, y)\left[1 - \left(\frac{z}{a}\right)^2\right]z \end{cases} \tag{5.14}$$

$$\varphi_4(x, y) = \psi_4(x, y) = \theta_5(x, y) = 0$$

As mentioned before in the previous section, the physical meaning of the higher order terms $\tilde{\varphi}_i(x, y)(i = 2, 4)$, $\tilde{\psi}_i(x, y)(i = 2, 4)$ and $\tilde{\theta}_i(x)(i = 3, 5)$ must always be associated with the form of the particular displacement field under consideration and results from certain mathematical manipulations that influence it. In the current approach these terms express higher order corrections to the initially assumed distributions of in-plane and transverse displacement components.

5.3 Displacement Fields of Structural Plate Elements

As mentioned before, the propagation of elastic flexural and shear waves in structural plate elements is associated with antisymmetric behaviour. Therefore, on the basis of Equations (5.1) and the considerations specified below the general form of the displacement field of two-dimensional plate spectral finite element for the needs of analysing propagation of elastic bending waves can be written in the following way:

$$\begin{cases} u_x(x, y, z) = \varphi_1(x, y)z + \varphi_3(x, y)z^3 + \varphi_5(x, y)z^5 \\ u_y(x, y, z) = \psi_1(x, y)z + \psi_3(x, y)z^3 + \psi_5(x, y)z^5 \\ u_z(x, y, z) = \theta_0(x, y) + \theta_2(x, y)z^2 + \theta_4(x, y)z^4 \end{cases} \tag{5.15}$$

where only nine terms of the Maclaurin series expansions of displacement functions $G_1(x, y, z)$, $G_2(x, y, z)$ and $G_3(x, y, z)$ are included.

The functions $\varphi_i(x, y)(i = 1, 3, 5)$, $\psi_i(x, y)(i = 1, 3, 5)$ and $\theta_i(x, y)(i = 0, 2, 4)$ defined in Equations (5.15) represent the independent nodal variables or degrees of freedom of the plate spectral finite element. Again it can be seen that in the current formulation the beam element has as many as 9 degrees of freedom in a single node. This number of independent nodal variables may be reduced, however, by taking into account the same zero traction condition, rewritten here:

$$\begin{gathered}\sigma_{zz}(x, y, z) = \tau_{yz}(x, y, z) = \tau_{zx}(x, y, z) = 0 \\ 0 \leq x \leq l, \quad 0 \leq y \leq b, \quad z = \pm a = \pm\frac{h}{2}\end{gathered} \tag{5.16}$$

where l and b now denote length and width, respectively, while h is the thickness of the plate spectral finite element.

Starting from the form of the displacement field given by Equation (5.15), the displacement fields for various one-mode, two-mode and other multi-mode theories can be constructed. Additionally, using Equation (5.16) representing the zero traction conditions on the lateral boundaries of the plate element allows one to enrich the displacement fields with some additional higher order terms. However, in most cases the resulting system of two differential equations is very complicated and cannot be solved analytically, but this can be avoided by means of simple substitutions and rearrangements of terms, thanks to which the zeroth-order terms for displacement functions $G_1(x, y, z)$, $G_2(x, y, z)$ and $G_3(x, y, z)$ can be represented as sums of all order terms in a manner similar to that presented in Equations (5.8).

Taking into account Equations (5.8), a new form of the displacement field of a two-dimensional plate spectral finite element for analysing the propagation of elastic bending (flexural) waves can be expressed as follows:

$$\begin{cases} u_x(x, y, z) = \varphi_1(x, y)z + \varphi_3(x, y)\left[1 - \left(\frac{z}{a}\right)^2\right]z + \varphi_5(x, y)\left[1 - \left(\frac{z}{a}\right)^4\right]z \\ u_y(x, y, z) = \psi_1(x, y)z + \psi_3(x, y)\left[1 - \left(\frac{z}{a}\right)^2\right]z + \psi_5(x, y)\left[1 - \left(\frac{z}{a}\right)^4\right]z \\ u_z(x, y, z) = \theta_0(x, y) + \theta_2(x, y)\left[1 - \left(\frac{z}{a}\right)^2\right] + \theta_4(x, y)\left[1 - \left(\frac{z}{a}\right)^4\right] \end{cases} \tag{5.17}$$

Specific theories of plate flexural behaviour known from the literature can be easily obtained starting from Equations (5.16) and (5.17). Different beam theories can be associated with different forms of the functions $\varphi_i(x, y)(i = 1, 3, 5)$, $\varphi_i(x, y)(i = 1, 3, 5)$ and $\theta_i(x)(i = 0, 2, 4)$. As an example, the dispersion curves for the Kirchoff–Love single-mode theory, the three-mode Mindlin–Reissner theory, the higher order three-mode Reddy theory, the higher-order three-mode theory, as well as the six-mode theory, are presented in

Figures 5.6 to 5.10. Their displacement fields can be presented conveniently in the following manner:

- *Kirchhoff–Love single-mode theory* [9]:

$$\begin{cases} u_x(x, y, z) = \varphi_1(x, y)z \\ u_y(x, y, z) = \psi_1(x, y)z \\ u_z(x, y, z) = \theta_0(x, y) \end{cases}$$
$$\varphi_1(x, y) = -\frac{\partial}{\partial x}\theta_0(x, y), \quad \psi_1(x, y) = -\frac{\partial}{\partial y}\theta_0(x, y) \tag{5.18}$$
$$\varphi_i(x, y) = \psi_i(x, y) = \theta_j(x, y) = 0, \quad i = 3, 5, \quad j = 2, 4$$

- *Three-mode Mindlin–Reissner theory* [10]:

$$\begin{cases} u_x(x, y, z) = \varphi_1(x, y)z \\ u_y(x, y, z) = \psi_1(x, y)z \\ u_z(x, y, z) = \theta_0(x, y) \end{cases} \tag{5.19}$$
$$\varphi_i(x, y) = \psi_i(x, y) = \theta_j(x, y) = 0, \quad i = 3, 5, \quad j = 2, 4$$

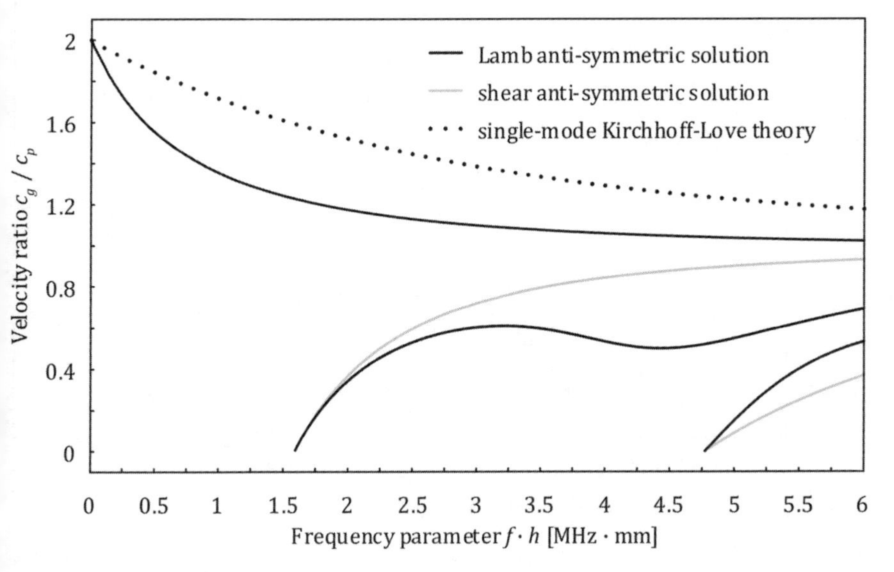

Figure 5.6 Dispersion curve for the velocity ratio c_g/c_p for the Kirchhoff–Love single-mode theory of plates ($c_l = 6.3$ km/s, $c_s = 3.2$ km/s)

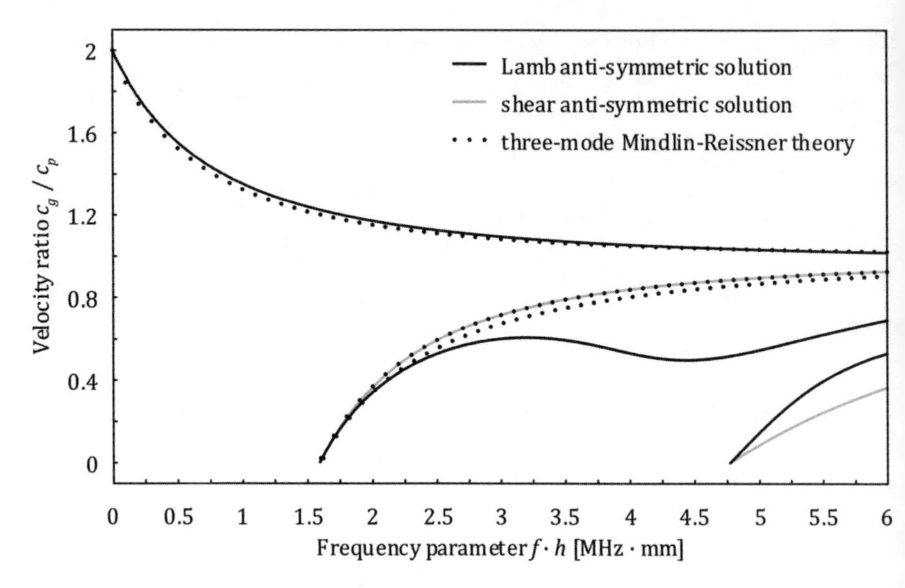

Figure 5.7 Dispersion curve for the velocity ratio c_g/c_p for the three-mode Mindlin–Reissner theory of plates ($c_l = 6.3$ km/s, $c_s = 3.2$ km/s)

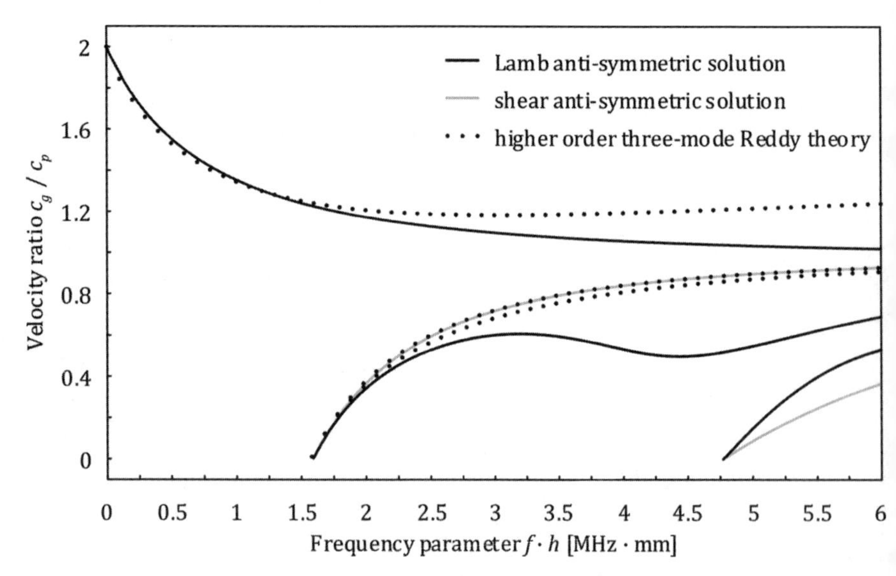

Figure 5.8 Dispersion curve for the velocity ratio c_g/c_p for the higher order three-mode Reddy theory of plates ($c_l = 6.3$ km/s, $c_s = 3.2$ km/s)

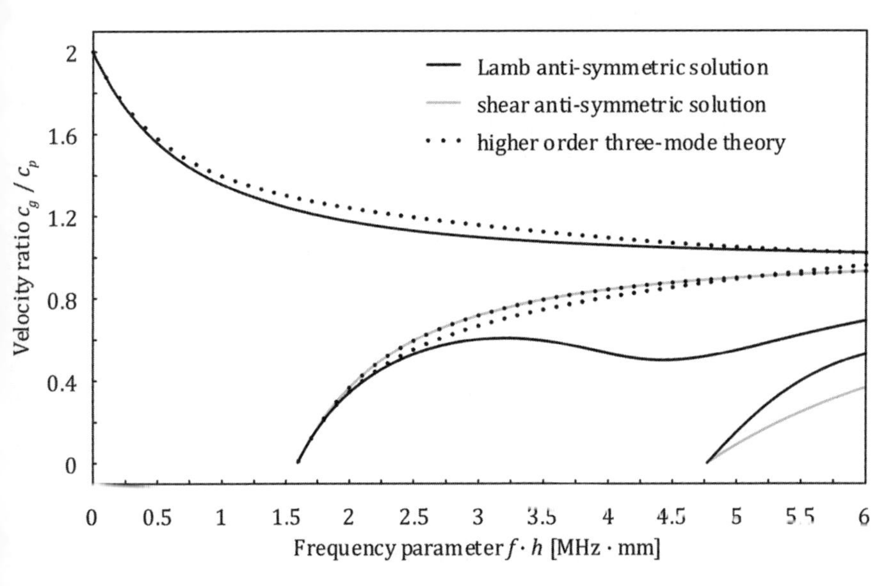

Figure 5.9 Dispersion curve for the velocity ratio c_g/c_p for the higher order three-mode theory of plates ($c_l = 6.3$ km/s, $c_s = 3.2$ km/s)

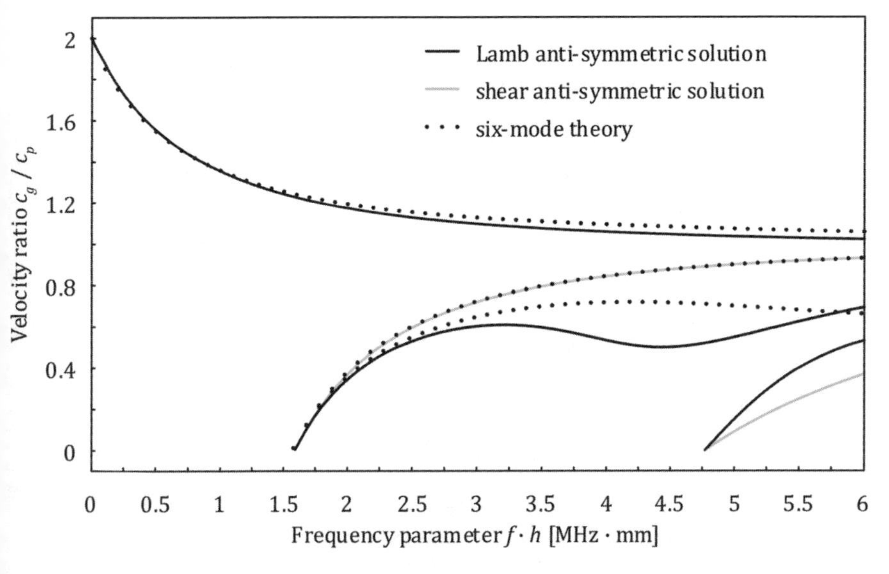

Figure 5.10 Dispersion curves for the velocity ratio c_g/c_p for the six-mode theory of plates ($c_l = 6.3$ km/s, $c_s = 3.2$ km/s)

- *Higher order three-mode Reddy theory* [11]:

$$\begin{cases} u_x(x,y,z) = \varphi_1(x,y)z + \varphi_3(x,y)\left[1-\left(\frac{z}{a}\right)^2\right]z \\ u_y(x,y,z) = \psi_1(x,y)z + \psi_3(x,y)\left[1-\left(\frac{z}{a}\right)^2\right]z \\ u_z(x,y,z) = \theta_0(x,y) \end{cases}$$

$$\varphi_3(x,y) = \frac{1}{2}\left[\varphi_1(x,y) + \frac{\partial}{\partial x}\theta_0(x,y)\right]$$

$$\psi_3(x,y) = \frac{1}{2}\left[\psi_1(x,y) + \frac{\partial}{\partial y}\theta_0(x,y)\right] \tag{5.20}$$

$$\varphi_5(x,y) = \psi_5(x,y) = \theta_2(x,y) = \theta_4(x,y) = 0$$

- *Higher order three-mode theory* [12]:

$$\begin{cases} u_x(x,y,z) = \varphi_1(x,y)z + \varphi_3(x,y)\left[1-\left(\frac{z}{a}\right)^2\right]z \\ u_y(x,y,z) = \psi_1(x,y)z + \psi_3(x,y)\left[1-\left(\frac{z}{a}\right)^2\right]z \\ u_z(x,y,z) = \theta_0(x,y) + \theta_2(x,y)\left[1-\left(\frac{z}{a}\right)^2\right] \end{cases}$$

$$\varphi_3(x,y) = \frac{1}{2}\left[\varphi_1(x,y) + \frac{\partial}{\partial x}\theta_0(x,y)\right] \tag{5.21}$$

$$\psi_3(x,y) = \frac{1}{2}\left[\psi_1(x,y) + \frac{\partial}{\partial y}\theta_0(x,y)\right]$$

$$\theta_2(x,y) = \frac{a^2}{2}\frac{\lambda}{\lambda+2\mu}\left[\frac{\partial}{\partial x}\varphi_1(x,y) + \frac{\partial}{\partial y}\psi_1(x,y)\right]$$

$$\varphi_5(x,y) = \psi_5(x,y) = \theta_4(x,y) = 0$$

- *Six-mode theory:*

$$\begin{cases} u_x(x, y, z) = \varphi_1(x, y)z + \varphi_3(x, y)\left[1 - \left(\frac{z}{a}\right)^2\right]z \\ u_y(x, y, z) = \psi_1(x, y)z + \psi_3(x, y)\left[1 - \left(\frac{z}{a}\right)^2\right]z \\ u_z(x, y, z) = \theta_0(x, y) + \theta_2(x, y)\left[1 - \left(\frac{z}{a}\right)^2\right] \end{cases} \tag{5.22}$$
$$\varphi_5(x, y) = \psi_5(x, y) = \theta_4(x, y) = 0$$

It should be emphasised that the physical meaning of the higher-order terms associated with the displacement functions $\varphi_i(x, y)(i = 3, 5)$, $\psi_i(x, y)(i = 3, 5)$ and $\theta_i(x)(i = 2, 4)$ must always be associated with the form of a particular displacement field under consideration and results from certain mathematical manipulations that influence it. In the current approach these terms express higher order corrections to the initially assumed distributions of in-plane and transverse displacement components.

5.4 Displacement Fields of Structural Shell Elements

Mechanical behaviour of shells as curved three-dimensional structural components couples the in-plane behaviour of membranes together with the out-of-plane behaviour of plates [13–16]. The coupling is geometrical and is the main reason for the complicated dynamical responses of these structures observed in practice. For the same reasons it is not possible in a general three-dimensional case to decouple the two types of behaviour observed in shell structures. However, it is very convenient to consider specific coordinate systems that greatly help and simplify the problems mentioned above.

Generally speaking, three different coordinate systems can be defined when considering a three-dimensional curved shell structure modelled by the spectral finite elements (see Figure 5.11):

1. The geometry of the element can be defined in a global coordinate system, which is a freely chosen well-known Cartesian coordinate system xyz. All nodal coordinates, nodal displacements, as well as the global stiffness and inertia, are expressed in this global coordinate system.
2. The second coordinate system is a curvilinear coordinate system $\xi\eta\zeta$. The ξ and η coordinates are defined on the mid-surface of the element, while

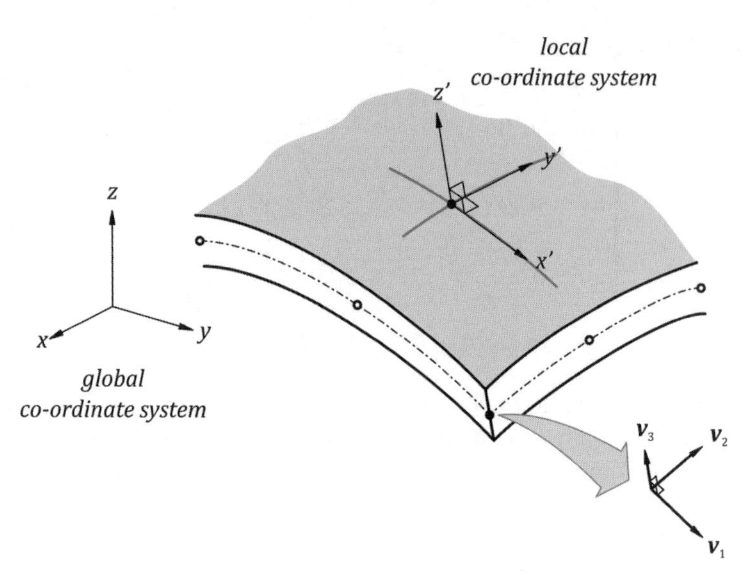

Figure 5.11 Global and local coordinate systems used for a curved shell element

the ζ coordinate is specified in the direction normal to the mid-surface. The curvilinear coordinates are normalised and usually vary between −1 and +1. The relations between the curvilinear coordinates ξ, η and ζ and the global coordinates x, y and z are presented in more details in Chapter 2 in Section 2.4, 'Computing characteristic matrices of spectral finite elements'.

3. Finally, a local coordinate system $x'y'z'$ is defined. This is also a Cartesian coordinate system related to the local normal directions at sampling points on the element surface (i.e. on the mid-surface of the element and where $\zeta = 0$). In these points the values of stresses and strains are calculated. The normal directions to the mid-surface of the element are determined by normal unit vectors v_3, while the remaining perpendicular directions (tangent to the mid-surface of the element in the $\xi\eta$ plane) are determined uniquely by two other perpendicular unit vectors v_1 and v_2. The relation between the curvilinear coordinates ξ, η and ζ and the local coordinates x', y' and z', the same as the relations necessary to determine the unit perpendicular vector v_3 as well as the normal unit vectors v_1 and v_2, are all well presented in References [14] and [15]. Very often various formulations of the curved shell elements assume full equivalence to the local and nodal coordinate systems.

Based on the above definitions of the global, normalised and local coordinate systems it can be noticed that the geometrical coupling of the in-plane and out-of-plane behaviour of curved three-dimensional shell structural components can be easily performed in the local coordinate system.

For example, in a specific case of the *elementary two-mode theory* of membranes, described by the displacement field given by Equations (5.10), and the *Kirchhoff–Love single-mode theory* of plates, described by the displacement field given by Equations (5.18), the coupled displacement field in the local coordinate system has the following form:

$$\begin{cases} u_x(x', y', z') = \varphi_0(x', y') + \varphi_1(x', y')z' \\ u_y(x', y', z') = \psi_0(x', y') + \psi_1(x', y')z' \\ u_z(x', y', z') = \theta_0(x', y') \end{cases}$$
$$\varphi_1(x', y') = -\frac{\partial}{\partial x'}\theta_0(x', y'), \quad \psi_1(x', y') = -\frac{\partial}{\partial y'}\theta_0(x', y') \tag{5.23}$$
$$\varphi_i(x', y') = \psi_i(x', y') = \theta_j(x', y') = 0, \quad i = 2, \ldots, 5, \quad j = 1, \ldots, 5$$

resulting in a *three-mode theory* of shell behaviour. It must be noted that in this theory the in-plane behaviour is represented by two symmetric modes enabling the propagation of longitudinal waves and symmetric shear waves, while the out-of-plane behaviour is represented by one antisymmetric mode enabling the propagation of flexural waves only. This is an undesired feature resulting in a nonbalanced mode (energy) conversion between the three modes; that is two symmetric modes can always convert into only one antisymmetric mode. Mode-balanced theories should allow for conversion of each symmetric and antisymmetric mode into its antisymmetric or symmetric counterpart. Theories of shell behaviour used for investigating wave propagation problems in shell structures that fulfil such a condition can be easily obtained from various theories of membranes and plates by taking an appropriate combination of equal-mode and same-order theories, such as: three-mode theories, higher order three-mode theories, six-mode theories, and so on. This can be well illustrated in the case of the *three-mode theories* of membranes (5.12) and plates (5.19) resulting in a *six-mode theory* of shells:

$$\begin{cases} u_x(x', y', z') = \varphi_0(x', y') + \varphi_1(x', y')z' \\ u_y(x', y', z') = \psi_0(x', y') + \psi_1(x', y')z' \\ u_z(x', y', z') = \theta_0(x', y') + \theta_1(x', y')z' \end{cases} \tag{5.24}$$
$$\varphi_i(x', y') = \psi_i(x', y') = \theta_i(x', y') = 0, \quad i = 2, \ldots, 5$$

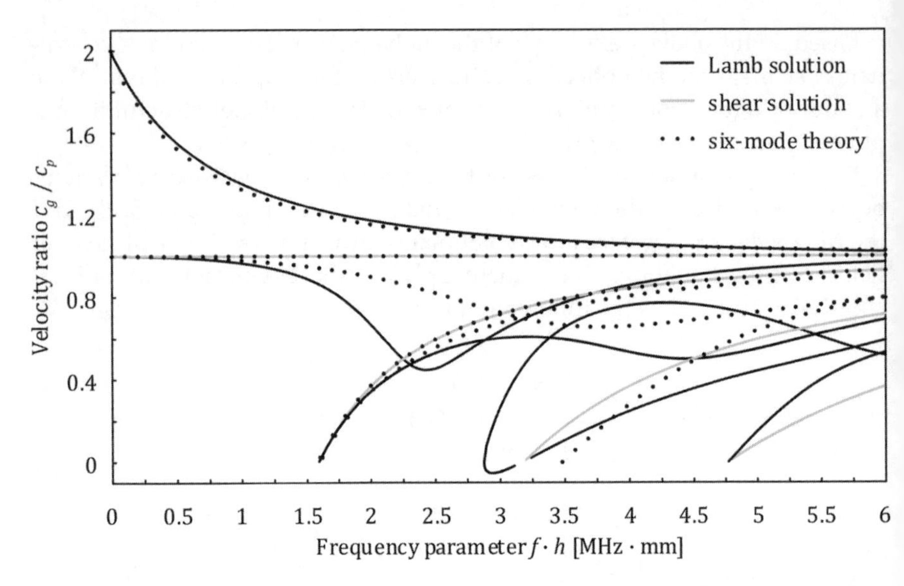

Figure 5.12 Dispersion curves for the velocity ratio c_g/c_p for the six-mode theory of shells ($c_l = 6.3$ km/s, $c_s = 3.2$ km/s)

Again it should be emphasised at this point that the dispersion curves of the resulting shell behaviour theory also represent combinations of dispersion curves of the constituent membrane and plate behaviour theories. This is illustrated by Figure 5.12.

5.5 Certain Numerical Considerations

It should be remembered that in engineering practice the investigation of wave-propagation-related phenomena in two- or three-dimensional structural membrane, plate and shell elements is very often limited only to the fundamental symmetrical or antisymmetrical modes [17–23]. In all such cases numerical simulations are required to provide computational results that remain not only in qualitative but more importantly in quantitative agreement with corresponding results of experimental observations. For that reason it is very important to know the accuracy level that can be reached by employing the numerical tools available, such as, for example, the spectral finite element method and the applications of the membrane and plate models presented.

The accuracy of the numerical results depends on many various factors [24, 25], but the most important one is the accuracy of the model used in the simulations. Its influence is very well illustrated in the case of the membrane and plate behaviour models discussed in the two previous sections. The accuracies of these models expressed as the relative error δ between the exact results provided by the theory of Lamb waves [26] (symmetrical solutions for membranes and anti-symmetrical for plates) are compared against the results provided by particular models and based on their dispersion curves, shown in Figures 5.1–5.10. The relative error is defined as follows:

$$\delta_T = \frac{f_T - f_L}{f_L} \times 100\%, \quad f_i = \frac{c_g}{c_p}, \quad i = T, L \tag{5.25}$$

where T denotes the ratio of the group velocity c_g and phase velocity c_p obtained on the grounds of the given behaviour theory (membrane or plate), while L denotes the ratio obtained on the basis of Lamb solutions (symmetric or antisymmetric). Furthermore, the average error has been defined in the following manner:

$$\Delta_T(fd) = \frac{1}{fd} \int_0^{fd} \left|\delta_T(s)\right| \mathrm{d}s \tag{5.26}$$

in order to estimate the broad frequency accuracy of the models under consideration.

It is very important to note at this point that the total relative error δ related to adopting particular membrane or plate behaviour theories is always higher and is increased by other numerical errors resulting from the order of approximation polynomials, mesh density, geometry approximation, solution algorithm, and so on.

The results presented in Table 5.1 and Figure 5.13 summarise the effectiveness of the investigated membrane behaviour theories and their accuracy is calculated against the Lamb analytical solution in the case of the fundamental symmetrical mode of wave propagation.

It can be clearly seen from Table 5.1 and Figure 5.13 that out of the five membrane behaviour theories investigated, the six-mode theory is characterised by the smallest values of both the relative error δ and average error Δ in the whole range of the frequency parameter fh. The maximum values of these errors are -0.3 % and 0.13 %, respectively. However, it should be noted that within the range of the frequency parameter fh up to 1.5 MHz·mm the most accurate results are obtained based on the higher order three-mode theory. The corresponding maximum values of the relative error δ and average error Δ are –0.12 % and 0.06 %, respectively. At the same time it

Table 5.1 Relative error δ for selected membrane behaviour theories calculated against the Lamb analytical solution for the fundamental symmetrical mode of wave propagation ($c_l = 6.3$ km/s, $c_t = 3.2$ km/s)

fd (MHz·mm)	δ_A (%)	δ_B (%)	δ_C (%)	δ_D (%)	δ_E (%)
0.20	0.11	0.05	0.03	−0.02	−0.02
0.40	0.44	0.24	0.13	−0.03	−0.03
0.60	1.05	0.60	0.36	−0.06	−0.06
0.80	2.02	1.22	0.76	−0.08	−0.08
1.00	3.53	2.26	1.47	−0.11	−0.12
1.20	5.87	4.02	2.74	−0.12	−0.17
1.40	9.62	7.02	5.01	−0.09	−0.22
1.60	15.94	12.35	9.25	0.10	−0.28
1.80	27.31	22.34	17.51	0.62	−0.30
2.00	49.14	41.96	34.14	1.61	−0.31
fd (MHz·mm)	Δ_A (%)	Δ_B (%)	Δ_C (%)	Δ_D (%)	Δ_E (%)
0.5	0.23	0.12	0.06	0.01	0.01
1.0	1.07	0.64	0.39	0.04	0.04
1.5	3.13	2.14	1.46	0.06	0.09
2.0	9.04	7.09	5.41	0.19	0.13

A – elementary two-mode, B – higher order two-mode, C – three-mode, D – higher order three-mode, E – six-mode.

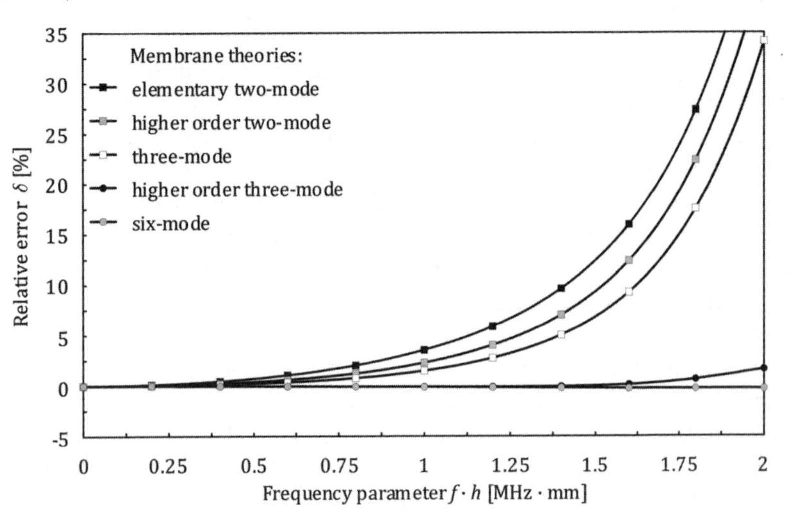

Figure 5.13 Relative error for selected membrane behaviour theories calculated against the Lamb analytical solution for the fundamental symmetrical wave propagation mode ($c_l = 6.3$ km/s, $c_s = 3.2$ km/s)

can be clearly seen that the usefulness of the elementary two-mode, the higher order two-mode and the three-mode theories of membrane behaviour for analysing wave propagation phenomena in two- or three-dimensional structural membrane, plate and shell elements caused by excitation signals in a broad range of the frequency parameter is very limited, due to the high values of both the relative error δ and average error Δ. On the other hand, these models of membrane behaviour can be successfully applied for solving slow or moderate-speed dynamic problems within the range of the frequency parameter fh up to 1.0 MHz·mm, when the values of both the relative error δ and average error Δ remain below 1 %.

In the case of plate behaviour theories the observed changes in the relative error δ as well as the average error Δ as a function of the frequency parameter fh, calculated against the Lamb analytical solution for the fundamental antisymmetrical wave propagation mode, are different and are presented in Table 5.2 and Figure 5.14.

As before, it can be seen from Table 5.2 and Figure 5.14 that out of the five plate behaviour theories investigated, the six-mode theory is characterised by the smallest values of both the relative error δ and average error Δ in the

Table 5.2 Relative error δ for selected plate behaviour theories calculated against the Lamb analytical solution for the fundamental antisymmetric wave propagation mode ($c_l = 6.3$ km/s, $c_t = 3.2$ km/s)

fd (MHz·mm)	δ_A (%)	δ_B (%)	δ_C (%)	δ_D (%)	δ_E (%)
0.20	9.48	−0.55	−0.49	0.12	0.04
0.40	16.29	−0.95	−0.73	0.44	0.13
0.60	21.18	−1.24	−0.76	0.91	0.27
0.80	24.65	−1.44	−0.60	1.52	0.46
1.00	27.05	−1.56	−0.29	2.20	0.67
1.20	28.63	−1.62	0.17	2.94	0.91
1.40	29.57	−1.63	0.75	3.67	1.16
1.60	30.01	−1.61	1.43	4.38	1.42
1.80	30.06	−1.55	2.20	5.03	1.67
2.00	29.83	−1.47	3.04	5.60	1.92
fd (MHz·mm)	Δ_A (%)	Δ_B (%)	Δ_C (%)	Δ_D (%)	Δ_E (%)
0.25	10.94	0.64	0.51	0.23	0.07
0.50	17.31	1.01	0.56	0.81	0.24
0.75	21.13	1.21	0.51	1.59	0.49
1.00	23.35	1.29	0.89	2.41	0.77

A – single-mode Kirchhoff–Love, B – three-mode Mindlin–Reissner, C – higher order three-mode Reddy, D – higher order three-mode, E – six-mode.

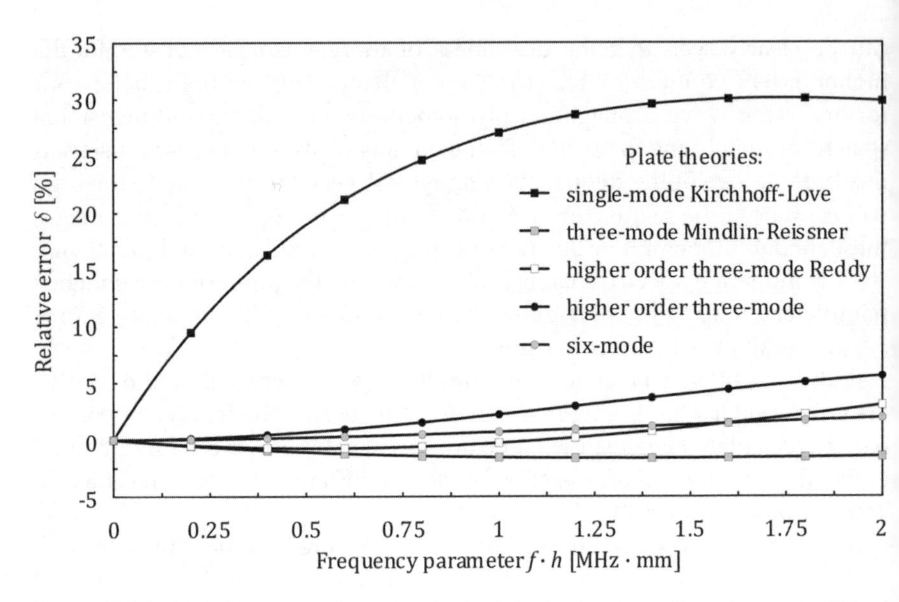

Figure 5.14 Relative error for selected plate behaviour theories calculated against the Lamb analytical solution for the fundamental antisymmetrical wave propagation mode ($c_l = 6.3$ km/s, $c_s = 3.2$ km/s)

whole range of the frequency parameter fh. The maximum values of these errors are 1.9 % and 0.7 %, respectively. Within the same range of the frequency parameter fh, the three-mode Mindlin–Reissner theory and the higher order three-mode Reddy theories also yield results of similar accuracies. However, the higher order three-mode Reddy theory is more accurate up to the first cut-off frequency, which corresponds to the value of the frequency parameter fh equal to 1.52 MHz·mm. It is interesting to note that the higher order three-mode theory is not as effective and its relative error δ as well as the average error Δ are small in a fairly short range of the frequency parameter fh up to approximately 0.5 MHz·mm. In that context the usefulness of the higher order three-mode theory is limited to slow or moderate-speed dynamic problems. On the other hand, the single-mode Kirchhoff–Love theory of plate behaviour appears to be appropriate only for analysing static problems due to very high values of both the relative error δ and average error Δ, which quickly exceed 10 % for relatively small values of the frequency parameter fh, that is for 0.22 MHz·mm.

Based on the results presented in Tables 5.1 and 5.2 and Figures 5.13 and 5.14 it can be concluded that the lower-mode theories of membrane and plate behaviour are badly suited for analysis of wave propagation phenomena. In general they can be applied for static or slow-to-moderate dynamic problems. For problems involving the propagation of elastic waves in two- or three-dimensional structural membrane, plate and shell elements generated by excitation signals in a broad range of the frequency parameter fh up to 1.0 MHz·mm, the use of three-mode theories is recommended. This also includes higher order higher-mode theories. Higher order higher-mode theories are generally superior to classical theories, but their superior performance can be narrowed to a relatively short range of the frequency parameter fh, where they can produce results of very high accuracies. Their application must always be proceeded by careful analysis of corresponding dispersion curves in order to avoid the regions of lower accuracies. If this cannot be done, the use of six-mode theories is always recommended.

However, it should be strongly emphasised that if a multi-mode behaviour of propagating waves is to be investigated, the lower-mode theories must be replaced by higher-mode (six-mode or more) or higher order higher-mode theories of membrane or plate behaviour that follow the behaviour of Lamb waves in a much broader range of the frequency parameter fh up to a few MHz·mm as well as in a broad range of modes.

5.6 Examples of Numerical Calculations

In this section the effectiveness of the spectral finite element method is demonstrated by selected numerical examples that illustrate wave propagation phenomena in two- and three-dimensional structural elements. Certain results of these simulations have also been compared with experimental measurements employing laser scanning vibrometry.

In general, aluminium alloy has been selected as the structural material. It exhibits the following mechanical properties: Young's modulus $E = 72.7$ GPa, Poisson's ratio $\nu = 0.33$, density $\rho = 2700\ \mathrm{kg/m^3}$. In all cases considered all edges of the elements under consideration have been assumed to be free, which corresponds to the free type of boundary conditions.

The excitation signal $F(t)$ acting over time $t_1 = 1/f_m$ has the form of a sine pulse modulated by a Hann window of the carrier frequency f_c. The carrier f_c and modulation f_m frequencies, as well as the total calculation time t_T, could vary in each simulation dependinging on the total number and type of spectral finite elements employed.

5.6.1 Propagation of Elastic Waves in an Angle Bar

It has been assumed for this simulation that the total length of the angle bar l is 1000 mm, the width of each bar section is 250 mm and the bar thickness h

is 10 mm. The carrier and modulation frequencies f_c and f_m of the excitation signal $F(t)$ of amplitude 1 N have been assumed to be $f_c = 75$ kHz and $f_m = 18.75$ kHz, respectively, which corresponds to four cycles of the carrier signal. This is equivalent to a range of the frequency parameter fh from $(f_c - 2f_m)h$ to $(f_c + 2f_m)h$, that is from 0.375 MHz·mm to 1.125 MHz·mm, in the dispersion curve for the velocity ratio c_g/c_p presented in Figure 5.12.

The calculation time t_T divided into 8000 time steps covers 0.5 ms in order to allow for multiple reflections of the propagating waves from both ends of the bar. The results of numerical simulation for the case of the aluminium bar element, in the form of three-dimensional snapshots of the displacement amplitudes $\sqrt{u_x^2 + u_y^2 + u_z^2}$, are presented in Figure 5.15.

In total 1800 spectral finite elements following the six-mode theory of shells have been used in this simulation (refer to Equation (5.24) for more details). It should be mentioned that the simulation employed the Lobatto distribution of nodes within spectral finite elements.

The obtained results of numerical simulation presented in Figure 5.16 are a very good example of mode conversion phenomena [26] at geometrical

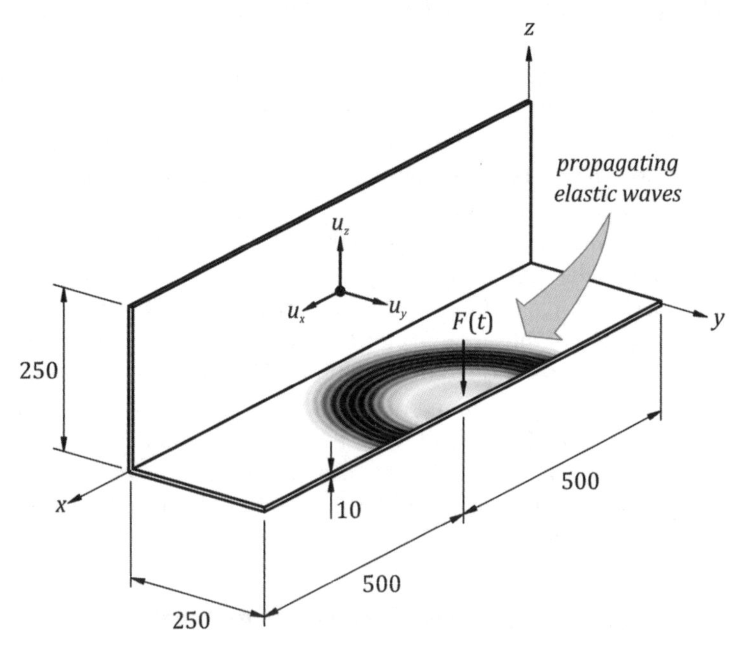

Figure 5.15 Geometry of an aluminium angle bar

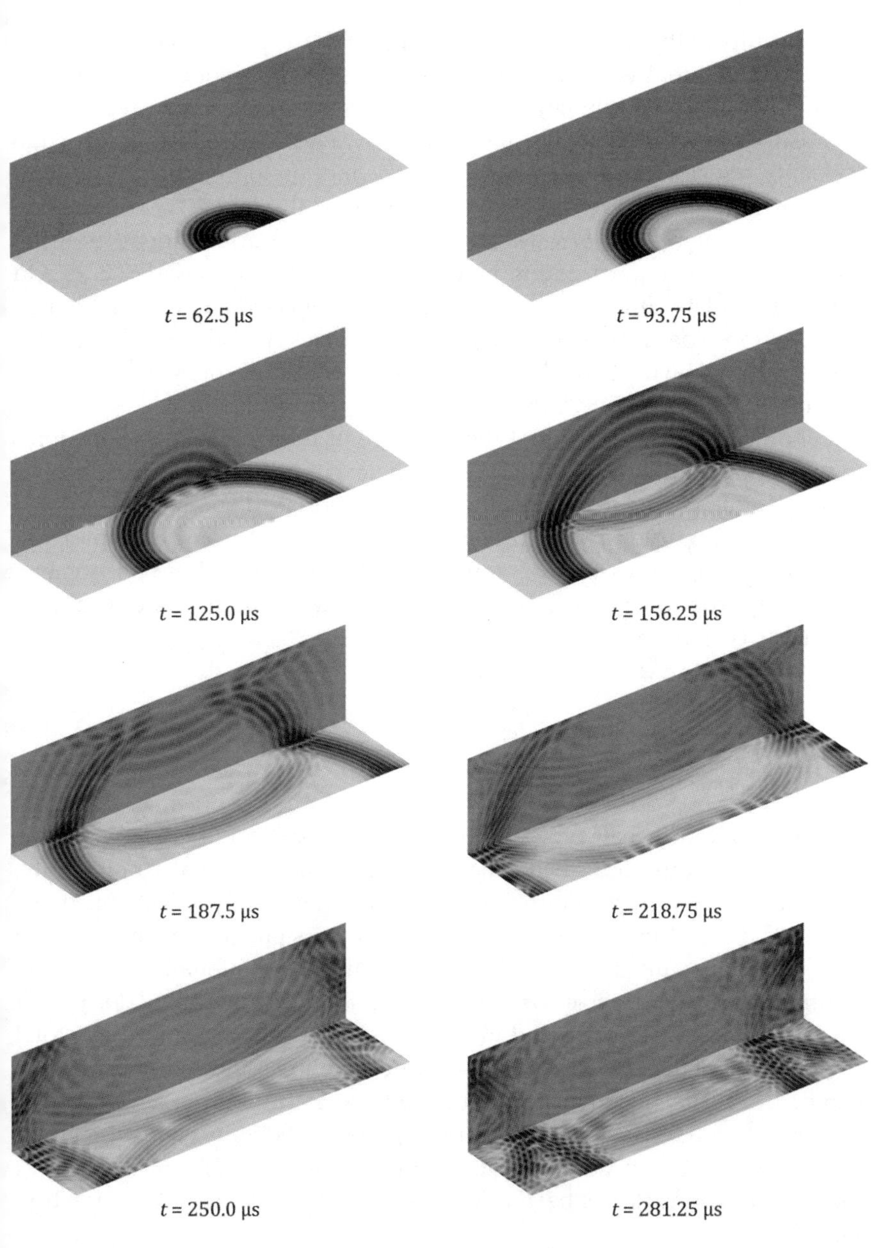

Figure 5.16 Snapshots of wave propagation patterns in an aluminium angle bar at various moments according to the six-mode theory of shells – results of a numerical simulation using the spectral finite element method

discontinuity in the form of the middle edge of the angle bar. It can be clearly seen that the initial wave propagating as the A_0 wave mode transmits and reflects from the middle edge of the angle bar as different wave modes. Because of the geometry of the angle bar, the wave motion at its middle edge is cleanly transformed from the out-of-plane motion to the in-plane motion, which results in a wave mode conversion from the initial A_0 wave mode to the resulting S_0 and SH_0 wave modes. This is clearly evident from the snapshots of wave propagation patterns taken at 125 μs and 156.25 μs. On the other hand, the initial A_0 wave mode is back-reflected without any change in its motion type, as two out-of-plane wave modes A_0 and SH_1. This is very well illustrated by the two following snapshots of wave propagation patterns taken at 187.5 μs and 218.75 μs.

It should be emphasised that the presence of mode conversion phenomena depends on the frequency range of the excitation signal, which should always be interpreted in conjunction with appropriate dispersion curves (see the dispersion curve presented in Figure 5.12 in the current example). On the other hand, successful and appropriate modelling of these phenomena requires appropriate numerical models that allow for multi-mode behaviour and mode conversion.

5.6.2 Propagation of Elastic Waves in a Half-Pipe Aluminium Shell

The geometry of a half-pipe aluminium shell used for this simulation is presented in Figure 5.17. It has been assumed that the total length of the element l is 1000 mm, its width measured circumferentially is 500 mm and its thickness h is 10 mm. Properties of the excitation signal and its modulation have been the same as in the previous example. Additionally it has been assumed that a transverse through-thickness and open crack is located at the half of the shell length and has a circumferential orientation. The total angular span of the crack is 3 degrees, while its centre is located at an angle of 17.5 degrees. The crack has been modelled using a technique developed by the authors in the past for various applications of the finite element method [27–29], an approach described in detail in Reference [30]. Shell stiffness loss due to the crack has been calculated according to the laws of fracture mechanics.

As before, the same calculation time t_T divided into 8000 time steps has been used in order to cover 0.5 ms and to allow for multiple reflections of propagating waves from all edges of the shell. The results of numerical

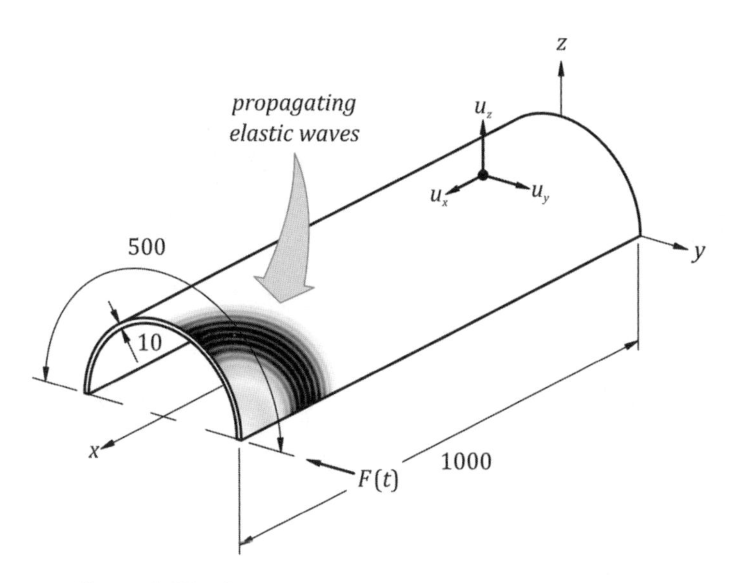

Figure 5.17 Geometry of a half-pipe aluminium shell

simulation, in the form of three-dimensional snapshots of the displacement amplitudes $\sqrt{u_x^2 + u_y^2 + u_z^2}$, are presented in Figure 5.18.

The same number of spectral finite elements following the six-mode theory of shells has been used as in the previous simulation (refer to Equation (5.24) for more details). It should be mentioned that the simulation employed the Lobatto distribution of nodes within spectral finite elements.

The results of numerical simulations presented in Figure 5.18 demonstrate that the excited wave propagates initially as the A_0 wave mode within the shell. As soon as the wave reaches the crack, it scatters and reflects at the discontinuity essentially as two modes: A_0 and SH_1; this is clearly visible in the snapshots of wave propagation patterns taken at 250 μs and 312.5 μs. In the same manner as was observed in the previous example, each time the incoming wave reaches any form of geometrical discontinuity, its scattering and reflection has a multi-mode nature. It should be said that due to the smooth geometry of the half-pipe shell the wave conversion phenomena are practically restricted to the crack site and the free edges of the shell. It is also interesting to point out that for the same reason (i.e. smooth geometry) the propagation of converted S_0 and SH_0 modes is not observed because of their very small amplitudes relative to the amplitudes of the A_0 and SH_1 modes.

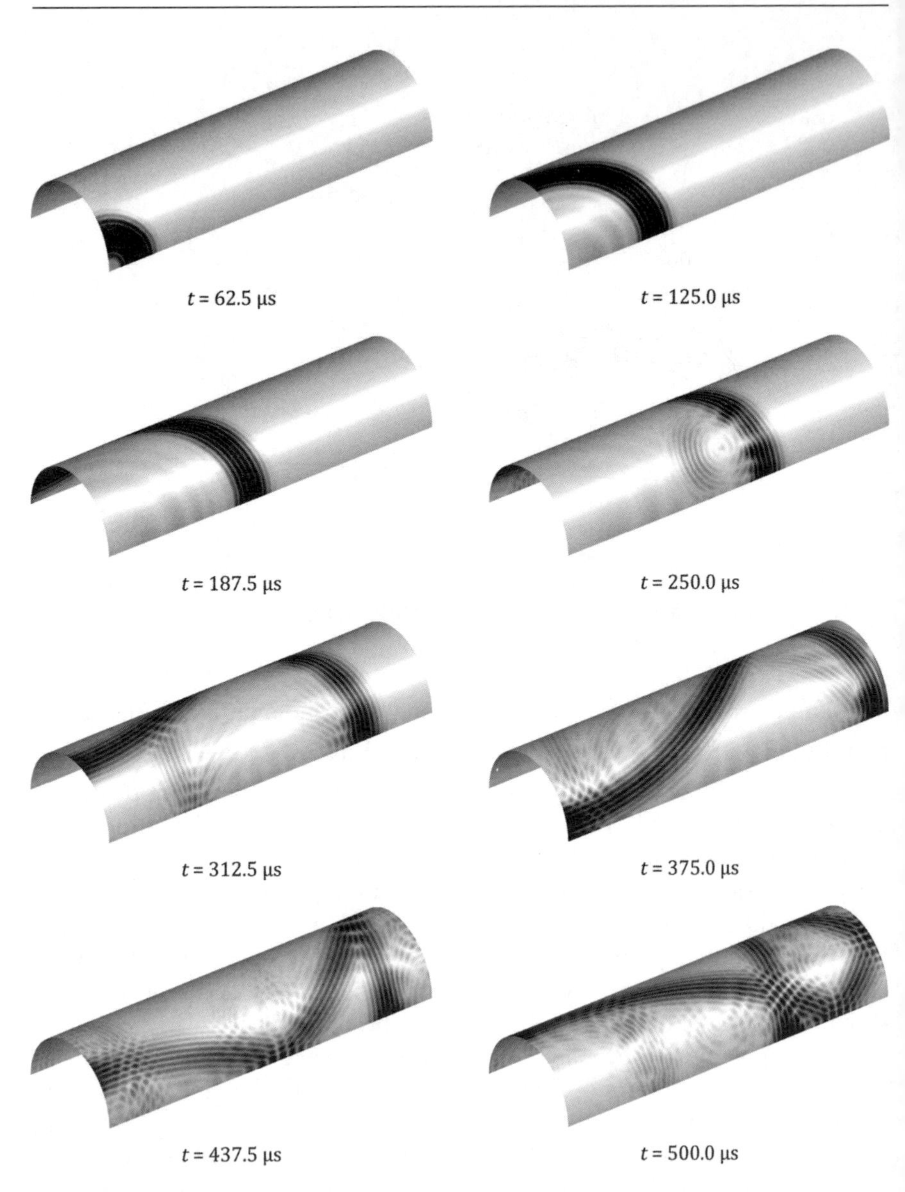

Figure 5.18 Snapshots of wave propagation patterns in a cracked aluminium half-pipe shell at various time instances according to the six-mode theory of shells – results of a numerical simulation using the spectral finite element method

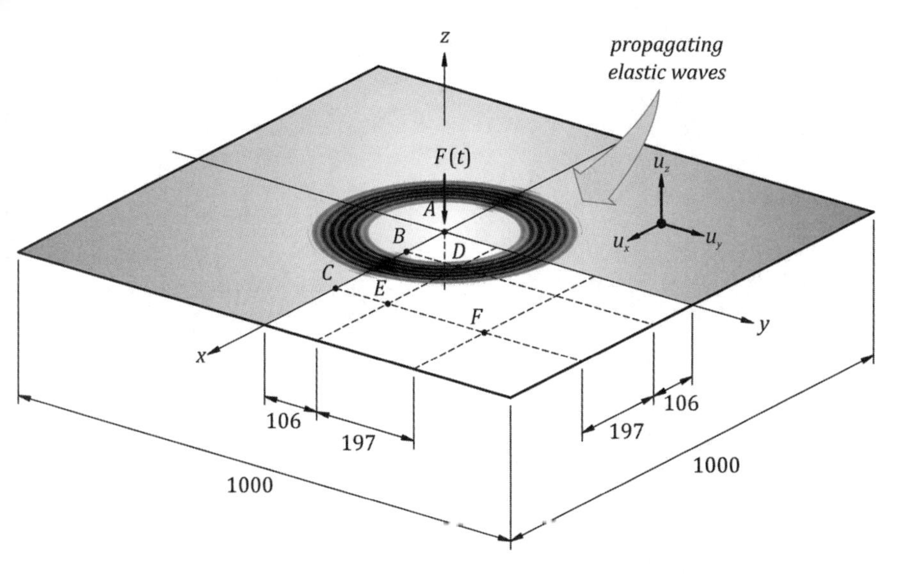

Figure 5.19 Geometry of a square aluminium plate

5.6.3 Propagation of Elastic Waves in an Aluminium Plate

The geometry of a square aluminium plate used for both numerical simulation and experimental measurements is presented in Figure 5.19. It has been assumed that the total length and width of the plate $l = b$ is 1000 mm, while its thickness h is 1 mm. Properties of the excitation signal and its modulation have been unchanged. For this particular case it has been assumed that the plate is composed of aluminium alloy of the following properties: Young's modulus $E = 68.0$ GPa, Poisson's ratio $\nu = 0.33$, density $\rho = 2660\ \text{kg/m}^3$.

In this case the results of numerical simulation, obtained by the use of the spectral finite element method, have been compared with the results of experimental measurements using laser scanning vibrometry. The comparison concerned wave propagation patterns for the A_0 mode, which is mostly associated with the transverse velocity component $\dot{u}_z$ at various time instances as well as time signals taken at selected points on the plate surface (points C and F), as presented in Figure 5.19.

Additionally, experimental wave propagation maps have been obtained from laser scanning measurements based on the gird of 225×227 points covering one-quarter of the plate and at 1024 time instances.

The calculation time t_T divided into 50 000 times steps covers 1 ms in order to allow reflections of the propagating waves from plate boundaries. In total

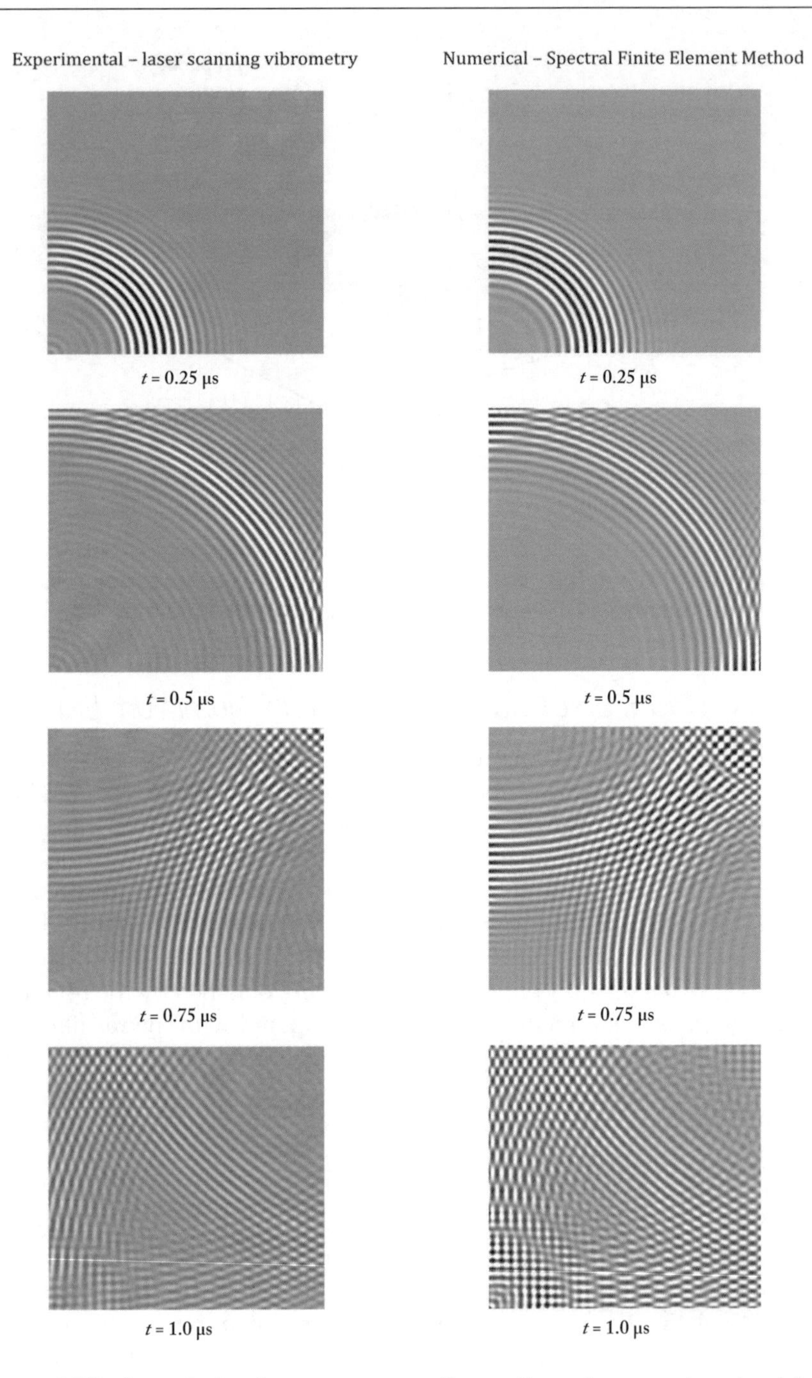

Figure 5.20 Snapshots of wave propagation patterns in a quarter aluminium plate at various time instances according to the three mode Mindlin–Reissner theory of plates – results of a numerical simulation using the spectral finite element method

10 420 spectral finite elements according to the three-mode Mindlin–Reissner theory of plate behaviour have been used in this simulation (refer to Equation (5.19) for more details). The Lobatto distribution of nodes within the spectral finite elements has been employed. In a similar manner as in the case of the coupled propagation of longitudinal and flexural elastic waves in a rod, the values of unknown parameters related to geometrical coupling due to the presence of a piezoelectric transducer, longitudinal and bending components of the excitation, together with material damping for the frequency range under investigation, have been estimated experimentally. In the numerical model, damping proportional only to the inertia matrix M^e has been considered with the value of the parameter $\alpha = 100$.

It can be clearly seen from Figure 5.20 as well as Figures 5.21 and 5.22 that very good agreement has been achieved between the results of numerical

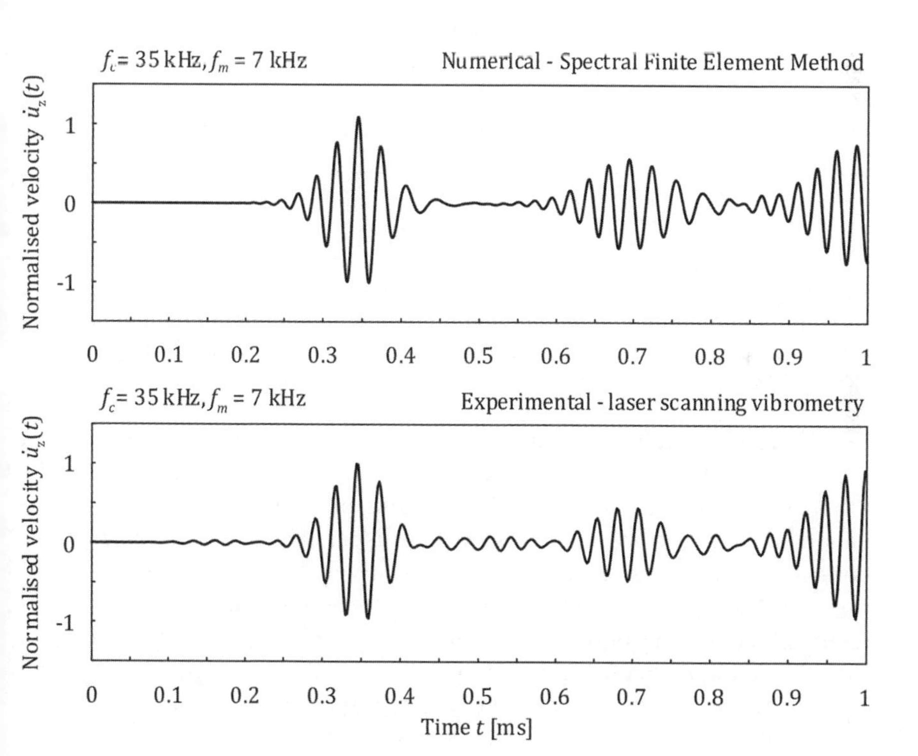

Figure 5.21 Time signals of the transverse velocity components $\dot{u}_z$ at point C on the plate surface – results of a numerical simulation using the spectral finite element method and experimental measurements by laser scanning vibrometry

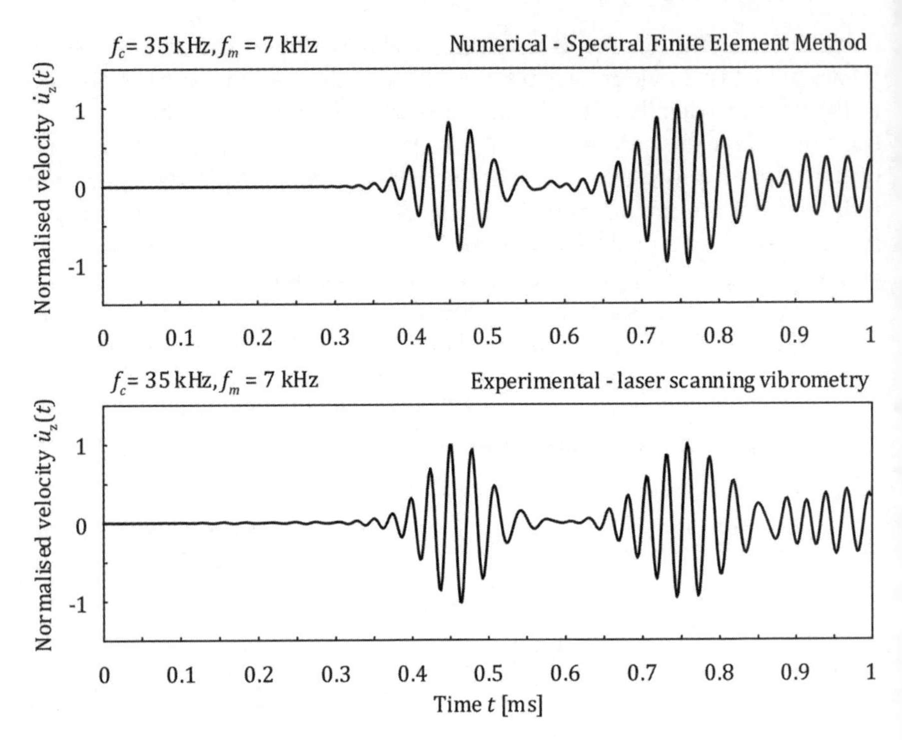

Figure 5.22 Time signals of the transverse velocity components $\dot{u}_z$ at point F on the plate surface – results of a numerical simulation using the spectral finite element method and experimental measurements by laser scanning vibrometry

simulations by the spectral finite element method and laser scanning vibrometry for the A_0 mode under investigation. This is true not only for the snapshots of the transverse velocity component $\dot{u}_z$ but also for the time signals of the same component. Again the dispersion of propagating waves observed in the experimental signals is relatively higher than in the case of numerical ones since the numerical model using the spectral finite element method takes into account the influence of the bond between the piezoelectric transducer and the plate [31] is a simplified manner. However, the remaining signal features observed have been successfully captured.

References

1. Rose, J.L. (1999) *Ultrasonic Waves in Solid Media*, Cambridge University Press, Cambridge.

2. Doyle, J.F. (1997) *Wave Propagation in Structures*, Springer-Verlag New York Inc, New York.
3. Achenbach, J.D. (1973) *Wave Propagation in Elastic Solids*, North-Holland Publishing Company, Amsterdam.
4. Żak, A. Krawczuk, M. and Ostachowicz, W. (2006) Propagation of in-plane waves in an isotropic panel with a crack. *Finite Elements in Analysis and Design*, **42**, 929–941.
5. Żak, A. Krawczuk, M. and Ostachowicz, W. (2006) Propagation of in-plane waves in a composite panel. *Finite Elements in Analysis and Design*, **43**, 154–154.
6. Kudela, P., Żak, A., Krawczuk, M. and Ostachowicz, W. (2007) Modelling of wave propagation in composite plates using the time domain spectral element method. *Journal of Sound and Vibration*, **302**, 728–754.
7. Żak, A. (2009) A novel formulation of a spectral plate element for wave propagation in isotropic structures. *Finite Elements in Analysis and Design*, **45**, 650–658.
8. Love, A.E. (1927) *A Treatise on the Mathematical Theory of Elasticity*, Dover Publications.
9. Reddy, J.N. (2007) *Theory and Analysis of Elastic Plates and Shells*, CRC Press.
10. Mindlin, R.D. (1951) Influence of rotatory inertia and shear on flexural motions of isotropic, elastic plates. *Trans. ASME, Journal of Applied Mechanics*, **18**, 31–38.
11. Ochoa, O.O. and Reddy, J.N. (1992) *Finite Element Analysis of Composite Laminates*, Kluwer Academic Publishers, Dordrecht.
12. Timoshenko, S. and Woinowsky-Krieger, S. (1959) *Theory of Plates and Shells*, McGraw-Hill Book Company, London.
13. Yang, J. and Guo, S. (2005) On using the Kane–Mindlin theory in the analysis of cracks in plates. *International Journal of Fracture*, **133**, L13–L17.
14. Rao, S.S. (1981) *The Finite Element Method in Engineering*, Pergamon Press.
15. Zienkiewicz, O.C. (1989) *The Finite Element Method*, McGraw-Hill Book Company, London.
16. Pozrikidis, C. (2005) *Introduction to Finite and Spectral Element Methods Using MATLAB*, Chapman & Hall/CRC.
17. Sharma, M.D. (2008) Propagation of inhomogeneous plane waves in anisotropic viscoelastic media. *Acta Mechanica*, **200**, 145–154.
18. Orrenius, U. and Finnveden, S. (1996) Calculation of wave propagation in rib-stiffened plate structures. *Journal of Sound and Vibration*, **198**, 203–224.
19. Williams, F.W., Ouyang, H.J., Kennedy, D. and York, C.B. (1995) Wave propagation along longitudinally periodically supported or stiffened prismatic plate assemblies. *Journal of Sound and Vibration*, **186**, 197–205.
20. Zhang, J., Jia, L. and Shu, Y. (2002) Wave propagation characteristics of thin shells of revolution by frequency–wave number spectrum method. *Journal of Sound and Vibration*, **251**, 367–372.
21. Haddow, J.B. and Jiang, L. (1999) Finite amplitude spherically symmetric wave propagation in a prestressed hyperelastic shell. *International Journal of Solids and Structures*, **36**, 2793–2805.

22. Wang, X., Lu, G. and Guillow, S.R. (2002) Stress wave propagation in orthotropic laminated thick-walled spherical shells. *International Journal of Solids and Structures*, **39**, 4027–4037.
23. Yuan, F.G. and Hsieh, C.C. (1998) Three-dimensional wave propagation in composite cylindrical shells. *Composite Structures*, **42**, 153–167.
24. Dauksher, W. and Emery, A.F. (1997) Accuracy in modelling the acoustic wave equation with Chebyshev spectral finite elements. *Finite Elements in Analysis and Design*, **26**, 115–128.
25. Christon, M.A. (1999) The influence of the mass matrix on the dispersive nature of the semi-discrete, second-order wave equation. *Computer Methods in Applied Mechanics and Engineering*, **173**, 147–166.
26. Rayleigh, J.W.S. (1945) *The Theory of Sound*, Dover Publications Inc.
27. Krawczuk, M. (1992) Modeling and identification of cracks in truss constructions. *Finite Elements in Analysis and Design*, **12**, 41–50.
28. Krawczuk, M. (1993) A rectangular plate finite element with an open crack. *Computers and Structures*, **46**, 487–493.
29. Krawczuk, M. (1994) Rectangular shell finite element with an open crack. *Finite Elements in Analysis and Design*, **15**, 233–253.
30. Ostachowicz, W. and Krawczuk, M. (2009) Modeling for detection of degraded zones in metallic and composite structures, in *Encyclopedia of Structural Health Monitoring* (eds C. Boller, F. Chang and Y. Fujino), John Wiley & Sons, Ltd, Chichester, pp. 851–866.
31. Lee, B.C., Palacz, M., Krawczuk, M. *et al.* (2004) Wave propagation in a sensor/actuator diffusion bond model. *Journal of Sound and Vibration*, **276**, 671–687.

6

Three-Dimensional Structural Elements

This chapter presents selected aspects of modelling elastic waves using solid spectral elements. Solid elements utilise the three-dimensional theory of elasticity as their foundation. Thanks to modelling simplicity, solid elements are particularly useful in situations where complex fields of variables (displacements, deformations, electric potential) need to be modelled. An undisputed disadvantage of solid elements is much larger number of degrees of freedom compared to two-dimensional elements, which directly results in significantly longer computation times.

As thin-walled structures are ubiquitous across today's technology, Chapter 5 focuses on modelling elastic wave propagation is such structures. For comparison, solid spectral elements are used in parallel to achieve the same goals. Dispersion curves are compared for both model types with respect to Lamb waves and shear waves. Moreover, the process of modelling piezoelectric transducers used for elastic wave generation is discussed, and also the effect of the thickness of bonding layer between the piezoelectric transducer and the structure on the shapes of propagating waves is investigated.

Guided Waves in Structures for SHM: The Time-Domain Spectral Element Method, First Edition.
Wieslaw Ostachowicz, Pawel Kudela, Marek Krawczuk and Arkadiusz Zak.

6.1 Solid Spectral Elements

As in curved shells, in three-dimensional structural elements longitudinal, shear and flexural waves are coupled one with another. The general form of the displacement field associated with a three-dimensional spectral element is given by the following formula:

$$\begin{cases} u(x, y, z) = \hat{u}^e(x, y, z) \\ v(x, y, z) = \hat{v}^e(x, y, z) \\ w(x, y, z) = \hat{w}^e(x, y, z) \end{cases} \tag{6.1}$$

where the terms $\hat{u}^e(x, y, z)$, $\hat{v}^e(x, y, z)$, $\hat{w}^e(x, y, z)$ are associated with degrees of freedom respectively in directions x, y and z. The discussed solid spectral elements are based on the linear three-dimensional theory of elasticity and use full stress and deformation fields.

In discussing elastic wave modelling, and particularly in the case of Lamb waves, it is important to choose an appropriate approximation of the field of displacements along the structural element thickness. As described in Chapter 1, the higher the frequency of the induced signal, the more complex the distribution of displacements along the structural element thickness. For solid spectral elements it is possible to adjust the degree of approximation of the field of displacements along the element thickness to the current modelling needs. Depending on the number of nodes along the thickness of the solid spectral element, one obtains the same number of terms of a Maclaurin series in the expansion of the displacement field for direction z as described in Chapter 5. Thus the general discussion of approximations of the relevant forms of elastic waves is analogous to the one presented in Chapter 5.

6.2 Displacement Fields of Solid Structural Elements

6.2.1 Six-Mode Theory

Using solid spectral elements with two nodes across the element thickness leads to the following form of the displacement field:

$$\begin{cases} u(x, y, z) = \boxed{u_0(x, y)} + u_1(x, y)z \\ v(x, y, z) = \boxed{v_0(x, y)} + v_1(x, y)z \\ w(x, y, z) = w_0(x, y) + \boxed{w_1(x, y)z} \end{cases} \tag{6.2}$$

where the framed terms are associated with symmetric modes of shear waves and Lamb waves, while the other ones are associated with antisymmetric modes of shear waves and Lamb waves. It can be seen that in the above formulation the solid spectral element has 6 degrees of freedom along its thickness, which are associated with the functions $u_i(x, y)$, $v_i(x, y)$, $w_i(x, y)$, $(i = 0, 1)$. Thus the theory is a six-mode one, exhibiting equilibrium in accuracy of approximation between the symmetric and antisymmetric modes of elastic waves.

One should note that the full field of stress and displacements in a solid spectral element does not fulfil all the assumptions of the theories of Lamb waves and SH shear waves (cf. Chapter 1). The three-dimensional six-mode theory does not satisfy the assumption of vanishing normal and tangential stress on the bounding surfaces:

$$\begin{gathered} \sigma_{zz}(x, y, z) - \tau_{yz}(x, y, z) - \tau_{zx}(x, y, z) = 0 \\ 0 \leq x \leq l, \quad 0 \leq y \leq b, \quad z = \pm a = \pm\frac{h}{2} \end{gathered} \tag{6.3}$$

Lack of bounds (6.3) results in a solution whose accuracy of representation of propagating Lamb waves and SH shear waves is lower than for the one produced by the six-mode theory of shells. Dispersion curves resulting from the discussed three-dimensional six-mode theory are presented in Figure 6.1.

6.2.2 Nine-Mode Theory

Using solid spectral elements with three nodes across the element thickness leads to the following form of the displacement field:

$$\begin{cases} u(x, y, z) = \boxed{u_0(x, y)} + u_1(x, y)z + \boxed{u_2(x, y)z^2} \\ v(x, y, z) = \boxed{v_0(x, y)} + v_1(x, y)z + \boxed{v_0(x, y)z^2} \\ w(x, y, z) = w_0(x, y) + \boxed{w_1(x, y)z} + w_2(x, y)z^2 \end{cases} \tag{6.4}$$

where the framed terms are associated with symmetric modes of shear waves and Lamb waves, while the other ones are associated with antisymmetric modes. One can note that in the above formulation the solid spectral element has 9 degrees of freedom along its thickness, and they are associated with functions $u_i(x, y)$, $v_i(x, y)$, $w_i(x, y)$, $\quad (i = 0, 1, 2)$. Thus, the theory is a nine-mode one.

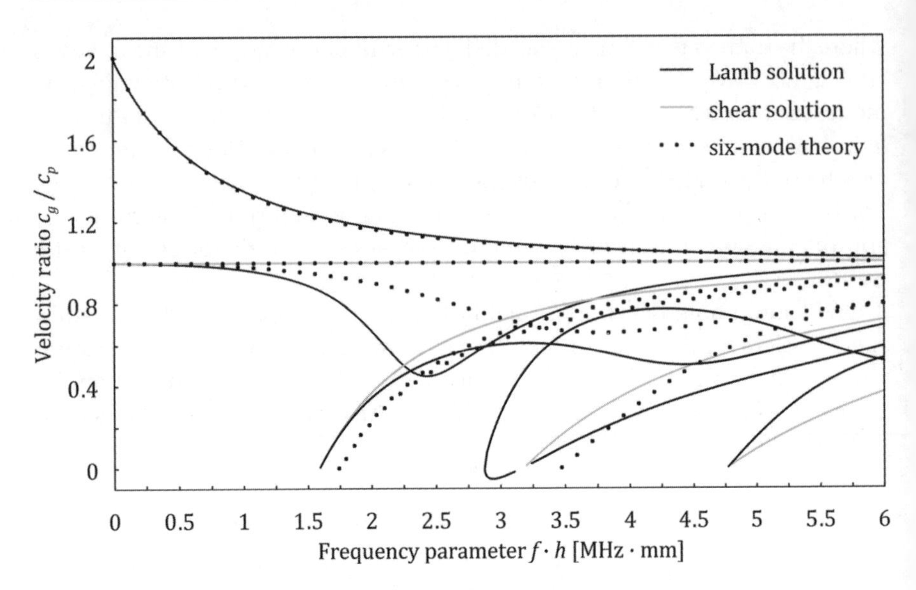

Figure 6.1 Dispersion curves for the velocity ratio c_g/c_p for the six-mode theory of 3D solids ($c_l = 6.3$ km/s, $c_s = 3.2$ km/s)

In the three-dimensional nine-mode theory the equilibrium of approximation accuracy between symmetric and antisymmetric modes is disturbed. Symmetric modes are represented by five terms, while antisymmetric modes are represented by four terms. This is an unwanted feature and unbalances the conversion between symmetric and antisymmetric modes. As for the six-mode theory, the assumed displacement field does not guarantee that the conditions of normal and tangential stress vanishing on bounding surfaces are met. As a result, a better approximation of symmetric modes is obtained, compared with the six-mode theory of solid elements described earlier. Dispersion curves for the three-dimensional nine-mode theory have been presented in Figure 6.2.

6.3 Certain Numerical Considerations

As in Chapter 5, the focus was put on the base modes of Lamb waves, which are most frequently employed in damage detection algorithms. In order to determine the accuracy and scope of applicability of the discussed theories of solid spectral finite elements, the relative error δ given by formula (5.25) and mean error Δ given by formula (5.26) were computed

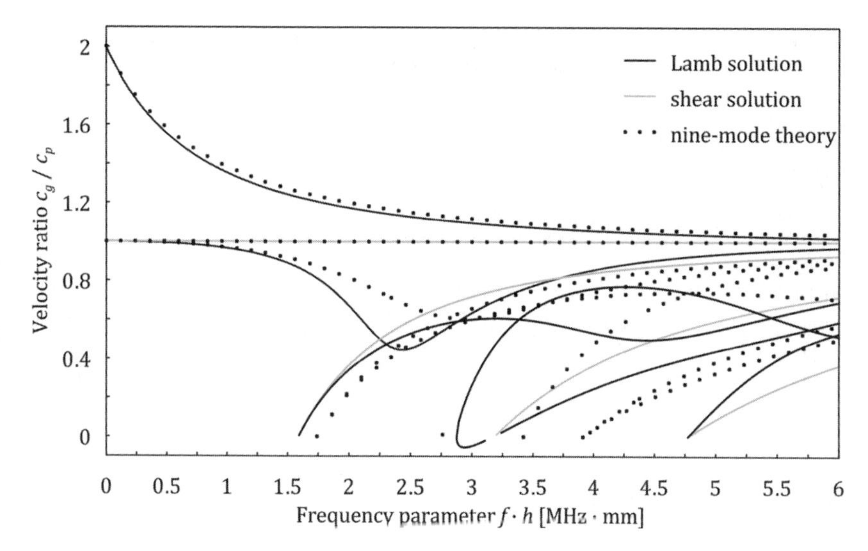

Figure 6.2 Dispersion curves for the velocity ratio c_g/c_p for the nine-mode theory of 3D solids ($c_l = 6.3$ km/s, $c_s = 3.2$ km/s)

again. These errors result from comparing the exact solution based on the Lamb wave theory with results obtained for both discussed theories of solid spectral elements, based on the dispersion curves presented in Figures 6.1 and 6.2.

Results presented in Table 6.1 and in Figure 6.3 summarise the effectiveness of the investigated theories of solid spectral finite elements and their accuracy of approximating the S_0 mode of Lamb waves. For comparison, results obtained for the chosen membrane theories in Chapter 5 were repeated.

Table 6.1 and Figure 6.3 demonstrate that the three-dimensional nine-mode theory provides much better accuracy than the three-dimensional six-mode theory. In the case of the three-dimensional nine-mode theory the relative error δ and mean error Δ do not exceed 1% for the frequency range of parameter fh up to 1.5 MHz·mm. The scope of applicability of the three-dimensional six-mode theory with regard to the frequency of parameter fh is much more limited, extending up to about 0.9 MHz·mm.

One should note that much more accurate results, even compared to the three-dimensional nine-mode theory, can be obtained by employing the three-mode theory of membranes of higher order. At the same time, computational costs are much lower as the ratio of degrees of freedom

Table 6.1 Relative error δ for selected 3D solid and membrane behaviour theories calculated against the Lamb analytical solution for the fundamental symmetric mode of wave propagation ($c_l = 6.3$ km/s, $c_p = 3.2$ km/s)

fd (MHz·mm)	δ_A (%)	δ_B (%)	δ_C (%)	δ_D (%)	δ_E (%)
0.20	0.03	0.00	0.03	−0.02	−0.02
0.40	0.15	0.01	0.13	−0.03	−0.03
0.60	0.39	0.03	0.36	−0.06	−0.06
0.80	0.81	0.12	0.76	−0.08	−0.08
1.00	1.54	0.35	1.47	−0.11	−0.12
1.20	2.82	0.86	2.74	−0.12	−0.17
1.40	5.12	2.00	5.01	−0.09	−0.22
1.60	9.40	4.45	9.25	0.10	−0.28
1.80	17.71	9.81	17.51	0.62	−0.30
2.00	34.41	21.40	34.14	1.61	−0.31
fd (MHz·mm)	Δ_A (%)	Δ_B (%)	Δ_C (%)	Δ_D (%)	Δ_E (%)
0.5	0.02	0.00	0.06	0.01	0.01
1.0	0.09	0.06	0.39	0.04	0.04
1.5	0.51	0.46	1.46	0.06	0.09
2.0	2.32	2.70	5.41	0.19	0.13

A – six-mode 3D solids, B – nine-mode 3D solids, C – three-mode theory of membranes, D – higher order three-mode theory of membranes, E – six-mode theory of membranes.

responsible for approximating the S_0 mode equals 3:5 for the three-mode theory of membranes of higher order and the three-dimensional nine-mode theory, respectively. The 3:5 ratio results from the fact that in the three-mode theory of membranes according to Equations (5.13) the S_0 mode is approximated with three independent terms (degrees of freedom), and in the three-dimensional nine-mode theory the S_0 mode is approximated with five terms (degrees of freedom), framed in Equations (6.4).

In the case of modelling of the A_0 mode of Lamb waves using the theory of solid spectral finite elements, the changes of both the relative error δ and mean error Δ, compared with Lamb's analytical solution, are presented in Table 6.2 and in Figure 6.4. For comparison, results obtained for chosen plate theories in Chapter 5 were repeated.

As discussed earlier, Table 6.2 and Figure 6.4 demonstrate clearly that the three-dimensional six-mode theory shows a lower relative error δ and mean

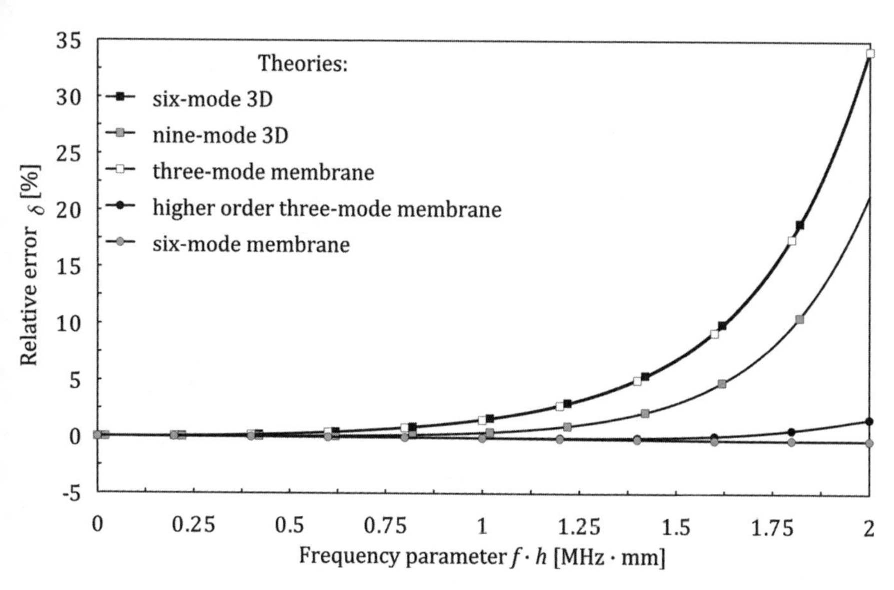

Figure 6.3 Relative error for selected 3D solid theories and membrane behaviour theories calculated against the Lamb analytical solution for the fundamental symmetric wave propagation mode ($c_l = 6.3$ km/s, $c_S = 3.2$ km/s)

error Δ than the three-dimensional nine-mode theory. This happens despite the fact that the three-dimensional nine-mode theory utilises one term more than the three-dimensional six-mode theory. Of course, this comparison only takes into account the terms responsible for approximating symmetric modes. In the whole frequency range of parameter fh up to 1.5 MHz·mm, the relative error for the six-mode theory remains below 1%. A similar accuracy can be obtained using the six-mode theory of plates.

Summarising, theories of solid spectral finite elements do not guarantee better representation of symmetric and antisymmetric modes of Lamb waves than the corresponding theories of shells, constructed by composing theories of plates and higher order theories of membranes. Moreover, theories of solid spectral finite elements can lead to numerical models of a much higher number of degrees of freedom than the corresponding theories of shells, which provide the same level of solution errors. Nevertheless, these theories allow for modelling certain three-dimensional structures that are difficult to model using shell-type spectral elements (e.g. structure joints, ribs, piezoelectric transducers).

Table 6.2 Relative error δ for selected 3D solid and plate behaviour theories calculated against the Lamb analytical solution for the fundamental antisymmetric wave propagation mode ($c_l = 6.3$ km/s, $c_p = 3.2$ km/s)

fd (MHz·mm)	δ_A (%)	δ_B (%)	δ_C (%)	δ_D (%)	δ_E (%)
0.20	0.05	1.00	−0.55	0.12	0.04
0.40	−0.08	1.62	−0.95	0.44	0.13
0.60	−0.28	1.98	−1.24	0.91	0.27
0.80	−0.47	2.20	−1.44	1.52	0.46
1.00	−0.63	2.32	−1.56	2.20	0.67
1.20	−0.75	2.38	−1.62	2.94	0.91
1.40	−0.83	2.39	−1.63	3.67	1.16
1.60	−0.87	2.37	−1.61	4.38	1.42
1.80	−0.88	2.33	−1.55	5.03	1.67
2.00	−0.86	2.27	−1.47	5.60	1.92
fd (MHz·mm)	Δ_A (%)	Δ_B (%)	Δ_C (%)	Δ_D (%)	Δ_E (%)
0.25	−0.02	1.08	0.64	0.23	0.07
0.50	−0.22	1.61	1.01	0.81	0.24
0.75	−0.40	1.86	1.21	1.59	0.49
1.00	−0.52	1.98	1.29	2.41	0.77

A – six-mode 3D solids, B – nine-mode 3D solids, C – three-mode Mindlin–Reissner, D – higher order three-mode theory of plates, E – six-mode theory of plates.

6.4 Modelling Electromechanical Coupling

Modelling electromechanical coupling constitutes an important element of modelling systems that utilise piezoelectric transducers integrated with the structure. Piezoelectric transducers can be parts of intelligent systems allowing for control of the underlying structure. Such systems are called *smart structures*. A smart structure is one that is able to respond adaptively. Such a structure should behave in an optimum fashion under diverse environment conditions and react to changes of the material it was made of (e.g. cracking).

In recent years many scientists have been drawn to researching smart structures because of their potential advantages in such applications as shape control, vibration reduction and diagnosing technical conditions. Using smart materials, such as piezoelectric elements manufactured as piezoelectric layers or tiles, embedded or bonded to composite structures, allows one to create

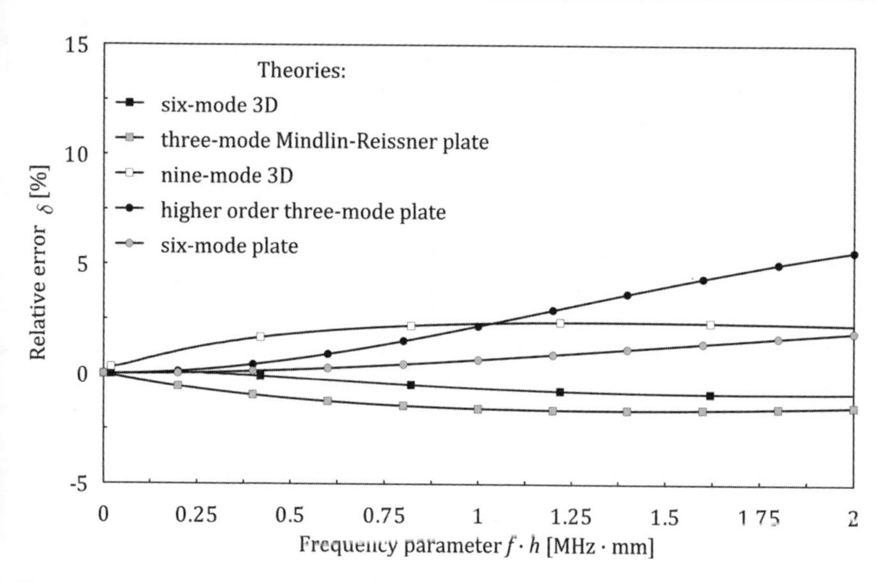

Figure 6.4 Relative error for selected 3D solid theories and plate behaviour theories calculated against the Lamb analytical solution for the fundamental antisymmetric wave propagation mode ($c_l = 6.3$ km/s, $c_s = 3.2$ km/s)

structures combining the great physical characteristics of composite materials with static and dynamic response capabilities.

Analysing composite structures like laminated plates or beams with embedded piezoelectric elements requires an efficient and accurate model capable of reproducing electromechanical coupling. The main aim is to represent both mechanical and electrical responses such as mechanical displacements and electric potentials. An accurate response of such structures can be obtained by solving piezoelectric constitutive equations while strictly fulfilling mechanical and electrical boundary conditions, as well as conditions of continuity between layers of composite laminates. Analytic solutions are only possible for three-dimensional problems in particular cases. These mostly apply to freely supported panels and rectangular plates made of piezoelectric materials [1–4]. Due to the fact that analytical solutions of three-dimensional piezoelectric problems are only available for regular shapes with simple, specific boundary conditions, applying approximation methods to problems of this kind is desirable. This applies particularly to composite structures with piezoelectric elements of complex shapes.

In the case of computing the mechanical displacement field, employing fully three-dimensional finite elements [5, 6] results in a very large problem size and high computational costs. Several two-dimensional models in which displacements are restricted to the central plane are available in the subject literature. These models incorporate additional assumptions concerning the distribution of the mechanical displacement field along the thickness [7]. The equivalent approximation of a single laminate layer combined with a *layer-wise* approximation constitutes one of the used computational techniques. Approximating a single layer on the basis of a first-order theory with account being taken for shear constitutes a good compromise between solution accuracy, computation economy and implementation simplicity. However, in the case of problems of elastic wave propagation such an approach may prove insufficient, as for higher signal frequencies of propagating waves the distribution of displacements along the element thickness is nonlinear (cf. Figure 1.14).

Also the distribution of an electric field potential along the element thickness can be approximated in different ways. In the simplest case one assumes a constant electric potential along the piezoelectric element thickness, that is in the direction that usually coincides with the polarisation direction. However, the above simplification may lead to significant errors. Mindlin [8] prepared a model of a piezoelectric plate assuming that the crosswise electric field is constant along the thickness. This assumption causes the electric potential function to vary linearly along the thickness of piezoelectric transducer layers. This model has found many uses. One should take into account, however, that such an approach is only sufficient for thin actuators. In the case of thick actuators higher order approximations need to be employed. Moreover, the Mindlin model neglects the piezoelectric potential induced in a piezoelectric sensor working in a closed circuit. For this reason, the local response in the form of an electric field potential cannot be correctly anticipated.

After analysing three-dimensional structures [3] it was noticed that the distribution of an electric potential along the thickness of the piezoelectric transducer is a higher-order function. Similar results are obtained when solid three-dimensional spectral elements are used (Figure 6.5). The nonlinear distribution of an electric potential is directly coupled with transducer bending and the effect is of the same order as transducer flexural stiffness [9].

Some researchers assumed a potential distribution in the form of a trigonometric function, for example 1 of the period of cosine function, taking into account the shear effect [7]. *Layerwise* theory [10] or multilayer modelling [11] are used to account more accurately for complex mechanical and electrical behaviour of laminates with piezoelectric elements. However, multilayer

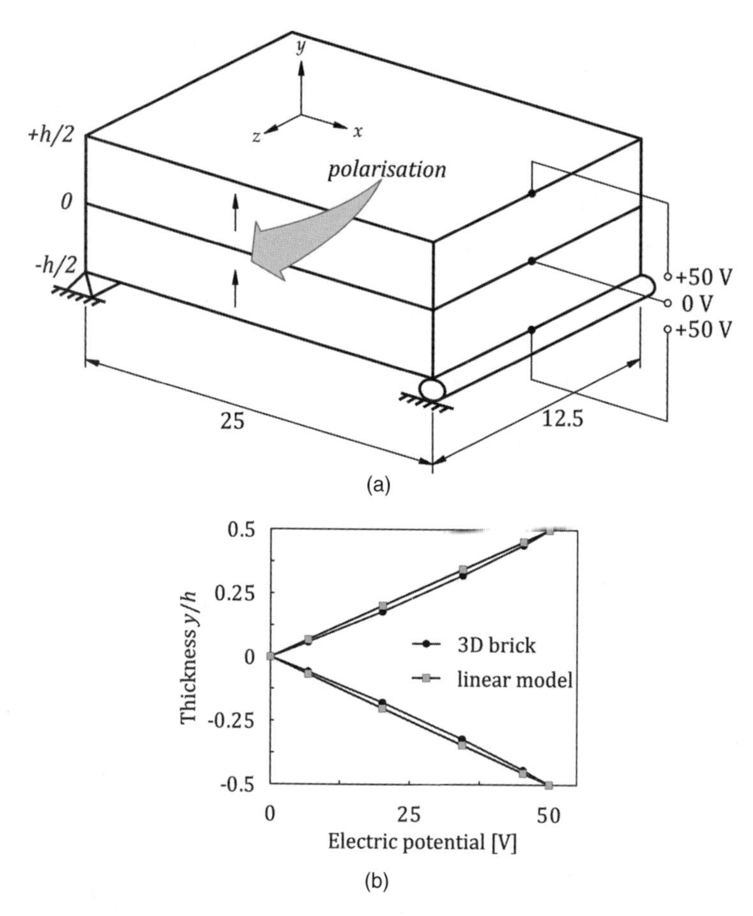

Figure 6.5 Piezoelectric bimorph (a) and resulting electric potential distribution (b) at the centre of the bimorph as a result of applied electric potential. Results are obtained by using 3D brick spectral elements for $h = 2.5$ mm

modelling can lead to very large numbers of electric field variables compared to the physical number of layers or the number of sensors and actuators. Kapuria [12] even proposed an electromechanical model combining an approximation of the displacement field by a third-order *zigzag* theory with an approximation of the electric potential by the *layerwise* theory. Also models incorporating a higher order approximation of the electric field have been proposed (including cubic terms), but these terms can be neglected for very thin plates. In other research [11] a biquadratic model was proposed for the electric potential distribution along the thickness. On the basis of this model it was concluded that in order to obtain a more accurate assessment of the stress field and electrical displacement field one has to take into account higher-order terms. Usually, for representing the electric potential, terms up to the fourth order are taken into account. Each of the approaches described above has advantages and disadvantages. Nevertheless, for problems of electric wave propagation, models with a higher order approximation of the displacement field in the direction of propagation and the static distribution of crosswise displacements depending on wave frequency and element thickness are preferred. The situation is similar with an electric potential field (when thick piezoelectric actuators are used, higher order discretisation is required).

Plate and membrane finite elements used for modelling composite laminates with built-in piezoelectric elements usually discretise mechanical and electrical displacement fields in the two-dimensional neutral plane. At the same time the electric potential is assumed to vary linearly across the element thickness (no discretisation along the thickness). Such assumptions can lead to discrepancies between the electric field and the mechanical displacement field, as the electric field can significantly affect the mechanical response by means of electromechanical coupling. For this reason there is no guarantee that both mechanical displacements and electric potential converge to the exact solution with mesh thickening. A numerical solution can heavily depend on the electric potential distribution along the thickness.

One should note that the existing analytical–numerical research concerning electromechanical coupling is largely restricted to free vibrations [13–15], while in wave problems this coupling is usually neglected and supplanted by forcing in the form of equivalent piezoelectric forces or displacements [16]. Equivalent piezoelectric forces are assumed to depend linearly on the applied voltage. Only rarely do analytical papers concerning Lamb wave induction take into account nonlinear stress distribution in piezoelectric actuators [17]. Nonlinear stress distribution depends on the frequency and mode of induced waves.

In the present chapter a three-dimensional solid spectral element with a high-order approximation is described. Such an approximation guarantees convergence of both the displacement field and potential field. The used approximation allows for modelling thick piezoelectric elements ($L/h < 10$, where L is the longest edge and h is thickness) of arbitrary shapes, providing accurate responses of both the piezoelectric element and the structure it is attached to.

6.4.1 Assumptions

The following modelling assumptions are taken henceforth:

- The piezoelectric element is made of an elastic, homogeneous material that is isotropic both dielectrically and mechanically.
- Piezoelectric actuator deformations take place under isothermal conditions.
- Electric potential is approximated by the same function as that of the displacement field, and the mesh of spectral finite elements overlaps the shape of the piezoelectric transducer.
- The main axes of the piezoelectric transducer are parallel to the orthotropic axes of the piezoelectric material and the direction of polarization is direction 3 (crosswise ζ).

6.4.2 Linear Constitutive Equations

In the case of the piezoelectric material, electrical and mechanical constitutive equations are coupled as follows:

$$\boldsymbol{\sigma} = \mathbf{C}^E \boldsymbol{\varepsilon} - \mathbf{e}^{\mathrm{T}} \mathbf{E} \tag{6.5}$$

$$\mathbf{D} = \mathbf{e}\boldsymbol{\varepsilon} + \mathbf{g}\mathbf{E} \tag{6.6}$$

where $\boldsymbol{\varepsilon}$ is the deformation vector, $\mathbf{D}$ is the piezoelectric displacement vector, $\boldsymbol{\sigma}$ is the stress vector, $\mathbf{C}^E$ is the matrix of elasticity coefficients under a constant electric field, $\mathbf{e}$ is the matrix of piezoelectric coupling constants, $\mathbf{E}$ is the electric field vector and $\mathbf{g}$ is the matrix of dielectric constants.

Equation (6.5) is the starting point for the operation of a piezoelectric transducer in the actuator mode, while Equation (6.6) is one for the operation in the sensor mode for recording signals of propagating waves.

The relationship between the electric field vector $\mathbf{E}$ and the electric potential field ϕ is given by the following formula:

$$\mathbf{E} = -\nabla\phi \tag{6.7}$$

Distribution of the electric potential in the spectral finite element is approximated using shape functions of the same order as the mechanical displacements. For example, for a spectral element with six nodes in the direction of axis ξ, six nodes in the direction of axis η and three nodes in the direction of axis ζ one obtains:

$$\phi^e(\xi,\eta,\zeta) = \mathbf{N}^e\hat{\phi}^e = \sum_{k=1}^{3}\sum_{j=1}^{6}\sum_{i=1}^{6} N_i^e(\xi)\, N_j^e(\eta)\, N_k^e(\zeta)\, \hat{\phi}^e(\xi_i,\eta_j,\zeta_k) \tag{6.8}$$

where $\hat{\phi}^e$ denotes the vector of the electric potential in the element nodes.

Leveraging Equation (6.7) and the approximation (6.8) one obtains the electric field vector expressed by the formula:

$$\boldsymbol{E} = -\boldsymbol{B}_\phi\hat{\phi} \tag{6.9}$$

where $\boldsymbol{B}_\phi$ is a matrix correlating the node values of the electric potential with the electric field, calculated as follows:

$$\boldsymbol{B}_\phi = \begin{bmatrix} \dfrac{\partial}{\partial x} \\ \dfrac{\partial}{\partial y} \\ \dfrac{\partial}{\partial z} \end{bmatrix} \boldsymbol{N}^e(\xi,\eta,\zeta) \tag{6.10}$$

6.4.3 Basic Equations of Motion

The equations of motion of a piezoelectric element can be derived by using Hamilton's variation principle, which takes into account the total kinetic and potential energies, as well as work performed by external mechanical and electrical forces. The final equations of motion of a piezoelectric element can be expressed in matrix form:

$$\begin{aligned} \boldsymbol{M}^e\hat{\ddot{\boldsymbol{u}}}^e + \boldsymbol{C}^e_{uu}\hat{\dot{\boldsymbol{u}}}^e + \boldsymbol{K}^e_{uu}\hat{\boldsymbol{u}}^e - \boldsymbol{K}^e_{u\phi}\hat{\boldsymbol{\phi}}^e &= \boldsymbol{f}^e \\ \boldsymbol{K}^e_{\phi u}\hat{\boldsymbol{u}}^e + \boldsymbol{K}^e_{\phi\phi}\hat{\boldsymbol{\phi}}^e &= \boldsymbol{q}^e \end{aligned} \tag{6.11}$$

where $\boldsymbol{M}^e$ denotes the element's matrix of inertia, $\boldsymbol{K}^e_{uu}$ denotes the matrix of mechanical stiffness, $\boldsymbol{K}^e_{u\phi}$ and $\boldsymbol{K}^e_{\phi u}$ denote matrices of the piezoelectric coupling, respectively, $\boldsymbol{K}^e_{\phi\phi}$ denotes the matrix of dielectric constants, $\hat{\boldsymbol{u}}^e$, $\hat{\dot{\boldsymbol{u}}}^e$ and $\hat{\ddot{\boldsymbol{u}}}^e$ denote vectors of displacements, velocities and node accelerations of the element, respectively, $\boldsymbol{f}^e$ denotes the vector of external forces and $\boldsymbol{q}^e$ denotes the vector of external electric charges applied in element nodes.

Individual characteristic matrices and vectors appearing in Equations (6.11) can be derived using the principles described in Chapter 2. For notation clarity, we shall now define the matrices resulting from electromechanical coupling that were not covered in Chapter 2:

$$\begin{cases} \boldsymbol{K}^e_{\phi\phi} = \int\limits_{V^e} (\boldsymbol{B}^e_u)^{\mathrm{T}}\, \mathbf{g}^e\, \boldsymbol{B}^e_\phi d V^e \\ \boldsymbol{K}^e_{u\phi} = -\int\limits_{V^e} (\boldsymbol{B}^e_u)^{\mathrm{T}} (\mathbf{e}^e)^{\mathrm{T}}\, \boldsymbol{B}^e_\phi d V^e \\ \boldsymbol{K}^e_{\phi u} = \left(\boldsymbol{K}^e_{u\phi}\right)^{\mathrm{T}} \\ \boldsymbol{q}^e = \int\limits_{V^e} (\boldsymbol{N}^e)^{\mathrm{T}} \boldsymbol{q}_0 d V^e \end{cases} \tag{6.12}$$

where $\mathbf{g}^e$ denotes the matrix of dielectric constants, $\mathbf{e}^e$ denotes the matrix of piezoelectric coupling constants and $\boldsymbol{q}_0$ denotes the vector of distribution of electric charge loads. Matrices and vectors from Equations (6.11) are subject to the same transformations in systems of coordinates that were described in Chapter 2 and therefore will not be repeated here.

6.4.4 Static Condensation

Characteristic matrices of elements are aggregated to the global form in a fashion typical for the finite element method. Vectors of unknown displacements and electric potentials could theoretically be computed directly from Equations (6.11). However, for piezoelectric materials typical values of the matrix $\boldsymbol{K}^e_{uu}$ are of the order of 10^8, while those of the matrix $\boldsymbol{K}^e_{\phi\phi}$ are of the order of 10^{-11}. This immense difference in absolute values can lead to bad conditioning of the global system of equations, when considered as a whole. In order to overcome these difficulties one can perform matrix static condensation of Equations (6.11). For this aim the part associated with the actuator is separated from the part associated with the sensor. In such a case Equations (6.11) are expressed in terms of displacements:

$$\begin{gathered} \boldsymbol{M}\ddot{\hat{\boldsymbol{u}}} + \boldsymbol{C}_{uu}\dot{\hat{\boldsymbol{u}}} + (\boldsymbol{K}_{uu} + \boldsymbol{K}_I)\,\hat{\boldsymbol{u}} = \boldsymbol{f} + \boldsymbol{f}_A \\ \boldsymbol{K}_I = \left(\boldsymbol{K}_{\phi u}\right)^{\mathrm{T}} \boldsymbol{K}^{-1}_{\phi\phi} \boldsymbol{K}_{\phi u} \end{gathered} \tag{6.13}$$

where $\boldsymbol{K}_I$ denotes the matrix of stiffness induced through electromechanical coupling and $\boldsymbol{f}_A$ denotes the vector of equivalent mechanical node forces generated by the piezoelectric actuator. One should add that induced stiffness in the form of matrix $\boldsymbol{K}_I$ is dependent on boundary conditions (open circuit, closed circuit, the transducer operating as an actuator). This stiffness can change in time (e.g. when after the transducer generates a signal it changes the operating mode to sensing).

6.4.5 Inducing Waves

Wave excitation is realised by means of a vector of equivalent piezoelectric forces generated by an actuator:

$$\boldsymbol{f}_A = \boldsymbol{K}_{u\phi}\hat{\boldsymbol{\phi}}_A \tag{6.14}$$

In other words, an electrical field applied to transducer electrodes affects the potential distribution $\hat{\phi}_A$, which in turn generates forces acting on the structure $\boldsymbol{f}_A$. It is relatively easy to control electrical voltage, and therefore it is possible to generate signals of diverse shapes. When the shape of the signal applied to an actuator is appropriate, elastic waves will be induced. It is also possible to reduce system vibrations by activating actuators spaced in such a fashion that their locations coincide with the largest displacement amplitudes for the given vibration form.

6.4.6 Recording Waves

Structure deformations result in induction of electric charges in the piezoelectric transducer attached to it. This is the piezoelectric effect, which can be leveraged for recording deformations, and in particular for recording propagating elastic waves. Wave recording by a piezoelectric sensor is associated with an electric field potential that can be computed according to the following formula:

$$\hat{\boldsymbol{\phi}}_S = -(\boldsymbol{K}_{\phi\phi})^{-1}\boldsymbol{K}_{\phi u}\hat{\boldsymbol{u}} \tag{6.15}$$

where $\boldsymbol{K}_{\phi\phi}$ and $\boldsymbol{K}_{\phi u}$ are the respective submatrices associated with the sensor considered as part of the open and closed circuits, respectively (imposed electrical boundary conditions).

6.4.7 Electrical Boundary Conditions

Electrical boundary conditions need to be imposed, because the matrix of dielectric constants $\boldsymbol{K}_{\phi\phi}$ is not positively definite. Depending on the type of electrical boundary conditions in Equations (6.13), respective submatrices of matrices $\mathbf{K}_{\phi\phi}$ and $\mathbf{K}_{\phi u}$ are taken into account.

A piezoelectric transducer was assumed to be attached to the structure surface and electrodes to be located on its upper and lower surfaces. In such a configuration the piezoelectric transducer can be used as a sensor in closed or open circuits or as an actuator. Thus, three types of electrical boundary conditions need to be considered:

1. *Piezoelectric sensor in closed circuit.* An electric charge gathers on electrodes of a piezoelectric material as a result of its deformations. Closing the electric circuit causes the charge to flow away from the top and bottom surfaces (electrodes) of the piezoelectric transducer. At the same time it is assumed that there is no free electric charge inside the piezoelectric material. In other words, the electric potential in nodes on the upper and lower surfaces of a solid spectral finite element can be assumed to be zero (grounding), while in the other nodes of the element the electric potential is induced. According to Equations (6.11), the charge equation for a sensor in a closed circuit can be written as:

$$\begin{bmatrix} \mathbf{K}_{\phi u}^{00} & \mathbf{K}_{\phi u}^{0i} & \mathbf{K}_{\phi u}^{0n} \\ \mathbf{K}_{\phi u}^{i0} & \mathbf{K}_{\phi u}^{ii} & \mathbf{K}_{\phi u}^{in} \\ \mathbf{K}_{\phi u}^{n0} & \mathbf{K}_{\phi u}^{nn} & \mathbf{K}_{\phi u}^{ni} \end{bmatrix} \begin{Bmatrix} \hat{\mathbf{u}}^0 \\ \hat{\mathbf{u}}^i \\ \hat{\mathbf{u}}^n \end{Bmatrix} + \begin{bmatrix} \mathbf{K}_{\phi\phi}^{00} & \mathbf{K}_{\phi\phi}^{0i} & \mathbf{K}_{\phi\phi}^{0n} \\ \mathbf{K}_{\phi\phi}^{i0} & \mathbf{K}_{\phi\phi}^{ii} & \mathbf{K}_{\phi\phi}^{in} \\ \mathbf{K}_{\phi\phi}^{n0} & \mathbf{K}_{\phi\phi}^{nn} & \mathbf{K}_{\phi\phi}^{ni} \end{bmatrix} \begin{Bmatrix} \mathbf{0} \\ \hat{\boldsymbol{\phi}}_c^i \\ \mathbf{0} \end{Bmatrix} = \begin{Bmatrix} \hat{\mathbf{q}}^0 \\ \mathbf{0} \\ \hat{\mathbf{q}}^n \end{Bmatrix} \tag{6.16}$$

where the upper indices denote degrees of freedom of a solid spectral element in nodes of, respectively:

- 0, bottom layer,
- i, middle layers and
- n, top layer,

as shown in Figure 6.6.

For notation clarity, in Equation (6.16) the index e marks the fact that matrices and vectors applying to an individual element have been omitted. One can calculate the induced potential by solving Equation (6.16):

$$\{\hat{\boldsymbol{\phi}}_c^i\} = -\left[\mathbf{K}_{\phi\phi}^{ii}\right]^{-1}\left[\mathbf{K}_{\phi u}^{i0} \quad \mathbf{K}_{\phi u}^{ii} \quad \mathbf{K}_{\phi u}^{in}\right] \begin{Bmatrix} \hat{\mathbf{u}}^0 \\ \hat{\mathbf{u}}^i \\ \hat{\mathbf{u}}^n \end{Bmatrix} \tag{6.17}$$

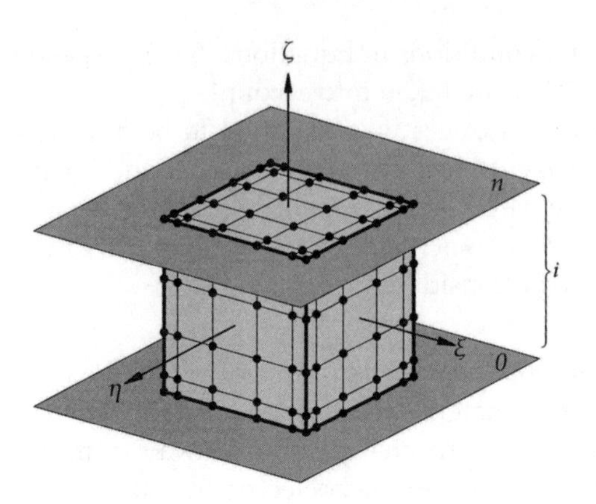

Figure 6.6 Numbering of node layers in a solid spectral finite element chosen for describing the electrical boundary conditions

According to Equations (6.13), the electrically induced stiffness of a sensor in the closed circuit $\mathbf{K}_I^c$ can be written as:

$$\mathbf{K}_I^c = \begin{bmatrix} \mathbf{K}_{\phi u}^{i0} \\ \mathbf{K}_{\phi u}^{ii} \\ \mathbf{K}_{\phi u}^{in} \end{bmatrix} \left[\mathbf{K}_{\phi\phi}^{ii}\right]^{-1} \left[\mathbf{K}_{\phi u}^{i0} \quad \mathbf{K}_{\phi u}^{ii} \quad \mathbf{K}_{\phi u}^{in}\right] \tag{6.18}$$

2. *Piezoelectric sensor in an open circuit.* An electric charge gathers on electrodes of a piezoelectric material as a result of its deformations. The electric potential on the bottom surface of a sensor is assumed to be zero (grounding). Thus, the electric potential is induced in the other nodes of a solid spectral finite element. According to Equations (6.11), the charge equation for a sensor working in an open circuit can be written as:

$$\begin{bmatrix} \mathbf{K}_{\phi u}^{00} & \mathbf{K}_{\phi u}^{0i} & \mathbf{K}_{\phi u}^{0n} \\ \mathbf{K}_{\phi u}^{i0} & \mathbf{K}_{\phi u}^{ii} & \mathbf{K}_{\phi u}^{in} \\ \mathbf{K}_{\phi u}^{n0} & \mathbf{K}_{\phi u}^{nn} & \mathbf{K}_{\phi u}^{ni} \end{bmatrix} \begin{Bmatrix} \hat{\mathbf{u}}^0 \\ \hat{\mathbf{u}}^i \\ \hat{\mathbf{u}}^n \end{Bmatrix} + \begin{bmatrix} \mathbf{K}_{\phi\phi}^{00} & \mathbf{K}_{\phi\phi}^{0i} & \mathbf{K}_{\phi\phi}^{0n} \\ \mathbf{K}_{\phi\phi}^{i0} & \mathbf{K}_{\phi\phi}^{ii} & \mathbf{K}_{\phi\phi}^{in} \\ \mathbf{K}_{\phi\phi}^{n0} & \mathbf{K}_{\phi\phi}^{nn} & \mathbf{K}_{\phi\phi}^{ni} \end{bmatrix} \begin{Bmatrix} \mathbf{0} \\ \hat{\boldsymbol{\phi}}_o^i \\ \hat{\boldsymbol{\phi}}_o^n \end{Bmatrix} = \begin{Bmatrix} \hat{\mathbf{q}}^0 \\ \mathbf{0} \\ \mathbf{0} \end{Bmatrix} \tag{6.19}$$

One can determine the induced potential by solving Equation (6.19):

$$\begin{Bmatrix} \hat{\boldsymbol{\phi}}_o^i \\ \hat{\boldsymbol{\phi}}_o^n \end{Bmatrix} = -\begin{bmatrix} \mathbf{K}_{\phi\phi}^{ii} & \mathbf{K}_{\phi\phi}^{in} \\ \mathbf{K}_{\phi\phi}^{ni} & \mathbf{K}_{\phi\phi}^{nn} \end{bmatrix}^{-1} \begin{bmatrix} \mathbf{K}_{\phi u}^{i0} & \mathbf{K}_{\phi u}^{ii} & \mathbf{K}_{\phi u}^{in} \\ \mathbf{K}_{\phi u}^{n0} & \mathbf{K}_{\phi u}^{ni} & \mathbf{K}_{\phi u}^{nn} \end{bmatrix} \begin{Bmatrix} \hat{\mathbf{u}}^0 \\ \hat{\mathbf{u}}^i \\ \hat{\mathbf{u}}^n \end{Bmatrix} \tag{6.20}$$

According to Equations (6.13), the electrically induced stiffness of a sensor in the open circuit K_I^o can be written as:

$$K_I^o = \begin{bmatrix} K_{\phi u}^{i0} & K_{\phi u}^{n0} \\ K_{\phi u}^{ii} & K_{\phi u}^{ni} \\ K_{\phi u}^{in} & K_{\phi u}^{nn} \end{bmatrix} \begin{bmatrix} K_{\phi\phi}^{ii} & K_{\phi\phi}^{in} \\ K_{\phi\phi}^{ni} & K_{\phi\phi}^{nn} \end{bmatrix}^{-1} \begin{bmatrix} K_{\phi u}^{i0} & K_{\phi u}^{ii} & K_{\phi u}^{in} \\ K_{\phi u}^{n0} & K_{\phi u}^{ni} & K_{\phi u}^{nn} \end{bmatrix} \tag{6.21}$$

3. *Actuator.* As we are dealing with a piezoelectric actuator, the electric charge is fed to electrodes. The voltage is assumed to be supplied to the top electrode, while the electric potential on the bottom surface of the piezoelectric transducer is assumed to be zero (grounding). Thus, the electric potential is induced in nodes of a solid spectral finite element between the top and bottom layers of nodes. According to Equations (6.11), the charge equation for the actuator can be written as:

$$\begin{bmatrix} K_{\phi u}^{00} & K_{\phi u}^{0i} & K_{\phi u}^{0n} \\ K_{\phi u}^{i0} & K_{\phi u}^{ii} & K_{\phi u}^{in} \\ K_{\phi u}^{n0} & K_{\phi u}^{nn} & K_{\phi u}^{ni} \end{bmatrix} \begin{Bmatrix} \hat{u}^0 \\ \hat{u}^i \\ \hat{u}^n \end{Bmatrix} + \begin{bmatrix} K_{\phi\phi}^{00} & K_{\phi\phi}^{0i} & K_{\phi\phi}^{0n} \\ K_{\phi\phi}^{i0} & K_{\phi\phi}^{ii} & K_{\phi\phi}^{in} \\ K_{\phi\phi}^{n0} & K_{\phi\phi}^{nn} & K_{\phi\phi}^{ni} \end{bmatrix} \begin{Bmatrix} 0 \\ \hat{\phi}_A^i \\ \hat{V}_A^n \end{Bmatrix} = \begin{Bmatrix} \hat{q}^0 \\ 0 \\ \hat{q}^n \end{Bmatrix} \tag{6.22}$$

where $\hat{V}_A^n$ denotes the voltage vector in nodes corresponding to the top surface of a solid spectral finite element. One can determine the induced potential by solving Equation (6.22):

$$\{\hat{\phi}_A^i\} = -\left[K_{\phi\phi}^{ii}\right]^{-1}\left[K_{\phi u}^{i0} \quad K_{\phi u}^{ii} \quad K_{\phi u}^{in}\right] \begin{Bmatrix} \hat{u}^0 \\ \hat{u}^i \\ \hat{u}^n \end{Bmatrix} - \left[K_{\phi\phi}^{ii}\right]^{-1}\left[K_{\phi\phi}^{in}\right]\left\{\hat{V}_A^n\right\} \tag{6.23}$$

According to Equations (6.13), the electrically induced stiffness of the actuator K_I^A can be written as:

$$K_I^A = \begin{bmatrix} K_{\phi u}^{i0} \\ K_{\phi u}^{ii} \\ K_{\phi u}^{in} \end{bmatrix} \left[K_{\phi\phi}^{ii}\right]^{-1}\left[K_{\phi u}^{i0} \quad K_{\phi u}^{ii} \quad K_{\phi u}^{in}\right] \tag{6.24}$$

After taking into account the electrical boundary conditions, Equations (6.13) are solved in the same fashion as described in Chapter 2.

6.5 Examples of Numerical Calculations

6.5.1 Propagation of Elastic Waves in a Half-Pipe Aluminium Shell

This part of the chapter presents the effectiveness of the spectral finite element method on an example of elastic wave propagation in a three-dimensional structure. The structure is a half-pipe made of aluminium alloy of properties identical to those in examples described in Chapter 5 (Young's modulus $E = 72.7$ GPa, Poisson's coefficient $\nu = 0.33$, density $\rho = 2700$ kg/m^3). All edges of the discussed structural element are assumed to be free, which corresponds to free boundary conditions.

An outline of the geometry of the aluminium half-pipe being simulated is presented in Figure 6.7. The total element length l is assumed as 1000 mm, its width measured circumferentially as 500 mm and its height h as 10 mm. Moreover, on the opposing ends of the half-pipe piezoelectric transducers made of PZT-5A material are located circumferentially. The simulation covers 16 actuators and 16 sensors. Transducer geometry has been adapted to a mesh of solid spectral elements with a size of 31 elements in the circumferential direction by 60 elements in the direction of axis x. Thus, transducer sizes

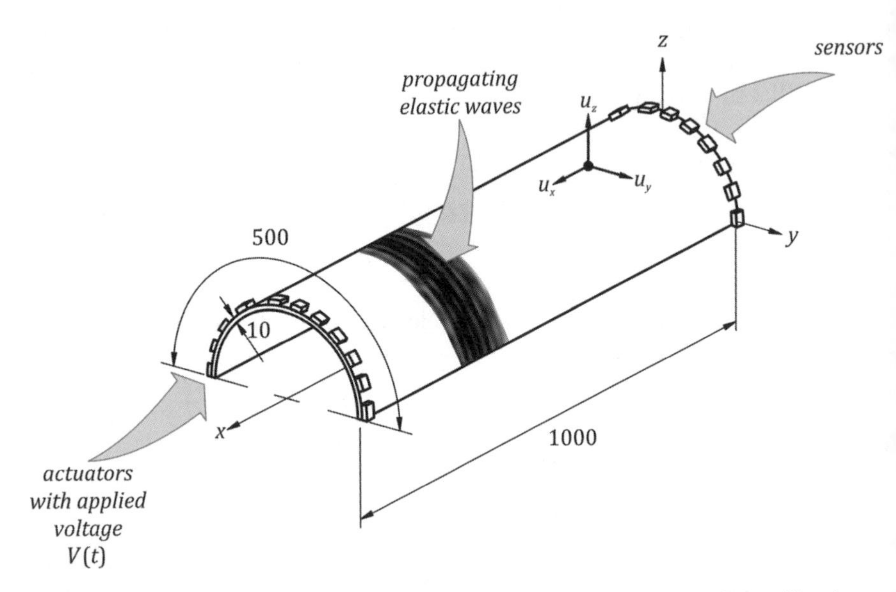

Figure 6.7 Geometry of a three-dimensional aluminium half-pipe with piezoelectric transducers

are as follows: the width in the circumferential direction is 16.12 mm, the length in the direction of axis x is 16.67 mm and the thickness is 1 mm. Both the main structure and piezoelectric transducers are modelled using solid spectral elements with three nodes in the radial direction (i.e. along the thickness), which corresponds to the three-dimensional nine-mode theory (cf. Equation (6.4)). Six nodes are used in each of the remaining directions, spaced according to the Lobatto node distribution.

Additionally, a crosswise, through-thickness, open crack has been placed at half the half-pipe's length, orientated circumferentially. The circumferential length of the crack equals about 15 mm and its centre is located at an angle of 19 degrees. The crack is modelled by disconnecting the appropriate nodes in adjacent solid spectral elements. In solid spectral elements, as in classical finite elements, the laws of fracture mechanics can be used for computing the loss of stiffness of the structural element at the crack location [18].

Elastic waves are generated in all actuators simultaneously, by applying the voltage $V(t)$ to their surfaces. The carrier frequency f_c and modulation frequency f_m of the excitation signal $V(t)$ with an amplitude of 100 V are assumed as $f_c = 75$ kHz and $f_m = 18.75$ kHz, respectively, corresponding to four cycles of the carrier signal. These values are equivalent to the range of frequency parameter $f\,h$ from $(f_c - 2f_m)\,h$ to $(f_c + 2f_m)\,h$, that is from 0.375 MHz·mm to 1.125 MHz·mm, in dispersion curves for the velocity ratio c_g/c_p shown in Figure 6.4.

For the needs of the numerical simulation, the time span covered has been chosen as $t_T = 0.5$ ms. The calculation time is divided into 8000 steps. This allows multiple reflections of propagating waves to be simulated from all edges of the three-dimensional shell. Results of numerical simulation in the form of three-dimensional frames of displacement amplitudes $\sqrt{u_x^2 + u_y^2 + u_z^2}$ are shown in Figure 6.8.

The results of numerical simulations shown in Figure 6.8 demonstrate that piezoelectric transducers induce a continuous front of the wave mode S_0, with the wave mode A_0 propagating behind. The nature of the induced wave front is consistent with rules for piezoelectric transducer systems, a so-called *phased array*. Continuity of the S_0 wave mode front guarantees that the condition $d < \lambda$, where d denotes the distance between the actuator centres and λ denotes the wave length, is met. An analogous condition is not met for the A_0 wave mode, which manifests as waves reflecting among actuators and causing additional interactions behind the main wavefront.

As for the shell used in the example in Chapter 5, each time an incoming wave encounters any kind of geometrical discontinuity of the pipe, its reflections are multi-modal in character. One needs to take into account, however,

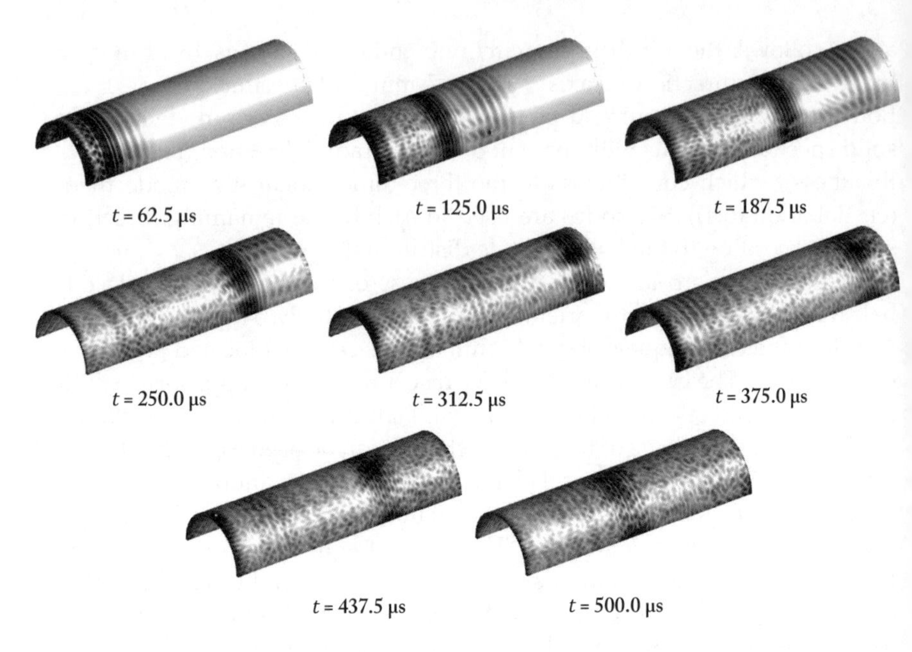

Figure 6.8 Snapshots of wave propagation patterns in a cracked aluminium half-pipe shell at various time instances according to the nine-mode theory of solids – results of the numerical simulation using the spectral finite element method

the fact that multiple wave reflections among piezoelectric transducers make mode conversions difficult to identify. Also wave reflections from damage sites are less noticeable than in the example in Chapter 5.

6.5.2 Propagation of Elastic Waves in an Isotropic Plate – Experimental Verification

The geometry of a square aluminium plate used for both numerical calculations and experimental measurements is shown in Figure 6.9. The aluminium plate is made of aluminium alloy 5754 of the following properties: Young's modulus $E = 68.0$ GPa, Poisson's coefficient $\nu = 0.33$ and density $\rho = 2660\ \mathrm{kg/m^3}$. A disc-shaped piezoelectric actuator with a diameter of 10 mm and thickness of 0.2 mm made of PZT-5A piezoelectric material has been bonded to the plate surface, in its geometric centre. The actuator is bonded with a very thin layer of wax. The actuator is used to induce a wave of carrier frequency $f_c = 35$ kHz and modulation frequency $f_m = 7$ kHz, which corresponds to five cycles of the carrier signal.

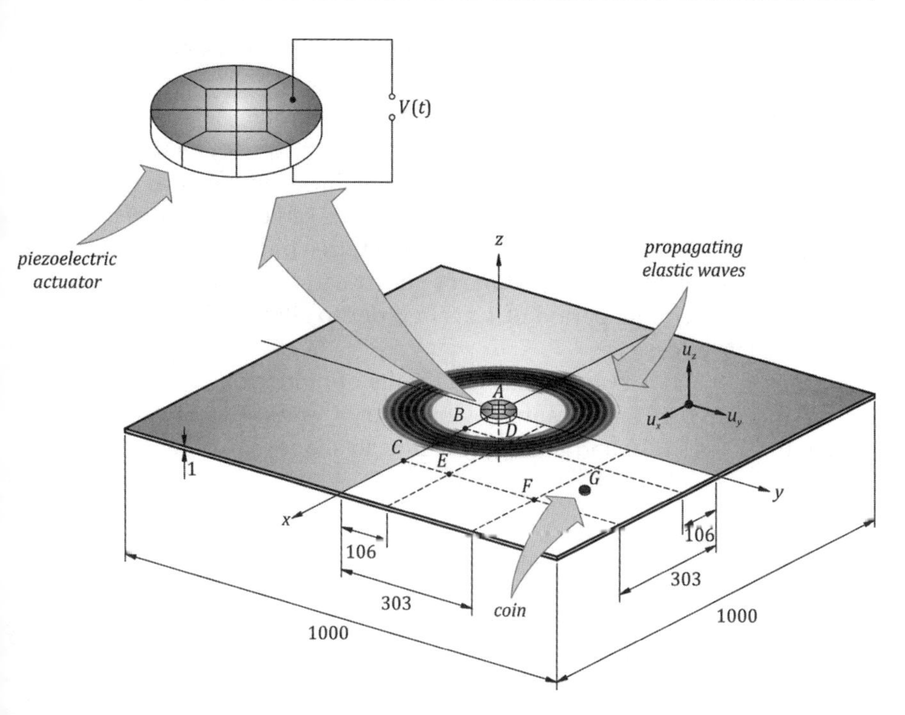

Figure 6.9 Geometry of a square aluminium plate showing the arrangement of measurement points, piezoelectric transducer and coin

In this case the results of numerical simulations utilising the spectral finite element method have been compared with results of experimental measurements obtained by means of laser vibrometry. The comparison concerns patterns of propagating waves at individual time instances for the mode A_0, which is associated chiefly with the transverse velocity component $\dot{u}_z$. The comparison also involves time signals in chosen points on the plate surface (points D and E), as shown in Figure 6.9.

Laser scanning vibrometry measurements have been performed on a single quarter of the plate. A 225-by-227-point mesh has been chosen, so that the propagating wave can be visualised. A single head of the Doppler laser vibrometer has been used for measuring velocities in the direction perpendicular to the plate surface. The recording time amounted to 1 ms and was synchronised with wave induction. The final results of laser scanning vibrometry measurements are the averaged results of 100 individual measurements.

For the needs of numerical calculations a mesh of solid spectral elements has been defined. It reflects the geometry of the whole plate and the piezoelectric transducer bonded to its surface. In all, 10 420 solid spectral elements are used to reproduce the plate geometry and 12 are used to represent the piezoelectric transducer. Solid spectral elements of $6 \times 6 \times 3$ nodes in directions ξ, η, ζ, respectively, with direction ζ being perpendicular to the plate, are used. Thus, the elements correspond to the three-dimensional nine-mode theory (cf. Equation (6.4)). The direction ζ is at the same time that of polarization of the piezoelectric transducer. Moreover, free boundary conditions are assumed on all plate edges. The time of numerical simulation t_T divided into 50 000 steps covers the period of 1 ms in order to allow for wave reflections from the plate edges.

Figures 6.10 and 6.11 juxtapose the results of numerical simulations with signals measured by laser scanning vibrometry. Consistency of the first wave packet in Figures 6.10 and 6.11 with regard to shape is excellent. The wave

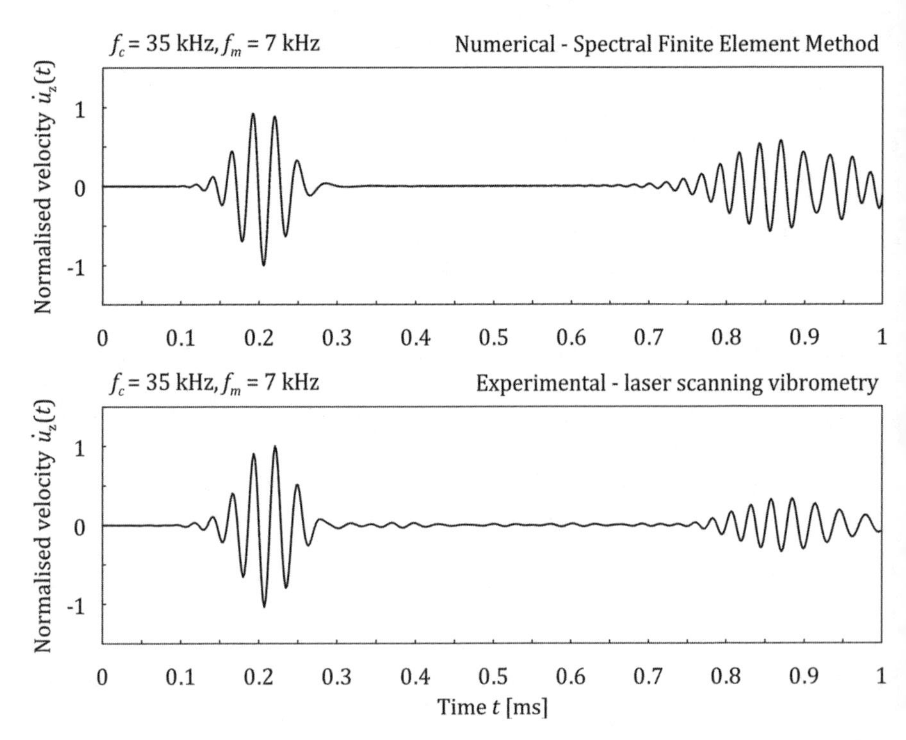

Figure 6.10 Time signals of transverse velocity components $\dot{u}_z$ at point D of the plate – results of the numerical simulation using the spectral finite element method (3D brick elements) and experimental measurements by laser scanning vibrometry

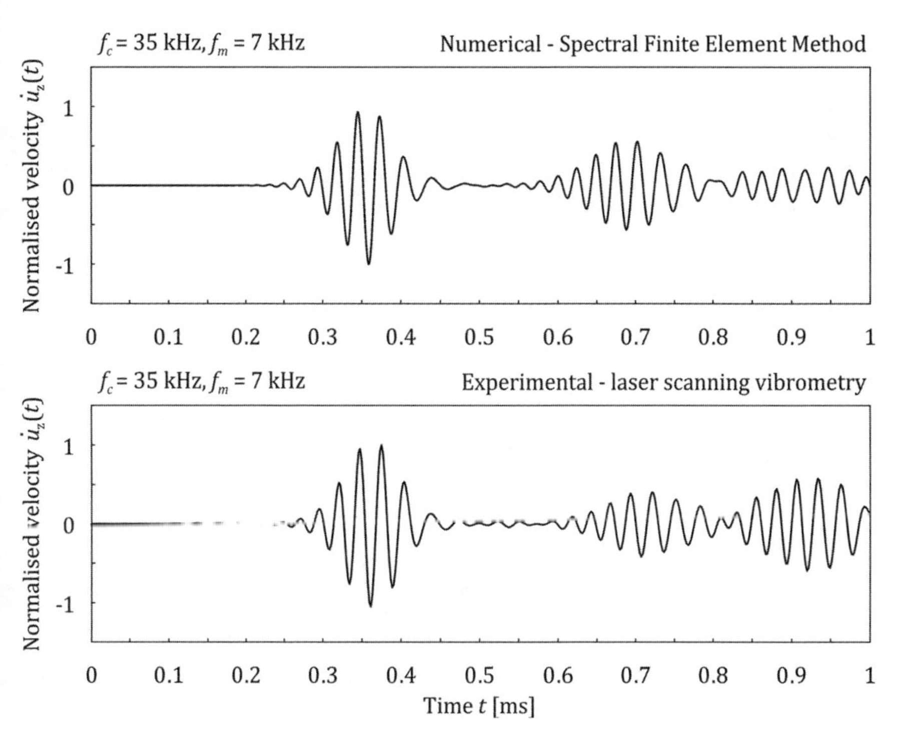

Figure 6.11 Time signals of transverse velocity components $\dot{u}_z$ at point D of the plate – results of the numerical simulation using the spectral finite element method (3D brick elements) and experimental measurements by laser scanning vibrometry

velocity in computed signals is slightly larger than the measured wave velocity, which is visible for subsequent wave packets reflected from the plate edges.

In the next stage of the experiment a coin with a diameter of 15.5 mm, thickness of 1.4 mm and mass of 1.64 g is bonded to the plate surface. The coin centre is located in point G of coordinates (0.2058 m, 0.3616 m) (cf. Figure 6.9). The coin can easily be located by means of analysing subsequent frames of propagating waves (Figure 6.12). The initially induced Lamb wave mode A_0, upon encountering the coin, reflects from its edge. Subsequently, waves reflected from the plate edge reach the coin again to reflect from its edges.

Figure 6.12 confirms the qualitative consistency of the numerical simulation with visualisation created on the basis of measured signals. In the numerical and experimental results being compared patterns of propagating

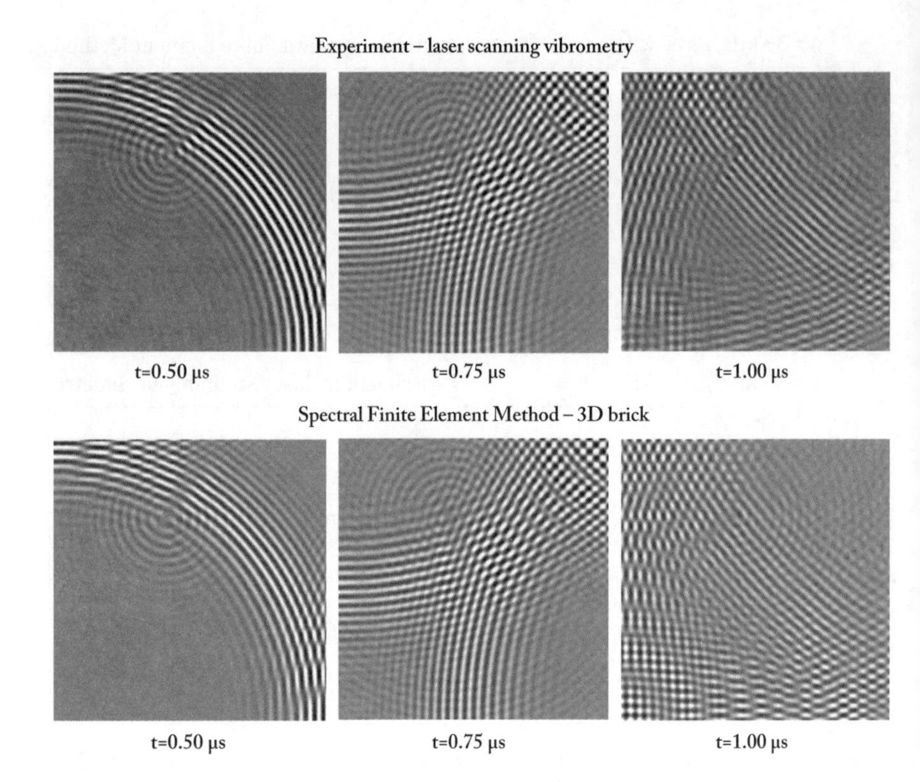

Figure 6.12 Frames of propagating waves and their interaction with a coin – comparison of experimental measurements with results of numerical calculations using solid spectral finite elements

and reflected waves are consistent. Dispersivities are similar and differences in the propagation velocity are minimal. An acute reader will notice differences in the first frames of wave propagation ($t = 0.50$ ms): in the experimental visualisation, waves of very low amplitude compared to the wavefront amplitude are visible between the wavefront and induction site (one can say that the tailing effect is noticeable). This effect is associated with the bonding layer between the piezoelectric transducer and the plate surface. Numerical calculations do not account for the bonding layer; therefore its effect is not visible in the respective visualisation. Moreover, an additional very small reflection appearing from behind the wavefront can be observed in the experimental visualisation. As no damage has been noticed on the surface of the investigated sample in the site of the additional wave reflection, one can conclude that it had been caused by an internal material defect.

6.6 Modelling the Bonding Layer

The numerical models described in Chapters 5 and 6 employed for investigating the phenomenon of elastic wave propagation do not account for the bonding layer between the piezoelectric transducer and the structure. Very good agreement between numerical results and experimental ones for forcing the carrier frequency of 35 kHz does not mean that the bonding layer can be neglected, however. The bonding layer affects the work characteristics of the piezoelectric element and modifies elastic wave dispersion relationships.

Despite much research into modelling of the bonding layer between the piezoelectric transducer and the structure, many aspects of the problem remain little understood. Significant differences in results observed between experimental research and numerical simulations of aspects of the electromechanical impedance [19] have caused researchers to focus on the so-called *shear lag effect*. This is a phenomenon involving piezoelectric element deformations differing from base structure deformations [20]. On the basis of numerical calculations for a system with one degree of freedom, Xu and Liu [21] demonstrated that if the quality of the bonding joint degrades, the resonance frequency of the system with the piezoelectric transducer increases. It was also determined that the system's dynamic stiffness matrix depends on the gluing process and bonding layer thickness.

The effects of the bonding layer thickness on elastic wave propagation are presented in Reference [22]. Spectral finite elements in the flat deformation state allow research to be undertaken of the effect of a 50 μm thick bonding layer on the shape of base modes of Lamb waves. Waves are induced in an aluminium plate about 1 mm thick. A sine-shaped inducing signal with a carrier frequency of $f_c = 400$ kHz modulated with a Hanning window is used. By intuition, one could expect the bonding layer to attenuate the wave in comparison with the case of perfect bonding. Simulations demonstrate, however, that increasing the bonding layer thickness results in increased amplitudes and peak counts in both the symmetric (S_0) and antisymmetric (A_0) wave modes. Numerical results are consistent with experimental observations. One should note that the scope of research reported in Reference [22] is limited and thus generalising the presented conclusions would be a mistake.

The bonding layer can be modelled using solid spectral finite elements. As the bonding layer is thin, a crosswise dimension of the finite element is also small, which results in the number of integration steps for equations of motion in the central differences method being very large. For this reason it is advantageous to use a three-dimensional spectral finite element with only three nodes along the thickness. Numerical simulations described in

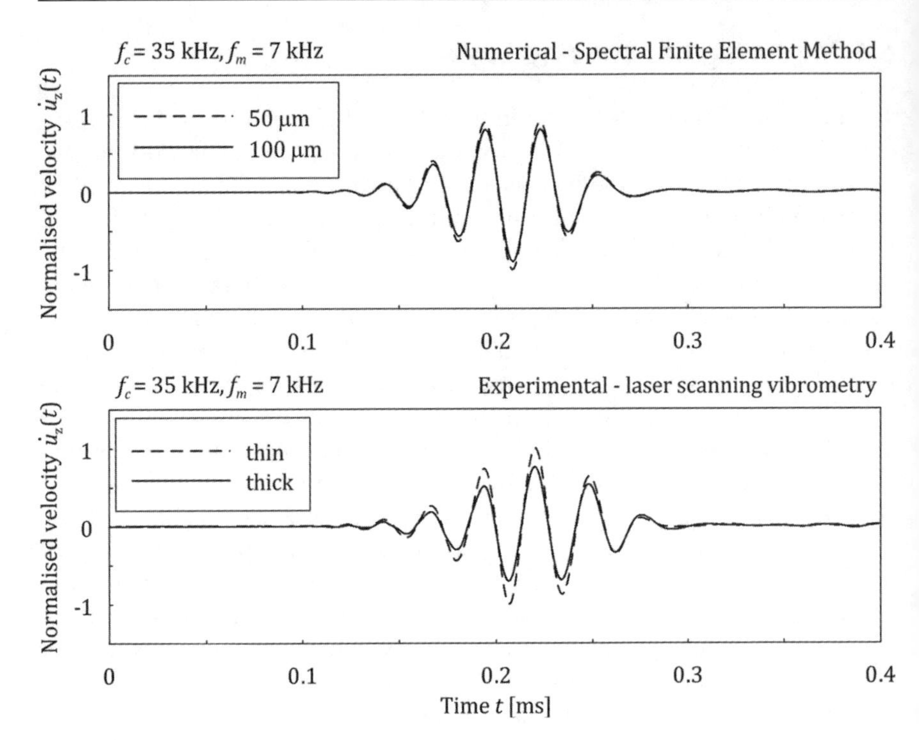

Figure 6.13 Time signals of transverse velocity components $\dot{u}_z$ at point D of the plate for various bonding layer thicknesses

Section 6.4 have been repeated with the bonding layer included. Bonding layer parameters have been assumed as follows: density of 1250 kg/m^3, Poisson's coefficient of 0.3, shear modulus of 2 GPa.

The results of numerical simulations presented in Figure 6.13 are consistent with measurements by laser scanning vibrometry. Figure 6.6 demonstrates that for a carrier frequency $f_c = 35$ kHz increasing the bonding layer thickness decreases the amplitude of the propagating wave. The wave packet shape remains virtually unaffected. The obtained result is completely different from the one described in paper [22] for inducing a signal of frequency $f_c =$ 400 kHz. Thus one can infer that the effect of the bonding layer on the shape of the induced elastic wave differs depending on the frequency of the induced signal.

Further research has been performed for the case of a composite panel made of epoxide resin reinforced with carbon fibre. The panel is a square with 439.5 mm long sides. The panel is composed of four layers of unidirectional

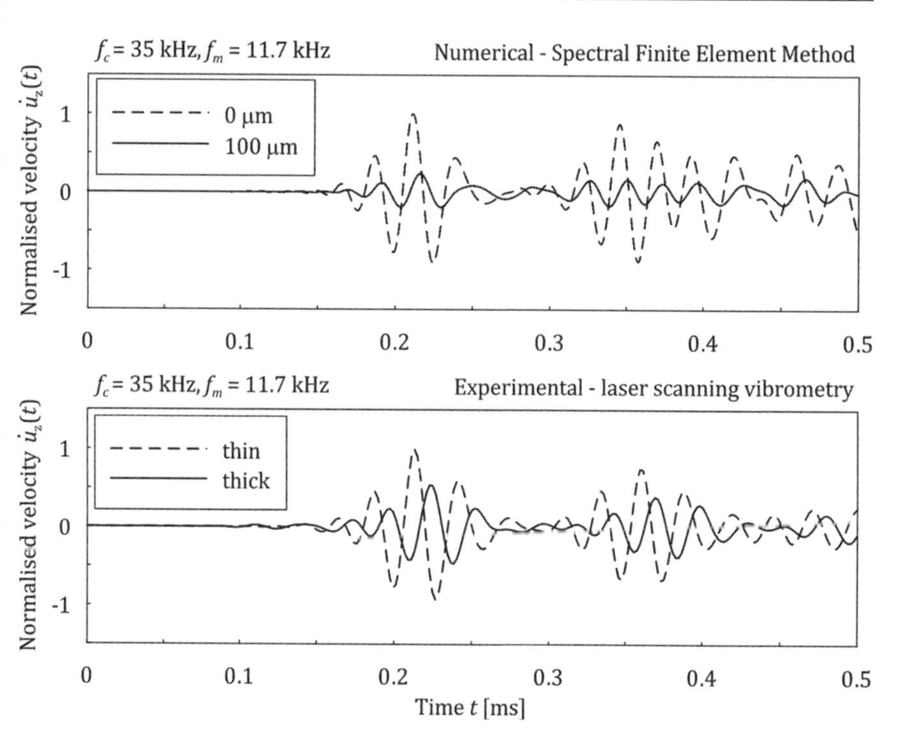

Figure 6.14 Time signals of transverse velocity components $\dot{u}_z$ calculated (top) and measured (bottom) on the surface of a composite panel for various bonding layer thicknesses

laminate with a total thickness of 2.8 mm. The volume content of the carbon fibre in each layer is identical and equal to 30%.

Figure 6.14 clearly demonstrates that the impact of the bonding layer on the shape of the induced wave packet is much larger than for the case of the aluminium alloy plate. The bonding layer decreases the wave amplitude. Moreover, bonding layer present locally between the piezoelectric transducer and the composite panel results in a global change of the propagating wave velocity.

What is interesting is that with increasing frequency of the induced signal, wave amplitudes for thin and thick bonding layers become almost equal (Figure 6.15, top) and then for even higher frequencies a thicker bonding layer results in a decreased wave amplitude (Figure 6.15, bottom).

Summarising, the bonding layer significantly affects the shape of the propagating elastic waves. This influence is directly related to dispersion relationships. The bonding layer should be accounted for as one of the laminate

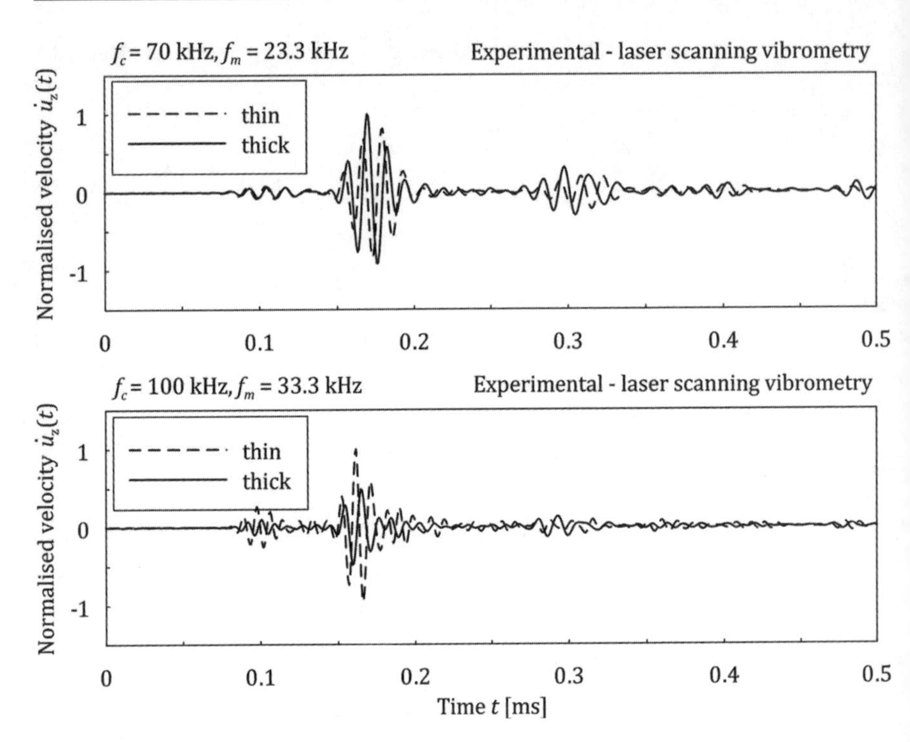

Figure 6.15 Time signals of transverse velocity components $\dot{u}_z$ measured on the surface of a composite panel by using laser scanning vibrometry for various bonding layer thicknesses and various excitation frequencies

layers. Wave velocity and dispersivity are altered depending on the bonding layer thickness and material properties of the bonding and of other layers of the structure. In other words, the effect of the bonding layer on elastic wave propagation depends on the frequency of the induced wave. Neglecting the bonding layer in numerical models can lead to significant errors in assessing wave velocity, especially for composite structures.

References

1. Ray, M.C., Bhattacharya, R. and Samanta, B. (1993) Exact solutions for static analysis of intelligent structures. *AIAA Journal*, **31**(9), 1684–1691.
2. Tzou, H.S. and Tiersten, H.F. (1994) Elastic analysis of laminated composite plates in cylindrical bending due to piezoelectric actuators. *Smart Materials and Structures*, **3**(3), 255–265.

3. Bisegna P. and Maceri, F. (1996) An exact three-dimensional solution for simply supported rectangular piezoelectric plate. *Trans. ASME, Journal of Applied Mechanics*, **63**, 628–638.
4. Ray, M.C., Bhattacharya, R. and Samanta, B. (1998) Exact solutions for dynamic analysis of composite plates with distributed piezoelectric layers. *Computers and Structures*, **66**(6), 737–743.
5. Tzou, H.S. and Tseng, C.I. (1990) Distributed piezoelectric sensor/actuator design for dynamic measurement/control of distributed parameter systems: A piezoelectric finite element approach. *Journal of Sound and Vibration*, **138**(1), 17–34.
6. Yao, L.Q. and Lu, L. (2003) Hybrid-stabilized solid-shell model of laminated composite piezoelectric structures under non-linear distribution of electric potential through thickness. *International Journal for Numerical Methods in Engineering*, **58**(10), 1499–1522.
7. Fernandes, A. and Pouget, J. (2002) An accurate modelling of piezoelectric multilayer plates. *European Journal of Mechanics A/Solids*, **21**, 629–651.
8. Mindlin, R.D. (1972) High frequency vibrations of piezoelectric crystal plates. *International Journal of Solids and Structures*, **8**(6), 895–906.
9. Yang, J.S. (1999) Equations for thick elastic plates with partially electrode piezoelectric actuators and higher order electric fields. *Smart Materials and Structures*, **8**(1), 73–82.
10. Saravanos, D.A., Heyliger, P.R. and Hopkins, D.A. (1997) Layerwise mechanics and finite element for the dynamic analysis of piezoelectric composite plates. *International Journal of Solids and Structures*, **34**(3), 359–378.
11. Bisegna, P. and Caruso, G. (2001) Evaluations of higher-order theories of piezoelectric plates in bending and in stretching. *Journal of Solids and Structures*, **38**(48–49), 8805–8830.
12. Kapuria, S. (2001) An efficient coupled theory for multilayered beams with embedded piezoelectric sensory and active layers. *International Journal of Solid and Structures*, **38**(50–51), 9179–9199.
13. Wang, S.Y. (2004) A finite element model for the static and dynamic analysis of a piezoelectric bimorph. *International Journal of Solids and Structures*, **41**, 4075–4096.
14. Mehrabadi, S.J., Kargarnovin, M.H. and Najafizadeh, M.M. (2009) Free vibration analysis of functionally graded coupled circular plate with piezoelectric layers. *Journal of Mechanical Science and Technology*, **23**, 2008–2021.
15. Wu, N., Wang, Q. and Quek, S.T. (2010) Free vibration analysis of piezoelectric coupled circular plate with open circuit. *Journal of Sound and Vibration*, **329**, 1126–1136.
16. Yang, Ch., Ye, L., Su, Z. and Bannister, M. (2006) Some aspects of numerical simulation for Lamb wave propagation in composite laminates. *Composite Structures*, **75**, 267–275.
17. Yu, L., Bottan-Santoni, G. and Giurgiutiu, V. (2010) Shear lag solution for tuning ultrasonic piezoelectric wafer active sensors with applications to Lamb wave array imaging. *International Journal of Engineering Science*, **48**, 848–861.

18. Ostachowicz, W. and Krawczuk, M. (2009) Modeling for detection of degraded zones in metallic and composite structures, in *Encyclopedia of Structural Health Monitoring* (eds C. Boller, F. Chang and Y. Fujino), John Wiley & Sons, Ltd, Chichester, pp. 851–866.
19. Abe, M., Park, G. and Inman, D.J. (2002) Impedance-based monitoring of stress in thin structural members. Proceedings of 11th International Conference on Adaptive Structures and Technologies, pp. 285–292.
20. Giurgiutiu, V. (2007) *Structural Health Monitoring with Piezoelectric Wafer Active Sensors*, Academic Press.
21. Xu, Y.G. and Liu, G.R. (2002) A modified electro-mechanical impedance model of piezoelectric actuator-sensors for debonding detection of composite patches. *Journal of Intelligent Material Systems and Structures*, **13**(6), 389–396.
22. Kim, Y., Ha, S. and Chang, F.-K. (2008) Time-domain spectral element method for built-in piezoelectric-actuator-induced Lamb wave propagation analysis. *AIAA Journal*, **3**, 591–600.

7

Detection, Localisation and Identification of Damage by Elastic Wave Propagation

The question of improving safety of operational structures has been investigated for many years in multiple research centres [1–10]. As is widely known, hidden damage in structures can develop under cyclic load, until it endangers the integrity of the whole structure. Undetected damage can be the cause of structure collapse. In order to prevent this, special diagnostic methods are being developed that allow the technical condition of structures to be assessed. Four levels of effectiveness of diagnosing the technical condition of structures can be distinguished: detection, location, identification and prediction [11]. Adding another level, classification, that would fit between location and identification, has also been proposed [12]. The first level provides information on whether damage exists in the structure. The second level answers the question concerning the damage location within the structure. The next level, identification, allows determination of the damage type and size (e.g. delamination, crack, matrix separation from the fibre in the composite material). The ultimate diagnostic level provides information regarding the remaining time of safe operation of the structure or the remaining time till the next repair.

Guided Waves in Structures for SHM: The Time-Domain Spectral Element Method, First Edition.
Wieslaw Ostachowicz, Pawel Kudela, Marek Krawczuk and Arkadiusz Zak.

Diagnostics of the technical condition of a structure utilises nondestructive testing (NDT) methods that allow the structure condition to be assessed without destroying it. Nondestructive testing methods can be divided into two groups: local and global ones. Local methods allow the technical condition of the structure to be determined locally, at a chosen point. Local methods include, among others, ultrasonic methods, eddy current methods, radiographic methods, penetration methods and magnetic dust methods. In recent years thermographic methods are being developed intensively.

In the case of using local methods, determining the technical condition of the whole structure requires repeating tests many times in order to check all the sensitive areas that are particularly prone to containing damage. Such an approach is time-consuming and in practice requires withdrawing the structure from operation for the time of the tests. Another disadvantage of local methods are the requirements concerning proper preparation of the structure and the need to employ highly skilled personnel.

Ultrasonic methods and eddy current methods combined with the visual inspection approach are universally used for testing aeronautic structures. For example, the cost of a visual inspection of a single Boeing 727 plane amounts to about 960 USD [13]. The average interval between subsequent visual inspections equals 15 days of use. Tests using the eddy current method are performed every 60 days of operation; the cost of such testing varies between 1900 and 3600 USD, depending on the scope [13]. The above information demonstrates how high the costs are that are associated with testing a technical condition of airships using local methods. For aeronautic structures, pauses in aeroplane operation caused by the need to perform diagnostic tests generate additional costs. A one-day-long pause in aeroplane operation costs the airline about 80 000 USD [14].

Because of such high economic costs, for many years researchers have been attempting to develop global methods that would allow the technical condition of the whole structure to be evaluated without removing it from operation or without the need for special preparation of the selected areas. Global methods also include vibration methods and methods employing elastic wave propagation. Developing effective global methods allows periodical inspections of a structure to be supplanted by continuous monitoring. Monitoring takes place during structure operation, without requiring pauses, and the obtained results can be analysed from any location in the world thanks to wireless data transfer. The structure requires no special preparation for tests and these no longer need to be performed by skilled personnel. A fundamental disadvantage of global methods is their lower sensitivity compared to local methods. Global methods are also more sensitive to noise and signal distortion. The obtained results are significantly affected by the way transducers are attached to the structure. In the case of the vibration method, accelerometers are bolted to the structure, and if this proves impossible, attached by a magnet located in the accelerometer base. Not all types of materials allow sensors to be attached using magnetic force, however. In such cases they can be glued or attached with wax. The latter approach is used mostly in laboratories. With the elastic wave propagation approach, piezoelectric transducers glued to the investigated structure are used most often. In such a case the thickness of the glue layer also affects the generated and registred signals, and therefore also the method's sensitivity [15, 16].

7.1 Elastic Waves in Structural Health Monitoring

In the recent years interest in the application of elastic wave propagation for structural health monitoring of structures has been significantly rising. This results from the fact that this method allows damage to be detected at early stages of its development, before it can endanger the safety of the structure. This capability in turn is a consequence of the method using high-frequency elastic waves, with frequencies reaching hundreds of kilohertz. Elastic waves can propagate in metal elements for long distances without a significant decrease in amplitude. This allows large areas of a structure to be tested during a single measurement.

The idea behind the elastic wave propagation method involves generating elastic waves that would propagate in the investigated structure and registering their amplitudes as a function of time. Waves are generated as 'packets' (see Figure 7.1). The discussed waves propagate until they encounter any discontinuities in the investigated element. These discontinuities can be edges or stiffening elements, but they can also be damage sites. After encountering discontinuities, waves reflect from them. Wave reflections provide the information on location, size and type of damage; this information is extracted from registered signals by appropriate algorithms.

At the moment, there are several commercial systems available that employ elastic wave propagation for monitoring the technical condition of pipelines. One of them was developed and implemented in the Imperial College [17] (see Figure 7.2). So far, a widely used system for monitoring the technical condition of the skin of structures containing stiffening elements and riveted, welded or bolted joints has not been developed. The reasons include the fact that in the case of pipes it is very easy to generate a flat wavefront propagating

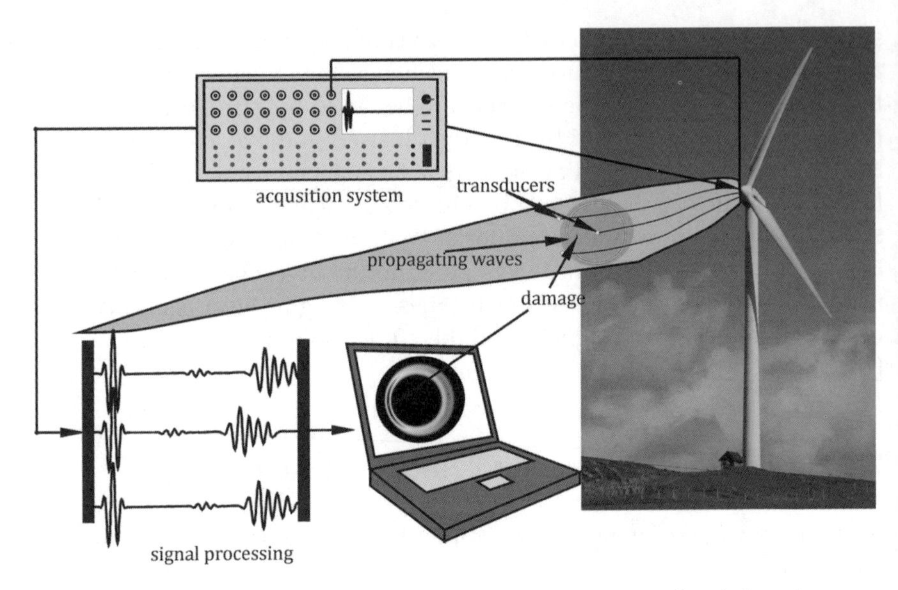

Figure 7.1 Idea of the elastic wave propagation method for damage detection

along the pipe length. However, in two-dimensional structural elements such as plates the problem of generation of a flat wavefront is much more complex. The known solutions usually employ the phased array approach [18–22]. This method allows the generation of a flat wavefront propagating in arbitrary direction. Such a solution allows the testing of the structure around the matrix

Figure 7.2 System used for locating damage in pipelines (courtesy of Guided Ultrasonics Ltd) (17)

of transducers generating and registering elastic waves. The number and distribution of transducers that generate and register elastic waves turn out to have an immense effect on this method's sensitivity. The directional recording of waves can also be achieved using MFC transducers, which allows the source of elastic waves to be located [23]. The directional generation of elastic waves can in turn be achieved by means of CLoVER transducers [24]. It should be emphasised that tests described here involved simple structures and do not solve problems present in real-life structures, as described earlier.

An important problem affecting attempts to employ elastic waves for damage detection is the dispersion phenomenon. Elastic wave dispersion manifests itself in the fact that their velocity depends heavily on the excitation frequency. This problem is discussed in detail in Chapter 1. Dispersion causes a wave packet propagating for significant distances to lengthen and change the original shape that was imposed by modulation. This phenomenon causes significant difficulties for locating damage, as it results in errors when determining the elastic wave velocity. For this reason, researchers strive to minimise dispersion by employing elastic waves of a specific frequency related to structure dimensions. One of the used methods, tone burst, involves inducing waves that propagate as wave packets. In practice, such a packet comprises several cycles of the sine function modulated with a time window (e.g. a triangular, Gauss or Hann window). A characteristic property of such elastic waves is a narrow frequency band. One should also note that for larger number of cycles the frequency band narrows but the signal duration increases (Figure 7.3). Longer elastic waves reduce the time resolution during the process of damage localisation. In such a case choosing the number of cycles involves a compromise between the precision of damage location and the reduction of the impact of dispersion [19]. The width of the frequency band and the distribution of harmonic components depend on the time window used for modulation. Figure 7.4 presents frequency spectra of wave packets with a carrier frequency of 300 kHz. These spectra result from modulation using rectangular, triangular and Hann windows, respectively. In the case of a rectangular window, characteristic side lobes around the carrier frequency can be seen. These side lobes are definitely smaller in the case of modulation with a triangular window and smallest when a Hann window is used for modulation. For this reason, in practice the Hann window is most often used for generating wave packets.

Another important problem involves choosing the carrier frequency of an elastic wave. It is well known that increasing the frequency enables damage of smaller dimensions to be detected. On the other hand, the number of elastic wave modes propagating simultaneously is known to rise with a frequency increase (cf. Chapter 1). The number of propagating wave modes

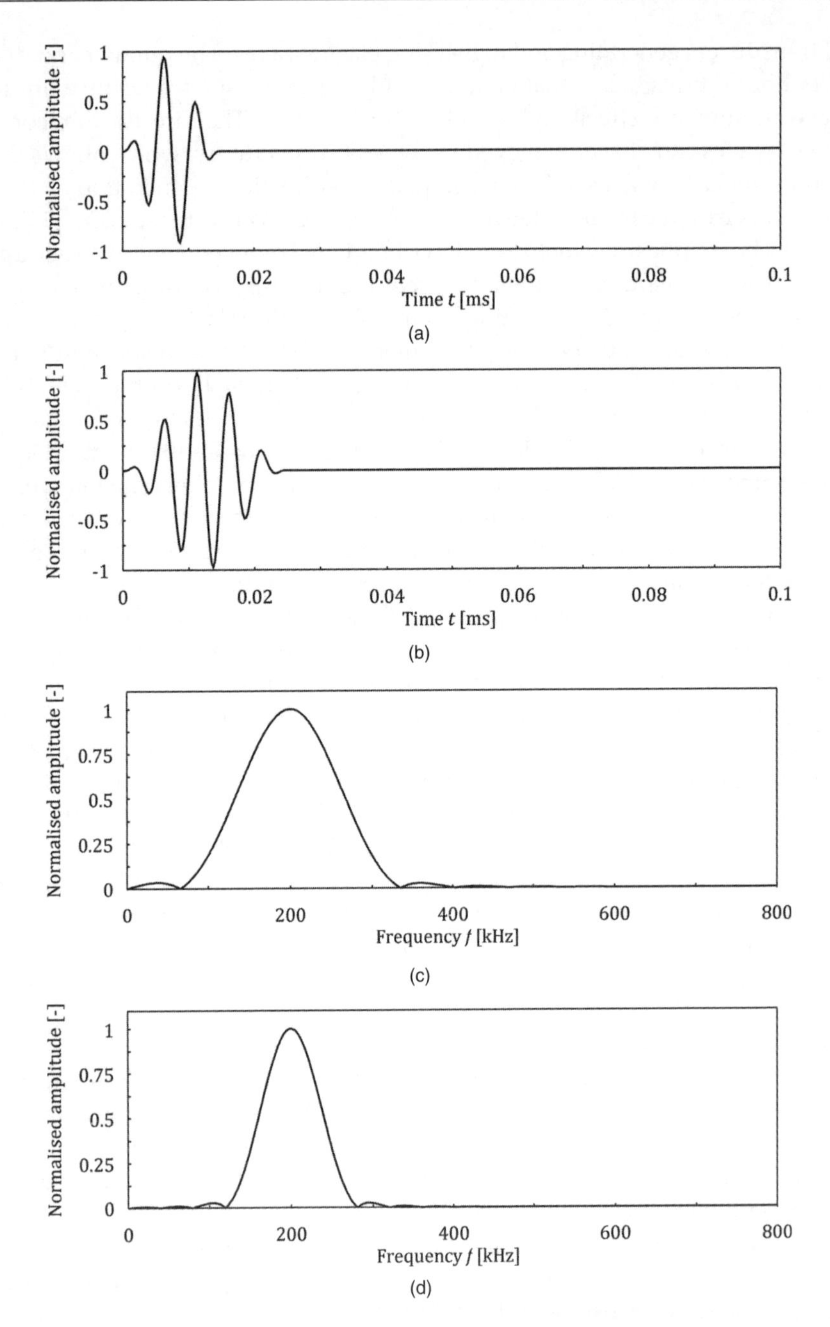

Figure 7.3 Forcing signal (200 kHz) modulated with a Hanning window in the time domain (a, b) and frequency domain (c, d) for three and five cycles, respectively

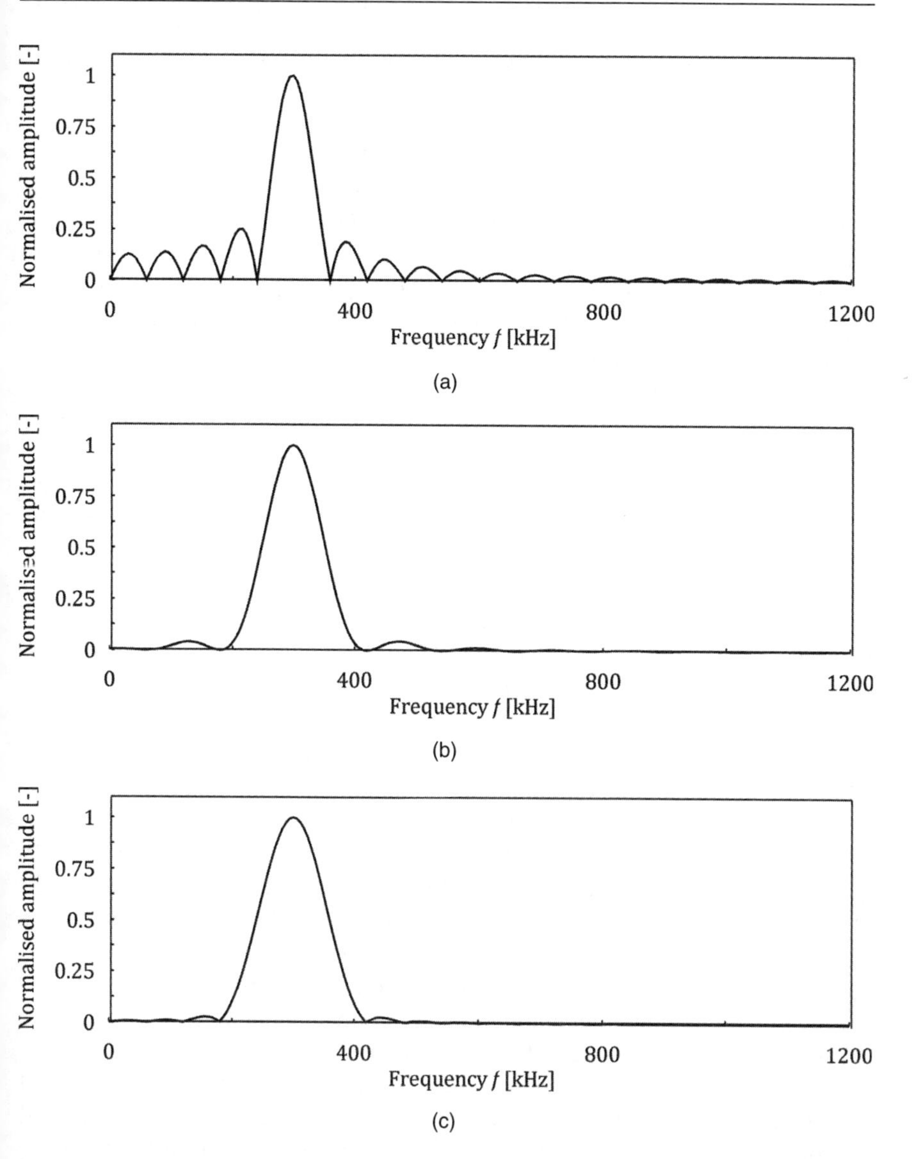

Figure 7.4 Frequency spectra of wave packets with a carrier frequency of 300 kHz modulated with: (a) rectangular window, (b) triangular window, (c) Hann window

depends on the excitation frequency and on the thickness of the element in which waves are generated. If dispersion curves for the given material and thickness of the element in which the wave will propagate are known, one can determine the number of induced wave modes. An important effect associated with the presence of different wave modes is their different propagation velocities. Reflections of various modes from damage and structure edges as well as their differing propagation velocities cause problems with reflection identification. Most importantly, it is difficult to identify where the given reflection comes from and which mode is involved. From the angle of effectiveness of damage location the most advantageous situation would involve only a single elastic wave mode propagating in the structure. For example, in 1 mm thick elements made of aluminium alloy, below the frequency of 1.5 MHz only two fundamental modes propagate. The two mentioned wave modes have different wavelengths, with the one for the S_0 mode being longer than the one for the A_0 mode. This results in the conclusion that for the same frequency mode A_0 should exhibit a higher sensitivity to damage than mode S_0.

One should also remember that besides wavelength, distribution of deformations across material thickness also significantly affects the method sensitivity. This distribution is associated with the type of propagating wave mode. For this reason, each wave mode exhibits a different sensitivity to individual damage types. The A_0 mode enables delaminations, transverse cracks and layer separation to be detected in composite elements [25–28]. The S_0 mode in turn enables the detection of cracks in metal elements [27, 28]. Sensitivity of the S_0 mode is independent of the depth at which discontinuity occurs, while in the case of the A_0 mode sensitivity is greatest on the element surface. For this reason the A_0 mode enables the detectiion of surface damage like scratches, cracks or corrosion [9, 29, 30]. The A_0 mode is also sensitive to simulated damage in the form of surface cuts on the investigated elements [31, 32]. The S_0 mode is better suited for damage detection in composite elements as it is attenuated to be weaker than the A_0 mode [29].

It remains to be determined whether the generation of a single mode of elastic waves only is possible. Recent research shows that for the defined geometry of a piezoelectric transducer one can determine plots of fundamental modes of elastic waves as functions of frequency [19]. It turns out that individual modes can be generated selectively (the other mode has an amplitude many times lower) [19]. Similarly, during signal registering sensitivity will be greater for the chosen mode for the specified elastic wave frequency. The transducer design turns out to affect significantly the type of generated and registered wave mode [33]. IDT transducers enable the generation and registration of the chosen mode of elastic wave (more precisely, the wavelength) [34, 35].

The process of elastic wave registration itself works as detailed below. The element in which elastic waves propagate is deformed. Structure deformations in turn cause deformations of the piezoelectric transducer glued to the discussed element. Deformations of the piezoelectric transducer cause the electric charge to appear on its electrodes. As the transducer exhibits some electric capacitance, at the moment of measurement the voltage difference between the piezoelectric transducer covers is read. It should be emphasised that in the case of transducers of high capacitance it is possible to connect the equipment registering voltage plots (e.g. an oscilloscope) directly to a transducer. However, if transducers of low capacitances, to the order of several to a dozen nanofarads, are involved, the use of charge amplifiers is necessary [36]. This approach allows for amplifying registered voltage signals as well as for eliminating the impact of lengths and paths of measurement cables on measurement results [36]. If charge amplifiers are not used, registered signals will not be reproducible. Figure 7.5 presents sample voltage plots registered on a piezoelectric transducer bonded to a structure in two cases: without damage and with damage. Elastic waves were generated and recorded by a pair of piezoelectric transducers located close to each other. Analysing the signal for the case of an undamaged structure (Figure 7.5(a)), one can easily distinguish two characteristic features. The first one is the wave packet associated with the direct wave propagation path from the excitation point to the registration point. The second one is associated with wave packets representing waves reflected from the edge of the structure element. In the case of a damaged structure (Figure 7.5(b)) there is also the characteristic third part, appearing between the wave packets described above, associated with wave reflection from damage.

Moving forward to problems appearing during elastic wave registration, one should emphasise the meaning of the amplitude of the wave reflected from damage. In the case of monitoring large areas of structure, propagation distances are large, which has effect on the wave amplitude. Amplitude is also affected by the way transducers are attached to the structure. The glue layer also can attenuate the elastic wave to some degree; therefore its thickness should be minimised. Varying the glue layer thickness or methods of constructing the glue joint cause wave amplitudes to differ between individual transducers, despite identical propagation paths. This results in a poor reproducibility of registered signals between individual transducers. Similar problems also affect actuators. A solution for wave registration involves contact-less measurements using laser vibrometry. The glue layer

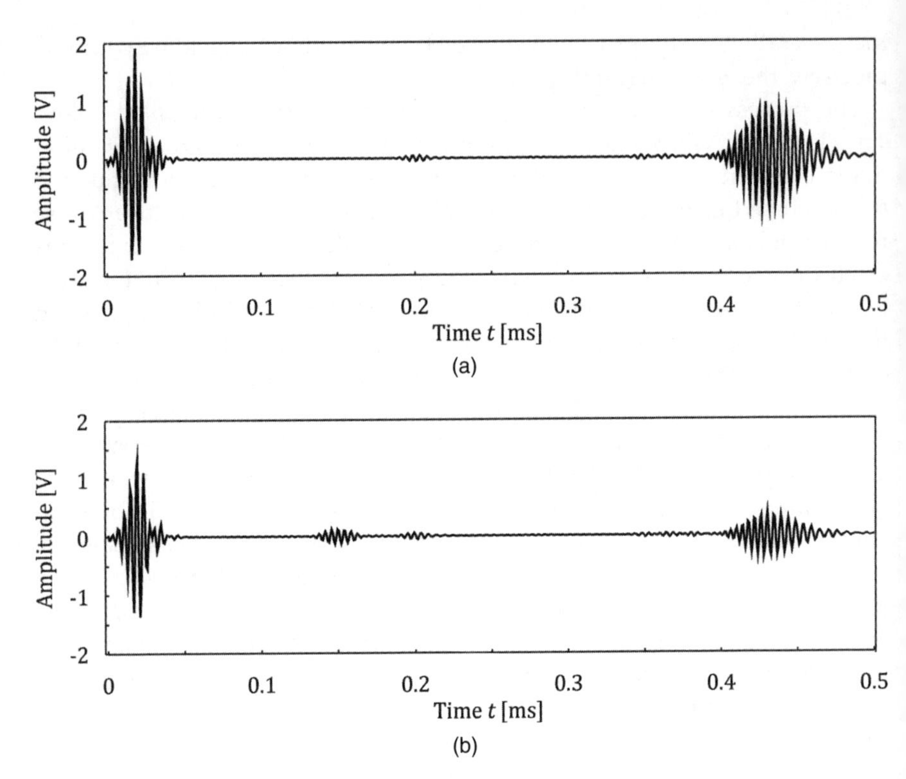

Figure 7.5 Sample signals registered for a structure: (a) without damage, (b) with damage

problem vanishes in this case. However, a problem of practical application of these methods in a real system for monitoring the technical condition of structures appears. It should be emphasised, however, that the approach described above is very useful in laboratory conditions, for research as well as for prototyping such systems.

Problems are also caused due to the design of the structure itself. As mentioned earlier, the elastic wave propagation method exhibits very high sensitivity. For this reason any discontinuities cause elastic wave reflections. One should remember that discontinuities are not limited to damage, but also include any changes in element thickness, stiffening elements, riveted joints, bolted joints, welded joints, and so on. For this reason, in the case of a fragment of a typical aeronautical structure, made of aluminium alloy with numerous stiffening elements joined with rivets, registered signals will contain multiple reflections. Stiffening elements cause almost total wave reflection [37, 38]. The wave amplitude beyond stiffening elements is very small, which virtually makes damage detection impossible.

In real-life cases, that is when elastic waves are employed in practice for damage detection and localisation, one should pay attention to operating conditions like temperature, stress in the structure or structure vibrations. According to research performed so far, a temperature increase results in a decrease in the wave propagation velocity [39, 40]. In such a case one needs a knowledge base containing measurements performed for different conditions in which the investigated structure is operated. Due to sensitivity of the elastic wave propagation phenomenon to temperature, procedures of optimal signal subtraction are being developed [41, 42]. Such a procedure is aimed at eliminating the need to collect many reference signals for temperatures to which the structure can be exposed during operation. Another approach to eliminating the need for collecting reference signals involves using a system of transducers installed on both sides of the investigated structure [43]. Subtracting signals registered on one surface of the specimen from the ones registered on the other side yields information about the presence of damage. This is a result of wave mode conversion in the presence of damage. Also stress present in the structure affects damage localisation in the case of using methods that utilise elastic wave propagation. Wave propagation velocity decreases with increasing stress. In isotropic materials the decrease in velocity of the propagating wave amounts to about 0.05 km/s per 1000 $\mu\epsilon$ [44]. In the case of composite materials the situation is more complex.

Researchers have also found noise appearing in registered signals. The reason for the noise is first and foremost electromagnetic interference. One should also note that the structure itself can generate noise during operation. Structure vibrations have rather low frequencies, that is they lie far away from the frequency band associated with propagating elastic waves. However, one should remember that impulse loads exhibit wide frequency spectra. Some of them can significantly contribute to the registered elastic waves. This fact carries particular weight in the case of monitoring large areas, when elastic waves propagate for large distances. In such cases, due to attenuation by the structure, amplitudes of registered waves can be very small and comparable to noise levels. Such a situation requires employing digital signal processing (DSP) techniques. It should be emphasised that this approach constitutes a very wide and rapidly advancing research direction.

There are many applications for DSP methods. The most important ones include: radar technology [45–47], medicine [47], acoustics [48], vibrations [49], processing of signals from measurement sensor matrices [50, 51] or general noise reduction in measurement systems [52]. In the context of the present discussion, the most important application is filtering the registered signals in order to eliminate noise.

As mentioned earlier, a characteristic property of elastic wave excitation employed for monitoring technical conditions of structures is that frequency components of such signals fit inside a narrow frequency band. Assuming that both the diagnostic system and tested structure operate within the linear range, registered signals must exhibit frequency components identical to those introduced into the investigated system by excitation. As the spectrum of the excitation signal, and in particular frequency bands with nonzero signal components, is known, the frequency spectrum of the registered signal is also known. One can thus state that all frequency components falling outside the signal band constitute noise. This noise interferes with diagnostic processes. Based on the provided information, it is obvious that a filter is required for removing the noise. The filter allows frequency components fitting inside a narrow frequency band to pass but blocks all other components. One should also note that methods taking into account nonlinear components of signals are used [53–55]. In such cases these would be present in the spectrum of the registered signal. For that reason using a band-pass filter designed to let through only the components associated with forcing is not possible.

If both the equipment and investigated structure operate within the linear range, band-pass filters are used. The most popular design of band-pass filters is a digital filter. An advantage of digital filters is the capability to almost arbitrarily shape the frequency characteristics, as opposed to analogue filters used earlier [56]. These filters have found application for filtering signals associated with elastic wave propagation [57]. Digital filters can be divided into low-pass filters, high-pass filters, band-pass filters and band-reject filters. As mentioned earlier, the best solution in the case of filtering signals associated with the propagation of elastic waves of a dispersive nature will involve band-pass filters. However, depending on the conditions of wave generation and registration a low-pass or high-pass filter may turn out to be sufficient for the needs of signal processing. Filter choice is associated with analysing the type of noise (its spectrum) present in signals.

Numerous research papers demonstrate that utilising time–frequency analysis in the form of a spectrogram constitutes a good approach to band-pass filtering. A spectrogram is constructed on the basis of a short-time Fourier transform (STFT). The spectrogram presents a time plot of amplitude changes of frequency components of the signal [57]. In the spectrogram one can isolate the part associated with the signal, in which components from a narrow frequency band appear. A very similar approach involves application of a continuous wavelet transform (CWT). The CWT enables a scalogram to be plotted, which is very similar to a signal spectrogram [58]. Upon comparing the wavelet transform with the short-time Fourier transform, one can easily spot the difference. The STFT uses a uniform division of the time scale for both low and high frequencies. The length of the time window used in the STFT is the same across all the frequency range. On the other hand, employing the continuous wavelet transform allows the time scale to be stretched for low frequencies and to be compressed for high frequencies. Such an approach is aimed at better visualisation of frequencies present in the signal being processed. Instead of frequency, the scalogram uses so-called scale, which is closely related to values of analysed frequencies. Filtering is possible in ways analogous to the spectrogram case. In such a case the signal for the narrow frequency band is also distilled from the scalogram.

Band-based signal filtering can also be applied for multilevel wavelet decomposition using a discrete wavelet transform (DWT) [57, 59]. Such an approach enables the signal to be divided into signals containing only frequency components from frequency bands of specified widths. By adjusting the decomposition level, one can choose the appropriate width of the frequency band.

Figure 7.6 presents a noisy measurement signal with a carrier frequency of 160 kHz. This signal was filtered using a digital band-pass filter, a scalogram obtained by means of CWT and wavelet decomposition.

In the case of a digital filter, frequency characteristics were adjusted to the spectrum of a generated signal. In the case of the scalogram, a signal of the frequency range overlapping the excitation frequency was chosen. In the case of wavelet decomposition, a second level of wavelet decomposition containing a signal with frequency components from the 156.25 to 312.5 kHz band was chosen.

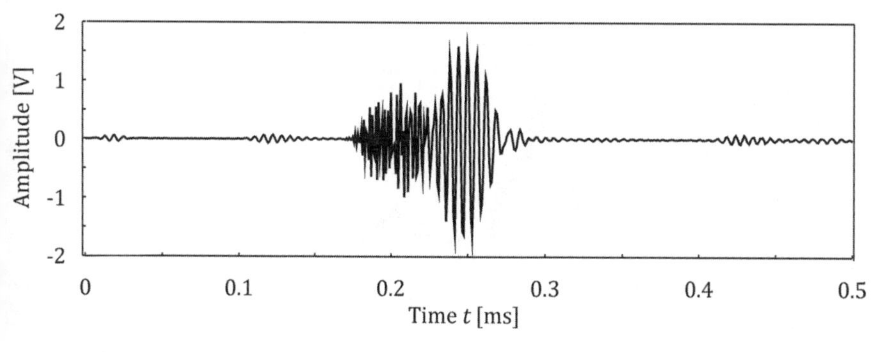

Figure 7.6 Registered signal of a carrier frequency of 160 kHz

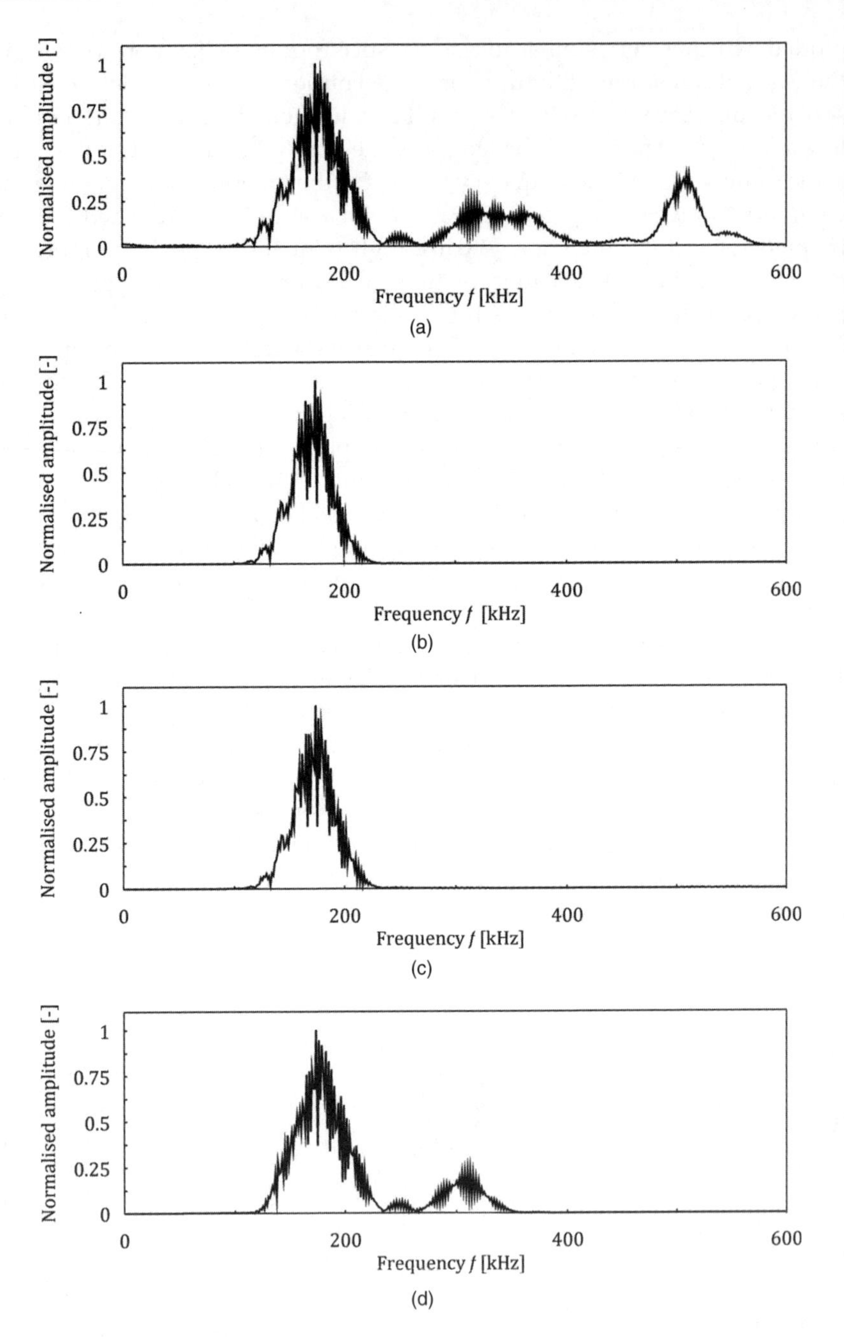

Figure 7.7 Spectrum of a noisy signal (a) and signals obtained after filtering using different methods: (b) digital filter, (c) CWT, (d) DWT

Figure 7.7 presents spectra of signals obtained for a raw signal as well as ones obtained after filtering using a digital filter, a continuous wavelet transform (CWT) and a wavelet decomposition transform (DWT). Figure 7.7(a) demonstrates that the registered signal beside the frequency of 160 kHz (frequency of a generated signal) also contains higher frequency components. After filtering, the band of signal components becomes narrower; analysis of individual cases leads to a conclusion that the widest band is produced by filtering using the DWT. Approaches using a digital filter or a continuous wavelet transform (CWT) yield very similar results, producing spectra of widths consistent with those of the forcing spectrum. A comparison of various filtering methods (Figure 7.7) does not take into account the approach using a spectrogram. This was caused by the fact that such filtering produces absolute values of signal samples.

7.2 Methods of Damage Detection, Localisation and Identification

A fundamental problem that appears when elastic waves are applied for monitoring the technical condition of a structure involves a spacing transducer, which will generate and register waves. Designing the correct layout for the mentioned transducers enables an effective damage location to be found. Such layouts require developing appropriate dedicated procedures for processing signals registered by sensors.

Methods of monitoring the technical condition utilising elastic wave propagation are classified as either of the pulse-echo type (Figure 7.8(a)) or of the pitch-catch type (Figure 7.8(b)). Methods of the first type utilise wave reflection from the damage site. A wave propagating from the generation site reflects from the damage site and is registered by the receiver. Pitch-catch methods utilise changes of wave characteristics (e.g. amplitude, propagation velocity, wave mode conversion) on the direct path between the transmitter and receiver. Pulse-echo methods are used successfully with both distributed and concentrated configurations. Conversely, pitch-catch methods require a distributed configuration.

The approach involving distributed configurations is based on covering a large area of the monitored structure with a network of sensors. Such an approach is represented by the system of 12 sensors designed in the course of research performed in the Department of Mechanics of Intelligent Structures, IFFM (Figure 7.9). Sensors are embedded in a dielectric film in order to guarantee their constant spacing. Moreover, the film protects the sensors

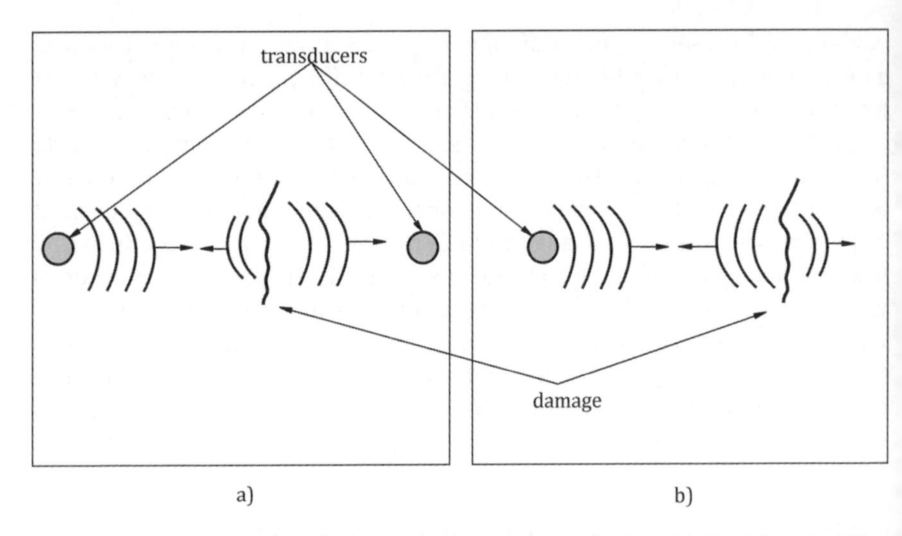

Figure 7.8 Illustration of the pitch-catch method (a) in which characteristics of the wave passing through the damage is analysed; illustration of the pulse-echo method (b) in which the wave reflection from damage is analysed

Figure 7.9 Example of the distributed sensor layout

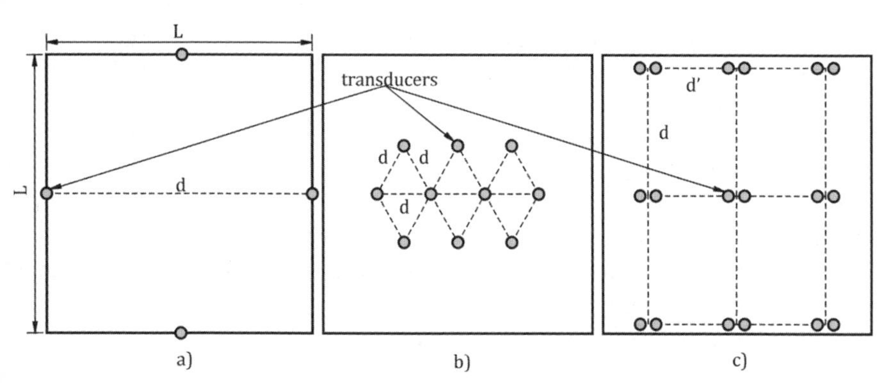

Figure 7.10 Selected examples of distributed sensor systems described in the literature: (a) four sensors located in the centres of edges of a monitored panel, (b) triangular sensor network, (c) rectangular network with double transducers in each node, external nodes near the plate edges

against external agents. One of the simplest solutions described in the literature is a monitoring system involving several sensors located along edges of a monitored square panel (Figure 7.10(a)). One of the approaches to signal processing aimed at damage detection involves utilising the symmetry of such a sensor layout [60]. Such systems are used for monitoring the condition of structures made of both isotropic [60] and anisotropic [61] materials. In the latter case, material properties require signal processing to take into account changes in the wave propagation velocity with the propagation direction. The discussed solution may be modified by altering the network geometry. Good results are obtained for triangular networks [62] (Figure 7.10(b)). Solutions based on such sensor networks use separate transducers for wave excitation and registration [61–63].

A characteristic property of damage localisation algorithms designed for such systems is that their indications are ellipse-shaped. Locations of generating and recording transducers correspond to ellipse foci. The circumference of this ellipse contains a discontinuity causing wave reflection. If there is no damage of other discontinuities, no reflection occurs. After considering all transducer pairs one obtains the overall result, which is the final result of the algorithm. For that reason, the type of transducer network determines the positions of ellipses. A different approach to the ideas of a sensor network involves placing two transducers in every network as a pair of sensor and acturator [64] (Figure 7.10(c)). Such a transducer layout causes the resulting ellipses to be virtually circular. Distributed systems are used not only for damage location but also for identification as well. The crack shape can be reconstructed with the help of a distributed system [65]. Distributed systems are also applied in tomography, where the area surrounded by sensors, for example a sensor ring, is monitored [66]. An image is generated using signals from all possible transducer pairs. In this case one of the transducers excites waves and the other registers them. In order to obtain a satisfactory result of a damage location, one should employ a large number of transducers. For this reason the method is best suited to monitoring chosen structure areas that are particularly susceptible to damage. The threat may involve delamination of stiffening elements from the skin [67].

A second group of sensor distribution layouts that find practical use consists of concentrated layouts. Concentrated layouts use a compact system of transducers designed to monitor the area around the system. An example of solutions used in practice is the system located in the Laboratory of Department of Mechanics of Intelligent Structures, IFFM (see Figure 7.11). Researchers analyse various sensor layouts in order to identify the optimum solutions for selected elements of monitored structures. One of the most popular solutions is the circular layout [68] (Figure 7.12(a)). Solutions used in practice differ with regard to the number of sensors constituting the system, for example 13 [68] or 16 [69]. The class of concentrated systems also involves

Figure 7.11 Example of a concentrated sensor layout

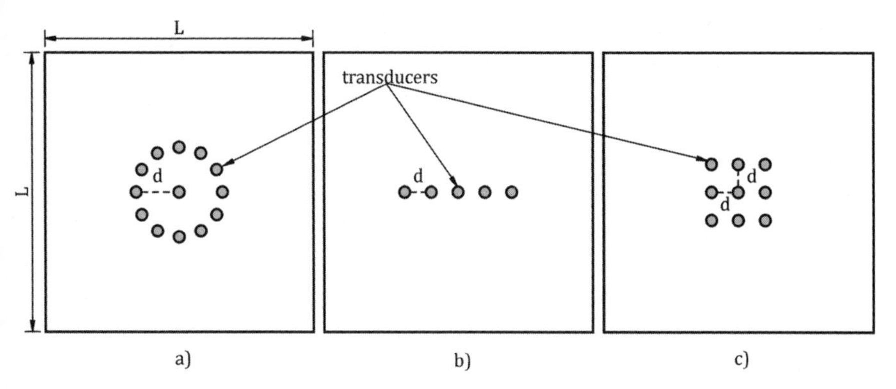

Figure 7.12 Selected examples of concentrated sensor layouts ($d \ll L$) described in the literature: (a) circular layout of 13 transducers, (b) linear layout, (c) square layout

phased array systems. Employing a special method of signal processing or specially tuned methods of wave generation in multiple transducers with precisely chosen phase delays allows wave interference to be obtained. This effect allows amplification of the waves reflected from damage [19]. Also, in the case of phased array systems different transducer layouts are investigated. The easiest of those is the linear layout (Figure 7.12(b)) used for locating damage in specimens; these are made of aluminium with curvature [70] or without curvature [71] as well as composites [71–73]. Linear layouts are used for damage identification, for example for visualising growth of fatigue cracks in an aluminium panel [74], amongst others. Attempts have been made to compare and choose from among traditional concentrated systems and phased array systems [75]. The main disadvantage of linear layouts is ambiguity of the damage location. The localisation result is symmetrical with regard to a straight line, along which sensors of the layout are located.

Two-dimensional systems constitute an improvement over linear (one-dimensional) systems, as they allow for unequivocal location of damage [76]. Their characteristics depend on a spatial configuration of sensors and the number of sensors in the system [19, 76, 77]. As for traditional concentrated systems, circular phased array layouts are used here. One should note the increase in damage location effectiveness when sensors cover the whole circle area as opposed to only its circumference [18]. Using more transducers in the latter case produces a significant improvement of localisation results. This is also confirmed by investigation of square systems, which are successfully used for detecting many types of damage [78] (Figure 7.12(c)).

The sensor layouts mentioned above utilise geometric shapes. Many research papers propose specific layouts and then test their effectiveness for monitoring the technical conditions of structures. Attempts to optimise sensor configurations are rare. The square layout guarantees unequivocal damage localisation within the whole angular range. For this reason it is often chosen and used [19]. There are attempts to automate the selection of sensor configurations by applying genetic algorithms to optimise sensor placement [29].

Both concentrated and distributed sensor systems have a common disadvantage, namely being unable to locate damage in the close vicinity of a transducer or just beneath it. For this reason, a so-called electromechanical (EM) impedance method is considered to be a natural compliment to methods utilising elastic wave propagation [19]. This method is based on measuring the electrical impedance of a piezoelectric transducer. The transducer couples an electric field with a mechanical response and thus allows mechanical changes to be registered in its vicinity [79–81].

Even wireless EM impedance sensors sensitive to temperature changes and detachment of the sensor itself are being developed [82]. Commercially available impedance analysers are not cheap, and, even more importantly, have a large size. For this reason, attempts to develop custom measurement systems are being made [83]. Summarising, the EM impedance method enables the detection of structure damage near transducers, as well as performing diagnostics of the sensor itself.

In recent years advances in laser vibrometry have enabled the development of methods that use data registered by scanning laser vibrometry. Their common property is that laser scanning allows measured amplitudes or velocities of propagating elastic waves to be used across the whole analysed structure area. Such methods include mapping maximum total amplitudes [84–86], mapping RMS values [87, 88] and mapping cumulative kinetic energy [89].

Besides damage localisation, a second important process of monitoring the technical condition of structures is damage identification. This process is aimed at determining the damage size, shape and type. Both the direct method and inverse methods are used for damage identification. Only several papers tackling this problem can be found in the literature on applications of elastic wave propagation for damage identification. One of the methods successfully employed for damage identification is wavelet decomposition. This method is being used for testing one-dimensional structures [90]. A thorough discussion of the problem of damage identification on the grounds of wavelet analysis can be found in paper [91]. Methods based on elastic wave propagation are used for identification of crack orientation. This procedure requires using a set of transducers surrounding the investigated crack. Transducers register waves propagating in various directions [57]. One should emphasise that damage identification constitutes a much more complex problem than damage localisation.

The rest of this chapter presents a number of methods enabling damage localisation and identification.

7.2.1 Energy Addition Method

The energy addition method assumes that damage or another discontinuity encountered by the propagating elastic wave is a source of reflection. For the needs of damage localisation the monitored area is covered with a network of measurement points P_i. The method assumes that N transducers (denoted R_n) register waves and one transducer (denoted G) generates them. Distances between the discussed transducers and an arbitrary point P, which can be identified by the damage position, are (Figure 7.13)

$$\begin{aligned} L &= |PG| \\ L_n &= |PR_n| \end{aligned} \tag{7.1}$$

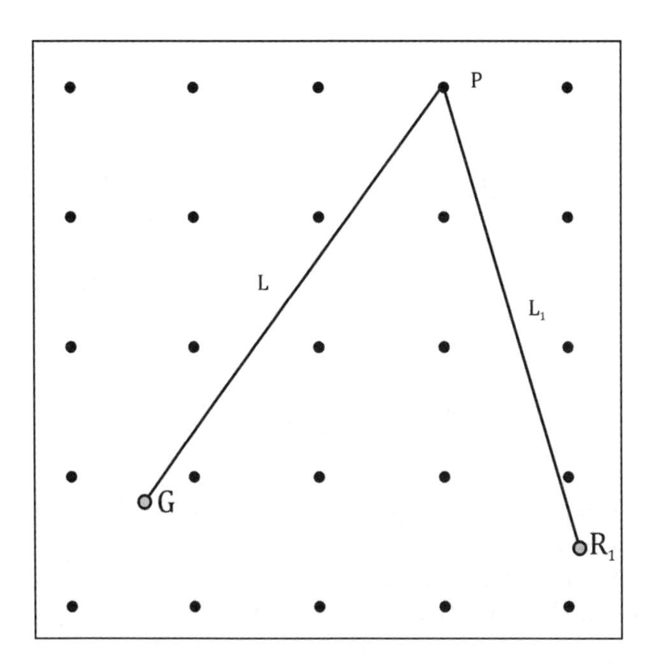

Figure 7.13 Graphical illustration of the localisation method: G – transducer generating waves, R_1 – transducer registering waves, P – network point, L and L_1 – respective distances

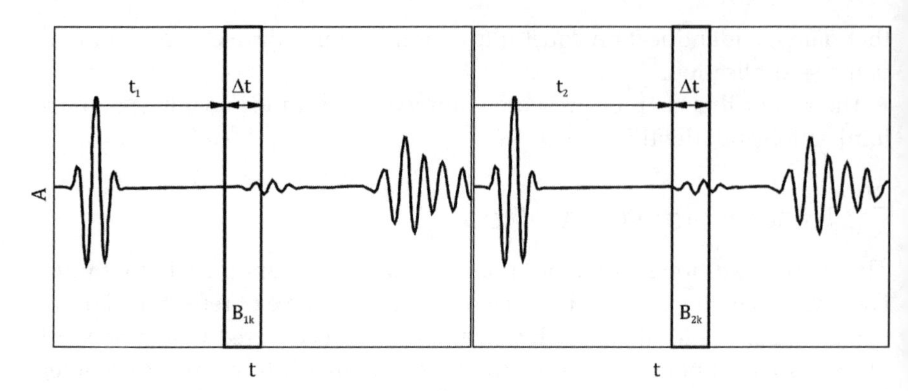

Figure 7.14 Illustration of the procedure of cutting out signal fragments in order to construct a damage indicator

Assuming that the time of the wave generation and the time of the starting wave registration are the same, the wave should be registered in transducers R_n after the time:

$$t_n = \frac{L + L_n}{c} \tag{7.2}$$

where c is the group velocity of the wave propagation.

Starting with time t_n, fragments of signals registered in transducers R_n are starting to cut out (Figure 7.14). The length of these fragments is denoted Δt. Cut out signal fragments are denoted B_{nk}, where k is an index changing from 1 to K (K describes the fragment length). Ultimately, the damage indicator associated with point P assumes the following form:

$$M(P) = \sum_{n=1}^{N} \sum_{k=1}^{K} (D_{nk} B_{nk})^2 \tag{7.3}$$

where D_{nk} are coefficients responsible for compensating the amplitude decay with the travelled distance.

If there are wave reflections in the cut out signals, the value of the indicator M for point P will be larger than for other points, thus indicating a potentially damaged area. The procedure described above is repeated for every point of the network and the results are plotted using coordinates associated with the monitored area.

A competing approach involves using a measure of damage present in the frequency domain. Using the fast Fourier transform algorithm, one obtains the signal fragment B_{nk} in the frequency domain:

$$\hat{B}_n(f) = |FFT(B_{nk})| \tag{7.4}$$

The frequency damage indicator is assumed to be the value of the above function for the frequency f equal to the excitation frequency f_0. This procedure is repeated for all exciter-reciver sensor pairs. The results are then summed:

$$M_f(P) = \sum_{n=1}^{N} \hat{B}_n(f)\big|_{f=f_0} \tag{7.5}$$

The obtained value is the frequency damage indicator and is associated with the discussed point P. As before, the indicator is computed for each point.

The main advantage of using the frequency damage indicator $M_f(P)$ (7.5) is its filtering nature. Selecting only the values for f_0 from the frequency response guarantees that the damage indicator value is not distorted by the measurement noise. Information contained in the signal B_{nk} for the selected excitation frequency is used.

A similar approach of mapping information contained in signals is used in paper [92]. Authors of this paper do not superimpose signal fragments, however, but only add values of signal amplitudes or envelopes for a single time sample corresponding to a point within the monitored area.

7.2.2 Phased Array Method

The phased array method utilises the physical phenomenon of elastic wave interference (Figure 7.15). The mathematical description that constitutes the grounds for signal processing based on the phenomenon of interference is presented below.

A wavefront at point $\boldsymbol{r}$ in the time instance t for a wave propagating with the velocity c can be expressed in the following form, facilitating the complex analysis:

$$F(\boldsymbol{r}, t) = \frac{A}{\sqrt{|\boldsymbol{r}|}} \mathrm{e}^{\mathrm{i}(\omega t - \boldsymbol{k}\cdot\boldsymbol{r})} \tag{7.6}$$

where A denotes real amplitude, ω denotes angular velocity and $\boldsymbol{k}$ is the wave vector.

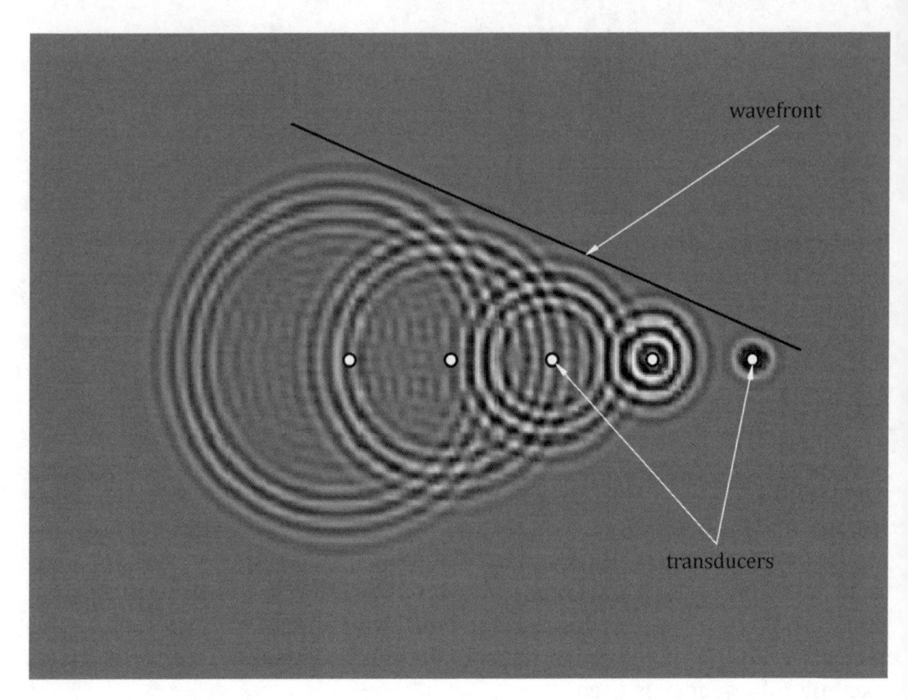

Figure 7.15 Illustration of the formation of a flat wavefront as a result of propagation of circular waves from five transducers placed in a line

In the case of N piezoelectric transducers, Equation (7.6) for the nth transducer assumes the form [19]:

$$F(\boldsymbol{r}_n, t) = \frac{A_n}{\sqrt{|\boldsymbol{r}_n|}} \mathrm{e}^{\mathrm{i}(\omega t - \boldsymbol{k}_n \cdot \boldsymbol{r}_n)} \tag{7.7}$$

Without losing generality, one can assume that:

$$A_n = A \tag{7.8}$$

because differing amplitudes will be accounted for in the further description. Expression (7.7) is reformulated by multiplying by unity [19] and using the relationship $\boldsymbol{k} \cdot \boldsymbol{r} = \frac{\omega}{c} |\boldsymbol{r}| \cos \alpha$:

$$F(\boldsymbol{r}_n, t) = \frac{A}{\sqrt{|\boldsymbol{r}|}} \mathrm{e}^{\mathrm{i}\omega(t - (|\boldsymbol{r}|)/c)} \frac{1}{\sqrt{|\boldsymbol{r}_n|}} \sqrt{|\boldsymbol{r}|}\, \mathrm{e}^{\mathrm{i}\omega((|\boldsymbol{r}| - |\boldsymbol{r}_n| \cos \alpha)/c)} \tag{7.9}$$

Assuming a circular wave propagation, it is assumed that $\mathbf{k}||\mathbf{r}$, that is $\alpha = 0,\ \cos\alpha = 1$, ultimately obtaining:

$$\begin{aligned} F(\mathbf{r}_n, t) &= F\left(\mathbf{r}, t - \frac{|\mathbf{r}|}{c}\right)\frac{1}{\sqrt{|\mathbf{r}_n|/|\mathbf{r}|}}\mathrm{e}^{\mathrm{i}\omega((|\mathbf{r}|-|\mathbf{r}_n|)/c)} \\ &= F\left(\mathbf{r}, t - \frac{|\mathbf{r}|}{c}\right)\frac{1}{\sqrt{r_1}}\mathrm{e}^{\mathrm{i}\omega((1-r_1)/(c/|\mathbf{r}|))} \end{aligned} \tag{7.10}$$

where $r_1 = |\mathbf{r}_n|/|\mathbf{r}|$. The overall signal for N transducers is expressed by the following formula:

$$S(\mathbf{r}, t) = F\left(\mathbf{r}, t - \frac{|\mathbf{r}|}{c}\right)\sum_{n=1}^{N}\frac{a_n}{\sqrt{r_1}}\mathrm{e}^{\mathrm{i}\omega((1-r_1)/c/|\mathbf{r}|)} \tag{7.11}$$

In order to account for arbitrary wave amplitudes, coefficients a_n are introduced. Multiplication by coefficients a_n is introduced with the aim of generalising the description. At the first stage identical amplitudes A have been assumed and later the introduction of coefficients a_n allowed for arbitrarily choosing amplitudes for the individual signals being summed. The first term of the product in formula (7.11) does not depend on transducer locations or coefficients a_n. What is important is the second term containing the sum, as it depends on transducer spacing and weights. This term is denoted I. As the aim is to obtain constructive interference of waves for the chosen direction φ (Figure 7.16), an additional expression (delay) is introduced into the exponent d_n:

$$I(a_n, \mathbf{r}_n) = \sum_{n=1}^{N}\frac{a_n}{\sqrt{r_1}}\mathrm{e}^{\mathrm{i}\omega((1-r_1)/c/|\mathbf{r}|-d_n)} \tag{7.12}$$

The maximum value of this expression is reached when the exponent is equal to zero:

$$\frac{1-r_1}{c/|\mathbf{r}|} - d_n = 0 \tag{7.13}$$

and thus:

$$I(a_n, \mathbf{r}_n) = \sum_{n=1}^{N}\frac{a_n}{\sqrt{r_1}} \tag{7.14}$$

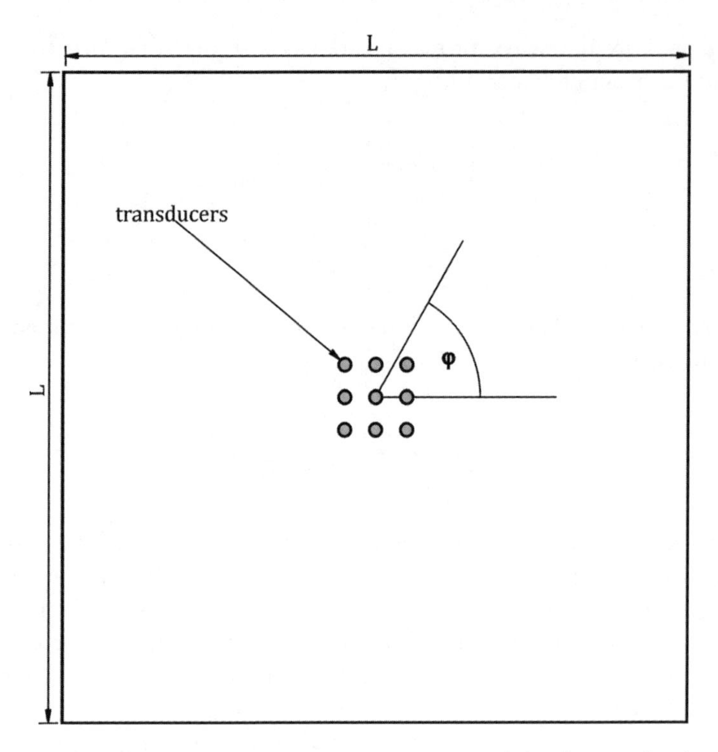

Figure 7.16 System of sensors placed on a square plate; the angle chosen for constructive interference is marked

Additionally, assuming coefficients:

$$a_n = \sqrt{r_1} \tag{7.15}$$

one obtains compensation caused by spatial configuration of the transducers. With the above assumptions, the overall signal expressed by formula (7.11) assumes the form:

$$S(\boldsymbol{r}, t) = NF\left(\boldsymbol{r}, t - \frac{|\boldsymbol{r}|}{c}\right) \tag{7.16}$$

This means an N-fold amplification of the signal that would be received from a single transducer in ordinary circumstances.

Recalling expression (7.13) and expanding this expression, one obtains:

$$d_n = \frac{1 - r_1}{c/|\boldsymbol{r}|} = \frac{|\boldsymbol{r}|}{c} - \frac{|\boldsymbol{r}_n|}{c} = t - t_n \tag{7.17}$$

where t denotes the time needed for the wave to propagate from/to the chosen reference point to/from the chosen point, while t_n is the time needed for propagation from/to the nth transducer to/from the same point. The obtained formula unequivocally determines the procedure for processing signals with the aim of simulating constructive interference.

The discussed procedure of signal addition is general in nature. The obtained formula is correct for an arbitrary point of location specified by the vector $\boldsymbol{r}$. However, if the wave source is located far away from the transducer system, or if the point in which the interference is desired is located far from the transducer system, one can introduce a simplification, assuming, after [19], that:

$$\frac{\boldsymbol{r}_n}{|\boldsymbol{r}_n|} \approx \frac{\boldsymbol{r}}{|\boldsymbol{r}|}, \quad \boldsymbol{k}_n \approx \boldsymbol{k}, \quad |\boldsymbol{r}_n| \approx |\boldsymbol{r}| \tag{7.18}$$

Expression (7.10) then assumes the following form:

$$\begin{aligned} F(\boldsymbol{r}_n, t) &= F\left(\boldsymbol{r}, t - \frac{|\boldsymbol{r}|}{c}\right) e^{i\boldsymbol{k}\cdot(|\boldsymbol{r}|-|\boldsymbol{r}_n|)} \\ &= F\left(\boldsymbol{r}, t - \frac{|\boldsymbol{r}|}{c}\right) e^{i(\omega/c)(\boldsymbol{r}\cdot\boldsymbol{s}_n)/|\boldsymbol{r}|} \end{aligned} \tag{7.19}$$

where $\boldsymbol{s}_n = \boldsymbol{r} - \boldsymbol{r}_n$. Ultimately, for the chosen approximation, one obtains the following expression, analogous to (7.12):

$$I'(w'_n, \boldsymbol{s}'_n) = \sum_{n=1}^{N} w'_n e^{i\omega((\boldsymbol{r}\cdot\boldsymbol{s}_n)/c|\boldsymbol{r}|-d'_n)} \tag{7.20}$$

The expression presented above reaches a maximum when:

$$d'_n = \frac{\boldsymbol{r}\cdot\boldsymbol{s}_n}{c|\boldsymbol{r}|}, \quad w'_n = 1 \tag{7.21}$$

This means that one only has to consider the angle:

$$\varphi = \angle(\boldsymbol{r}, \boldsymbol{s}_n) \tag{7.22}$$

in which interference is forced and the sensor location s_n is:

$$\frac{\boldsymbol{r}}{|\boldsymbol{r}|} \cdot \boldsymbol{s}_n = |\boldsymbol{s}_n| \cos(\varphi) \tag{7.23}$$

After the author of [19], one can specify the conditions that must be met for one to be able to use the approximation introduced. In such a case two characteristic distances can be distinguished. The first one describes the distance from the transducer system, beyond which one can use the approximation described by Equation (7.20):

$$R_2 = \frac{2D^2}{\lambda} \tag{7.24}$$

The second limit distance:

$$R_1 = 0.62\sqrt{\frac{D^3}{\lambda}} \tag{7.25}$$

is one below which the theory of phased array systems is no longer applicable. Thus, the scope of applicability of the nonsimplified formula (7.14) fits between the limits:

$$R_1 < R \leq R_2 \tag{7.26}$$

while approximation (7.20) can be used when:

$$R > R_2 \tag{7.27}$$

where R is the distance from the phased array and D is the system span, that is the distance between the outer transducers.

An important question regarding phased arrays is the choice of spacing d between the array elements [19]. If the condition $d \leq 0.5\lambda$ is not met, the array characteristics start to exhibit ambiguities, causing wave amplification, not only for the chosen angle φ but also for other angles. For example, Figure 7.17(a) presents characteristics of a 5×5 array for the angle $\varphi = 60°$ and sensor spacing $d = 1.15\ \lambda$. For the listed parameters the array response contains the chosen angle φ, the symmetric angle $360° - \varphi$, as well as angles $114°$ and

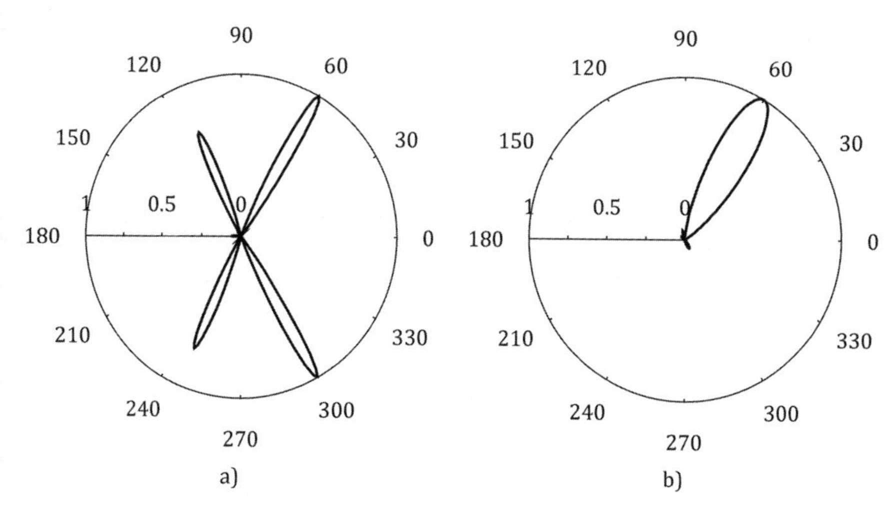

Figure 7.17 Characteristic of a 5 × 5 phased array for the chosen angle $\varphi = 60^\circ$ for spacing between sensors of: (a) $d = 1.15\,\lambda$, (b) $d = 0.4\,\lambda$

$360 - 114^\circ$. Decreasing the spacing to $d = 0.4\,\lambda$ yields an unequivocal selection of the angle $\varphi = 60^\circ$ (Figure 7.17(b)).

7.2.2.1 Damage Localisation Procedure

According to the discussion presented above, the signal can be amplified N times. This amplification results from choosing appropriate weights and delay. The point defined by the vector $\boldsymbol{r}$ is positioned relative to the reference point, which was chosen at the position of the central transducer of the phased array for convenience. The vector $\boldsymbol{r}_n$ in turn indicates the direction from the mth transducer to the wave focusing point. Thus, t denotes the time needed for the wave to propagate from the chosen reference point to the chosen focusing point, while t_n is the time needed for propagation from the nth transducer to the same focusing point. One should emphasise that a single application of the interference algorithm, that is multiplying by weights and shifting signals, is not sufficient. The reason is that the phased array system is aimed at emitting a wavefront and reading the wave after some time. For this reason one should take into account the different times needed for the wave to return to individual transducers. In practice this procedure has been implemented as follows:

1. All signals recorded in the nth transducer for excitation in all the other $N - 1$ transducers are multiplied by appropriate weights, shifted in phase

and summed:

$$\mathrm{sig}_n(t) = \sum_{k=1}^{N-1} a_k \mathrm{sig}(t - d_k) \tag{7.28}$$

This procedure has yielded N signals corresponding to interference after a wave emission.

2. Signals obtained in the previous step are again multiplied, phase-shifted and summed:

$$SF(t) = \sum_{n=1}^{N} a_n \mathrm{sig}_n(t - d_n) \tag{7.29}$$

Finally, a single signal is obtained, the reference point for which is the central transducer (as assumed earlier).

The range of the time variable t depends on the location of the transducer pair being considered, as well as of point P, that is:

$$t_1 \leq t \leq t_2 \tag{7.30}$$

where:

$$t_1 = \frac{|PQ| + |PW|}{c} \tag{7.31}$$

Here Q and W denote the locations of the transducers being considered. The value t_2 can be defined most generally as:

$$t_2 = t_1 + \Delta t \tag{7.32}$$

The signal obtained in such a fashion corresponds to waves for the chosen angle φ. Selecting subsequent points of the analysed plane and repeating steps 1 and 2, one obtains diagnostic information in the time domain, which can be correlated with geometrical coordinates, as the wave velocity is known (Figure 7.18).

Practical implementation of the algorithm can take the following shape:

1. For the chosen angle and radius, coordinates of the wave focusing point P are computed.

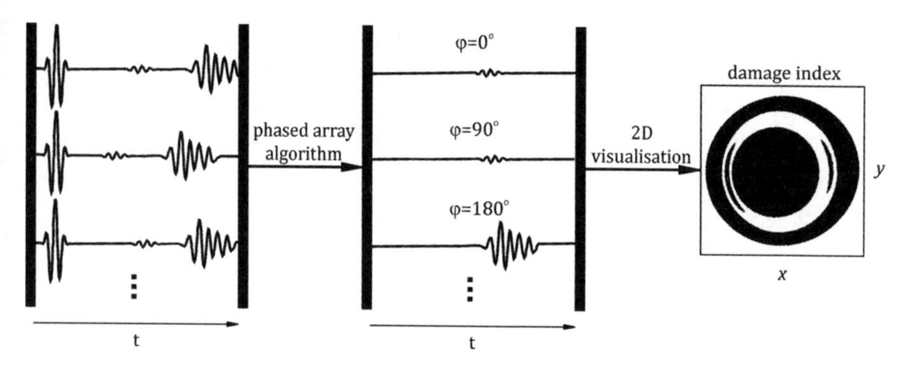

Figure 7.18 Illustration of the damage location procedure based on signal processing algorithms for phased array systems

2. All actuator–sensor pairs are considered. Their distances from point P are computed, as well as the weights and phase shift. Subsequently, signals are multiplied by those and shifted.
3. In further steps, signal fragments corresponding to point P are added. These operations are performed for each recording transducer.
4. Step 2 is repeated, and then combinations of each transducer with the central one are considered.
5. The signal for the analysed point P is constructed.
6. The square of the signal is computed.
7. Coordinates corresponding to the obtained signal are computed.
8. One obtains a damage indicator in the form of the square of the signal assigned to the discrete coordinates x and y. This indicator is denoted $DI(x,y)$.

This algorithm can be speeded up if potential damage is known to lie further away from the transducer system (7.20). One only has to fix one of the values and perform all computations for the assumption of a varying angle.

7.2.3 Methods Employing Continuous Registration of Elastic Waves within the Analysed Area

7.2.3.1 Maximum Total Amplitudes

This method is based on the assumption that propagating elastic waves reflect from discontinuities and interferes with the original induced waves. For this reason an area of increased maximum amplitude is present near the damage site [84, 86]. When this method is employed, one creates a map, for which in every measurement point the maximum total amplitude of velocity of vibrations perpendicular to the plate surface is determined. The presented method has been verified numerically and experimentally using the example of an incised aluminium plate as well as of delamination in the composite plate.

7.2.3.2 Root Mean Squared (RMS) Value

The root mean squared (RMS) value within the chosen time interval $T_1 \leq t \leq T_2$ for the continuous function $f(t)$ specified in the time domain is defined as:

$$f_{rms} = \left[\frac{1}{T_2 - T_1} \int_{T_1}^{T_2} f(t)^2 \, dt \right]^{1/2} \tag{7.33}$$

If the measured velocity $f(t)$ is taken as the function $v(t)$ in the measurement point of coordinates, the RMS value is proportional to the square root of the normalised (with regard to mass) mean kinetic energy associated with the investigated point within the specified time interval:

$$v(x, y)_{rms} = \left[\frac{1}{T_2 - T_1} \int_{T_1}^{T_2} v(x, y, t)^2 \, dt \right]^{1/2} \tag{7.34}$$

For discrete signals v_i of N samples the RMS value is determined as the square root of the arithmetic mean of squared values:

$$v_{rms} = \left(\frac{1}{N} \sum_{i=1}^{N} v_i^2 \right)^{1/2} \tag{7.35}$$

Applying Parseval's theorem, one can demonstrate that RMS values can also be determined in the frequency domain. For a discrete signal the following equality is valid:

$$\sum_{i=1}^{N} v(t)_i^2 = \frac{1}{N} \sum_{i=1}^{N} |V(f)_i|^2 \tag{7.36}$$

where $V(f)$ is the discrete Fourier transform (DFT) of the discrete signal determined in the time domain $v(t)$, having N samples:

$$V(f) = FT\left\{v(t)\right\} \tag{7.37}$$

Thus, the root mean squared value can also be computed from the relationship:

$$v_{rms} = \frac{1}{N}\left(\sum_{i=1}^{N} |V(f)_i|^2\right)^{1/2} \tag{7.38}$$

The RMS value is used as a tool for damage localisation [87]. This value is determined separately for every measurement point. The obtained results constitute an RMS map that forms the basis for locating damage.

In order to improve the legibility of the RMS value, one can filter measured signals with the aim of selecting only the conceivable reflections from damage [88]. For this aim the Fourier transform is used, enabling transition from the space–time domain to the wavenumber–frequency domain. Original wave filtering is performed in the wavenumber–frequency domain. Then, one returns to the time domain employing the inverse Fourier transform. Signals prepared in this fashion are developed into an RMS map, on which areas of the largest values indicate possible locations of damage.

7.2.3.3 Cumulative Kinetic Energy

Another indicator similar to the RMS is the cumulative kinetic energy [89]. This value is computed from the following relationship:

$$E(x, y) = \frac{1}{2}\int_{T_1}^{T_2} v^2(x, y, t)\mathrm{d}t \tag{7.39}$$

where $E(x, y)$ is a measure of the normalised cumulative kinetic energy of the propagating elastic wave for the time interval $T_1 \le t \le T_2$. For measured discrete signals this coefficient takes the following form:

$$E = \frac{1}{2}\sum_{i=1}^{N} v_i^2 \tag{7.40}$$

As this indicator is proportional to the square of RMS, maps obtained using both these indicators are similar.

7.2.4 Damage Identification Algorithms

7.2.4.1 Direct Methods

In one-dimensional problems the wave transition and reflection coefficients can be used as measures of damage size. However, more often the energy of a wave packet reflected from damage e_R compared to the energy of the induced packet e_I is used. Usually this is cumulated energy, computed on the basis of signals registered in multiple sensors and related to the geometry of the investigated element by means of the elastic wave propagation velocity. In other words, one can simultaneously assess damage location and size. Assessing damage size requires knowing the dependency of e_R/e_I on the damage size (Figure 7.19). This dependency can be obtained by means of either numerical computations or experiments [93]. It should be added that due to the multi-modal nature of propagating elastic waves, extracting the relevant packets carrying information about damage is difficult.

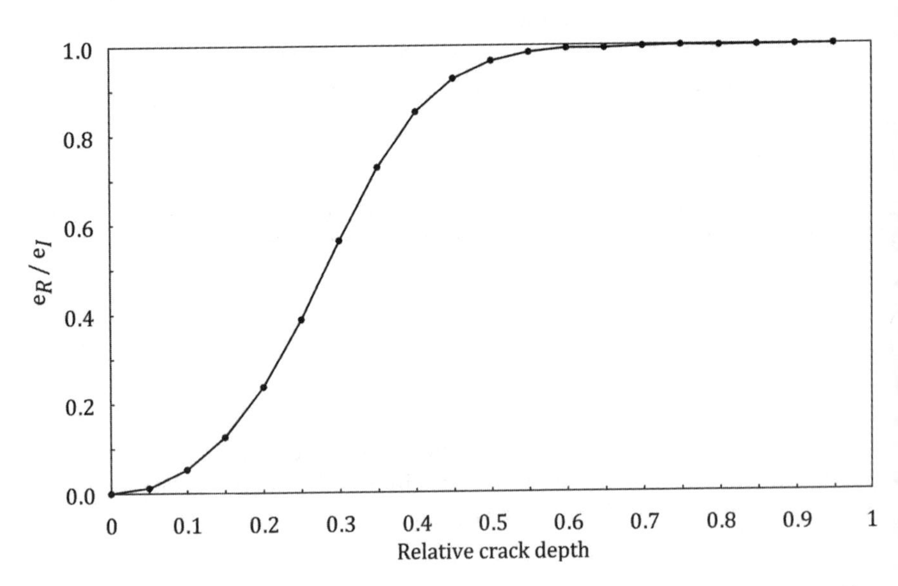

Figure 7.19 Indicator of the relative crack depth in isotropic bar obtained by means of numerical computations

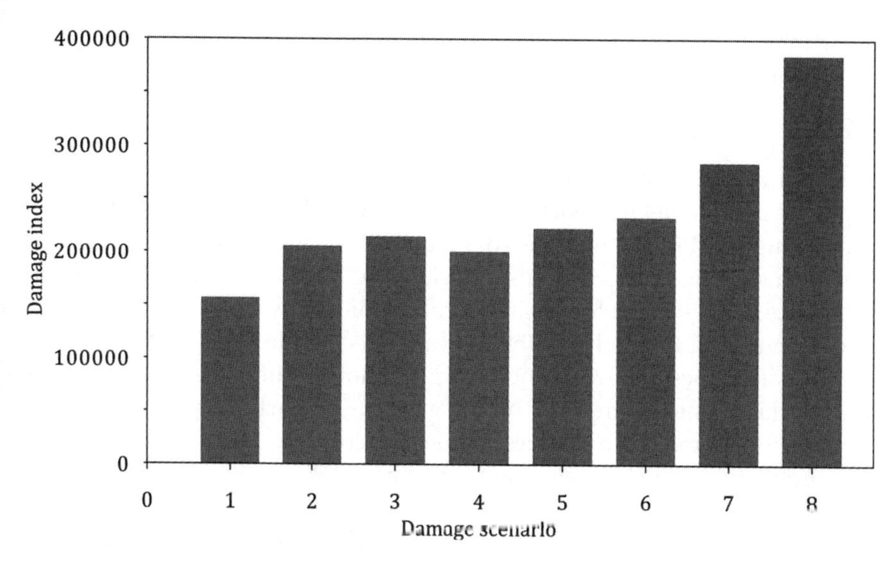

Figure 7.20 Damage indicator based on the cumulated energy for a 1 mm thick composite plate with damage in the form of (1) a 4 mm long, 0.2 mm deep crack, (2) a 4 mm long, 1 mm deep crack, (3) an 8 mm long, 1 mm deep crack, (4) a 14 mm long, 1 mm deep crack, (5) an 18 mm long, 1 mm deep crack, (6) two 1 mm deep cracks with lengths of 18 mm and 4 mm, respectively, (7) two 1 mm deep cracks with lengths of 18 mm and 8 mm, respectively, (8) two 1 mm deep cracks with lengths of 18 mm and 15 mm, respectively

Analysing single wave packets reflected from damage also allows a method to be developed that identifies such discontinuities in bar and beam structures [94]. This method investigates the correlation between the induced and reflected wave packets. Such an approach distinguishes discontinuities in the form of a fatigue crack from a change in stiffness of the structure element caused by a change in the cross-section or a decrease of Young's modulus.

In the case of two-dimensional structures it is virtually impossible to construct a meaningful damage indicator based on signals registered by piezoelectric sensors. In the two-dimensional case the difficulty is that waves reflect from structure damage not necessarily perpendicularly to the wavefront, that is towards the wave exciting transducer. This prevents unequivocally correlating the energy of reflected waves with a crack depth in the pulse-echo method (Figure 7.20). In the pitch-catch method, in turn, perturbation of the propagating wave caused by damage present in the structure need not happen on the path from the actuator to the sensor. This mandates using measurement techniques enabling measuring the full field of displacements associated with wave propagation, such as laser vibrometry.

7.2.4.2 Inverse Methods – Genetic Algorithms

In the case of inverse methods used for damage identification, the key part is played by a model that enables simulating the phenomena of elastic wave propagation and interaction with damage with good approximation. Models based on spectral finite elements described in this book can be used for this purpose. Damage in the form of material discontinuity is modelled by means of disconnecting the nodes of the relevant spectral elements and modifying the global stiffness matrix. Modifying the global stiffness matrix involves adding the stiffness computed on the grounds of fracture mechanics to the global stiffness matrix in degrees of freedom corresponding to damage [93]. The problem is that infinitely many damage cases are possible and in practice one can only parameterise a finite set of locations and damage sizes. As computations are time-consuming, problems concerning damage location can only be solved for one-dimensional structural elements like bars or beams.

If one has both measurement signals and a numerical model allowing for simulating elastic wave propagation, for example in the case of a bar with a transverse crack, one can simultaneously identify the crack location and depth. For the needs of employing a genetic algorithm both decision variables, the crack location m and the crack depth n, can be assumed to be encoded in the chromosome divided into two parts. Decision variables are represented by integers from the range of $[0, 2^p - 1]$, where p denotes the number of bits constituting the chromosome. This allows easy modelling of damage, as its location is unequivocally defined and overlaps with mesh of spectral finite elements (e.g. the crack number m is located beyond the element number m). In optimisation tools using genetic algorithms the chromosome is coded automatically [95].

The most important step of optimisation methods is formulating the objective function. In the problem of identifying the crack depth and location one can search for an objective function of the following form:

$$f(m, n) = \sum_{i=1}^{N} \sum_{j=1}^{M} |(R_{ij} - S_{ij}(m, n))| \tag{7.41}$$

where N denotes the number of points in the recorded signal, M denotes the number of signals, R_{ij} denotes the ith amplitude of the jth measurement signal and $S_{ij}(m, n)$ denotes amplitudes of signals obtained from the numerical model for parameters defined by the decision variables m and n. Signals $S_{ij}(m, n)$ are usually computed on the fly, because the number of possible crack locations and sizes can be relatively high (depending on the required computation accuracy) and the genetic algorithm does not process every possible combination. On the other hand, one-time construction of a sizable database of numerical computation results for combinations of crack locations and depths makes the genetic algorithm run very fast. It should be noted that in the discussed case only one damage site appears in the bar. The simultaneous presence of multiple damage sites complicates the algorithm and requires modification of the objective function. Results of a genetic algorithm employing a nine-bit chromosome lead to an assessment of a crack location on a 1 m long bar with an accuracy of about 2 % and a crack depth with an accuracy of about 6 %.

7.3 Examples of Damage Localisation Methods

7.3.1 Localisation Algorithms Employing Sensor Networks

Damage localisation algorithms described in Section 7.2 were applied to experimental results. A $1000 \times 1000 \times 1\ \text{mm}^3$ aluminium panel was considered. Additional mass was installed on its surface (Figure 7.21). A piezoelectric actuator with a diameter of 10 mm was glued in the centre of the element area. A transducer was used for exciting elastic waves in the investigated element (see Figure 7.21). The excitation frequency was equal to 16.5 kHz. The excitation signal comprised five periods of a sine function of the given frequency modulated with a Hann window. Signals were measured in contactless fashion using laser scanning vibrometry (Chapter 3). Laser measurements were performed on the other surface of the element and 13 653 measurement points forming a rectangular grid were defined. Signals measured in these points represent changes of the surface vibration velocity as a function of time. From the set of points the part constituting the desired sensor configuration was selected. The central point of configuration was assumed to correspond to coordinates of the piezoelectric transducer. A circular grid of angular spacing between points of $\Delta\alpha = 2^\circ$ and radial spacing of $\Delta r = 1$ mm was chosen for the needs of location result visualisation. Both concentrated and distributed configurations were considered. A square grid (Figure 7.22(a)) was chosen as a distributed configuration. Concentrated square-type systems (Figure 7.22(a)) were also investigated. The last type of system considered in this section

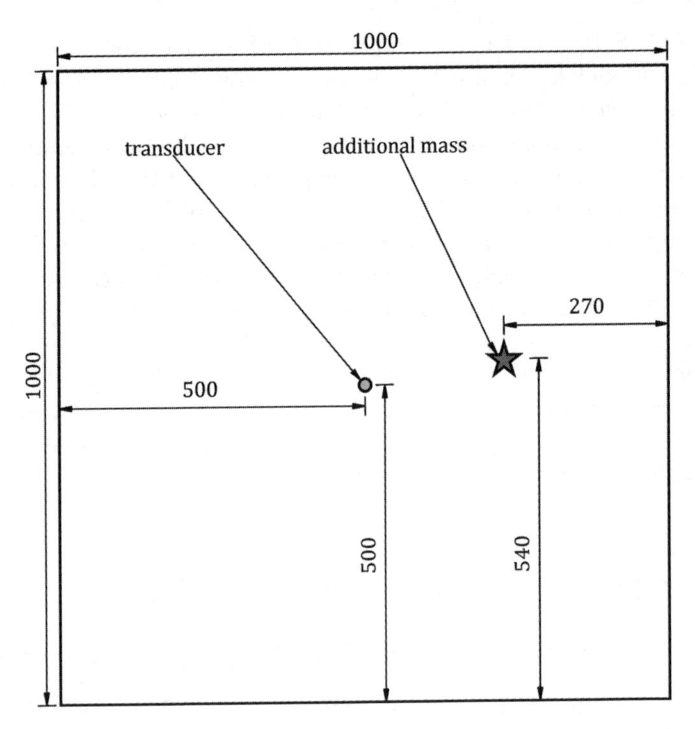

Figure 7.21 Aluminium plate with a piezoelectric transducer located in the centre and additional mass

were phased array systems in a square configuration (Figure 7.22(a)) and a star configuration (Figure 7.22(b)).

A uniform method of presenting results was chosen for all damage location methods by representing the damage indicator using a colour scale mapped on the surface of the investigated element. The maximum damage value was presented in black, the minimum value was presented in white and 256 levels of grey were chosen for presenting intermediate values. Images presented according to these conditions map the damage indicator values on to specimen coordinates.

A concentrated square sensor layout of spacing between sensors of $d =$ 9 mm containing $N = 9$ elements (Figure 7.22(a)) was chosen as the first example. The value of the damage indicator is visualised by a damage influence map, presented in Figure 7.23(a). Individual measurement points constituting the system cannot be distinguished in the illustration due to the distance d being small. Significant increases of the damage indicator are visible in the form of a black circle. The radius of this circle corresponds

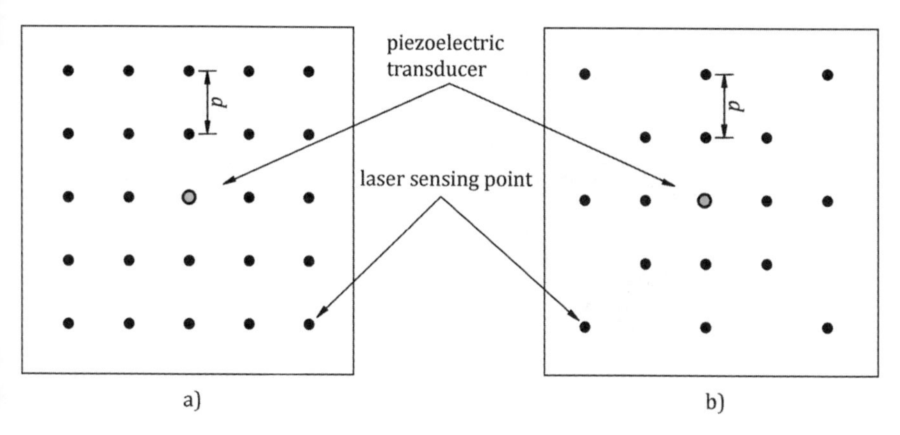

Figure 7.22 Considered sensor layouts for $N = 25$: (a) square layout, (b) star layout. Depending on spacing d, these layouts can be considered either as concentrated or distributed. The wave was excited by a single piezoelectric transducer located in the layout centre, while measurements in the other points were performed by laser vibrometry

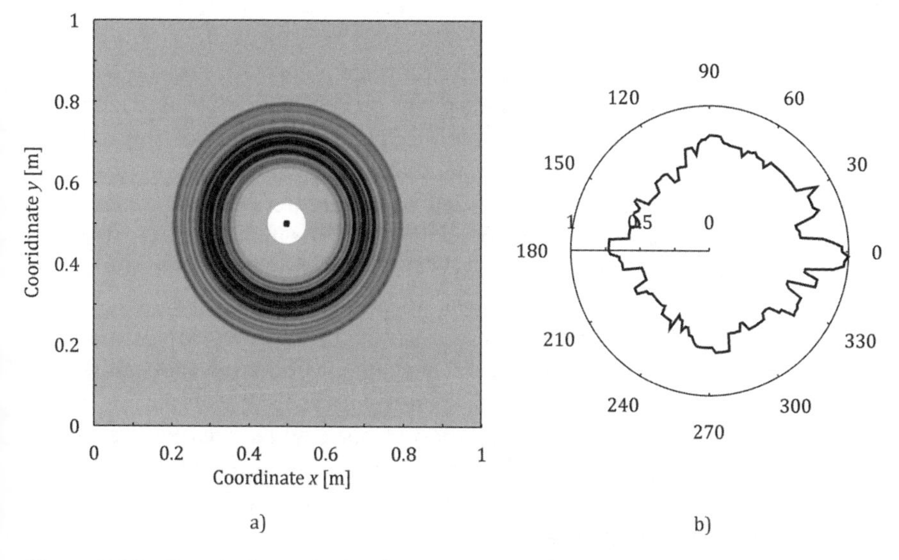

Figure 7.23 Square layout ($d = 3$ mm, $N = 9$): (a) result of locating discontinuity, (b) maximum value of the damage indicator for various angles

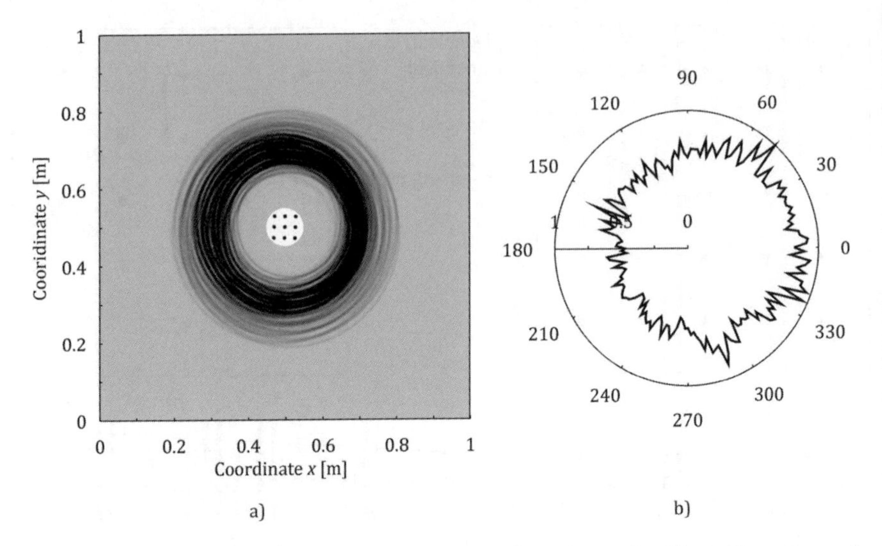

Figure 7.24 Square layout ($d = 30$ mm, $N = 9$): (a) result of locating discontinuity, (b) maximum value of the damage indicator for various angles

to the distance from the centre of the measurement system to the location of the sought discontinuity. However, the obtained result does not indicate to which side of the sensor system the damage is located. For the needs of diagnostic classification the obtained result can be called damage detection, but not damage localisation. In order to visualise values of the damage indicator around the sensor system, its maximum values were investigated across the whole angular range (0–360°) (see Figure 7.23(b)). It turns out that a partial location was achieved, as the indicator maximum corresponds to the damage location (angles around 0°). However, for other angles the damage indicator exhibits large values, up to 0.8.

In the course of further research the same concentrated layout ($N = 9$) was revisited for larger spacing of measurement points ($d = 30$ mm). The result of the damage location can be seen in Figure 7.24. Comparing this to the result presented in Figure 7.23, one notices a wider area with elevated damage indicator values. According to the used numerical algorithm (Section 7.3.1), the excitation point and measurement point constitute foci of the ellipse on which damage lies. In the previous case these foci were located so close together that the indication was practically a circle. Increasing the distance d results in a clearer appearance of the elliptic shapes. As the angular graph indicates (Figure 7. 24(b)), an indication of the damage location did not improve. The next layout considered was a square phased array. To

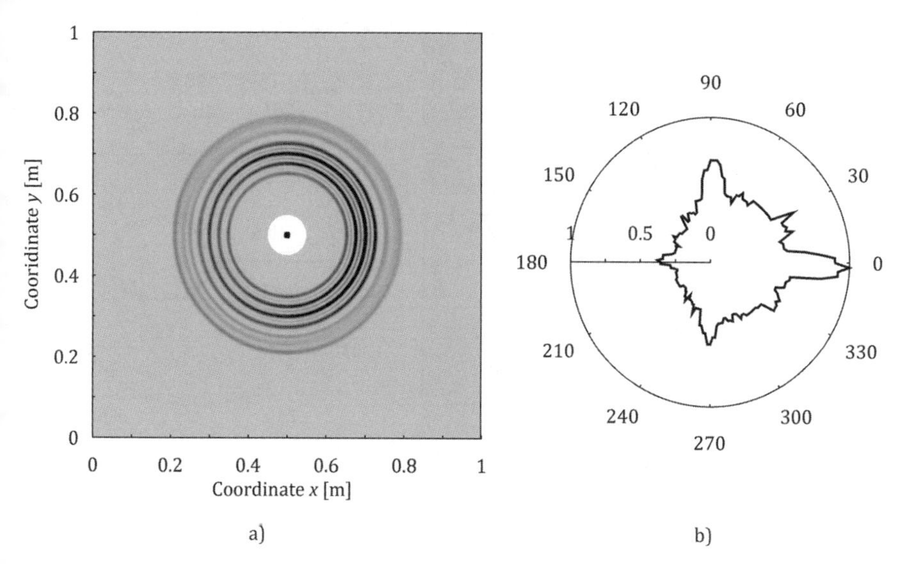

Figure 7.25 Square phased array layout ($d = 3$ mm, $N = 9$): (a) result of locating discontinuity, (b) maximum value of the damage indicator for various angles

facilitate comparison, the layout parameters (d and N) were identical to those presented in Figure 7.23. The location result shows significant improvement of the damage location indication (Figure 7.25(a)). This is visible particularly in the angular graph (Figure 7.25(b)). The location algorithm enabled an increase of the damage indicator for the angle corresponding to the damage location. Advantages of the phased array layout are visible even more clearly when the number of measurement points is increased to $N = 25$. The result for that case can be seen in Figure 7.26. The last of the phased array layouts considered was a star-shaped one (Figure 7.22(b)). Compared to the square layout, an advantage of this one is the number of sensors reduced eight times. As the results demonstrate (Figure 7.27), the location result is generally identical. After having analysed concentrated systems, it is worthwhile to analyse a distributed system using the same example. Experiments were performed for a square layout of the parameters $d = 163$ mm and $N = 5$. The result of the damage location is presented in Figure 7.28. One can distinguish characteristic ellipses that are visible even better thanks to larger distances between their foci (compare with Figure 7.24). Due to multiple indications, the result is difficult to interpret. One cannot be sure that only a single damage site is present.

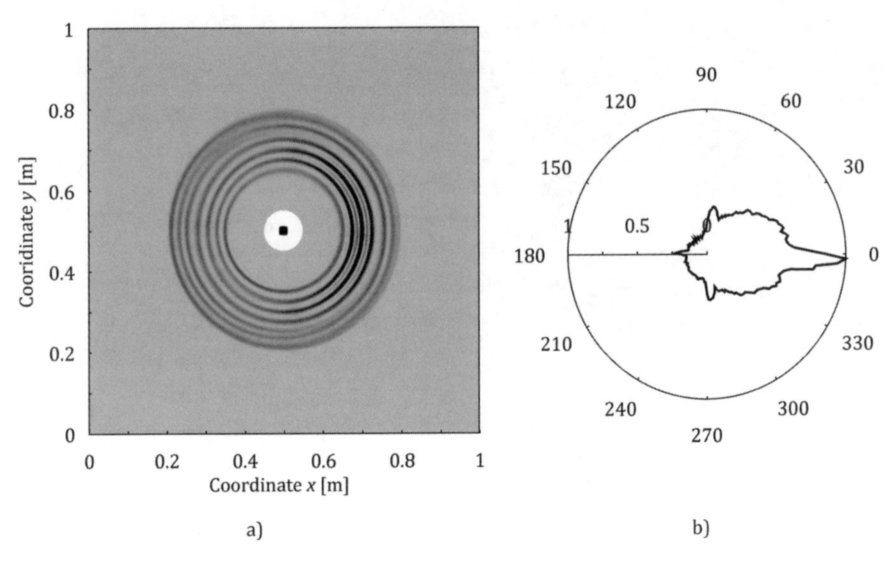

Figure 7.26 Square phased array layout (d = 3 mm, N = 25): (a) result of locating discontinuity, (b) maximum value of the damage indicator for various angles

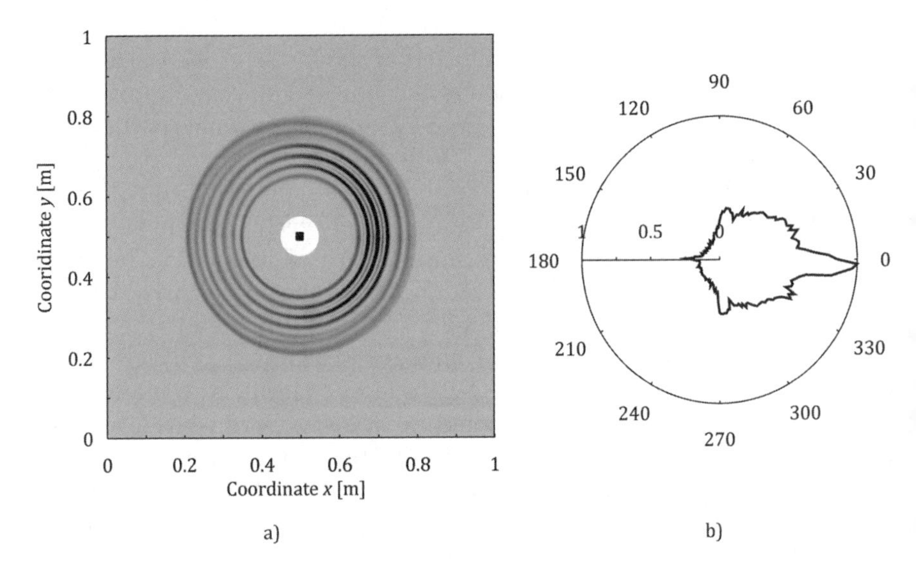

Figure 7.27 Star-shaped phased array layout (d = 3 mm, N = 25): (a) result of locating discontinuity, (b) maximum value of the damage indicator for various angles

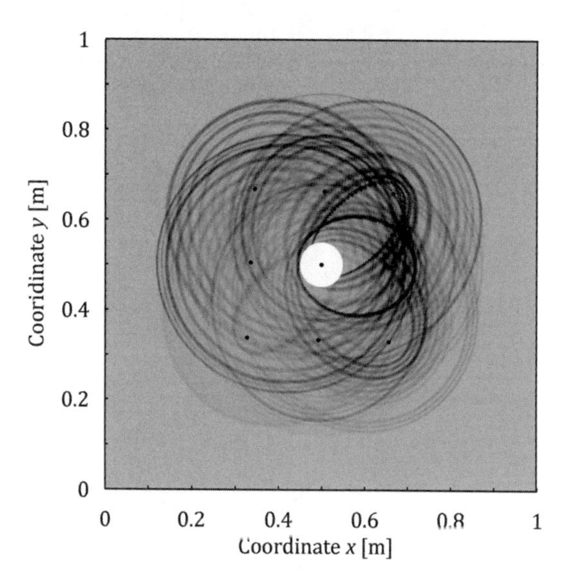

Figure 7.28 Square distributed layout ($d = 163$ mm, $N = 9$): result of locating discontinuity

7.3.2 Algorithms Based on Full Field Measurements of Elastic Wave Propagation

In order to verify the effectiveness of described damage detection methods employing measurements of a full velocity field representing the phenomenon of elastic wave propagation, numerical and experimental research was performed on identical specimens. A $1000 \times 1000 \times 1$ mm^3 aluminium plate with a piezoelectric transducer located in the centre of its surface was considered. Discontinuity was modelled by attaching a small disc (of mass of 1.6 g and diameter of 15 mm) in the upper right quarter of the specimen (see Figure 7.29).

For the needs of numerical simulation this disc was modelled as an additional circular mass. Only the quarter containing the mass was chosen for a result presentation. Measurements were performed by laser scanning vibrometry, recording the vibration velocity in the direction perpendicular to the measuring beam. The measurement grid consisted of 226×226 uniformly spaced points. In order to provide the best possible consistency, velocities instead of displacements were read from spectral grid nodes of the numerical

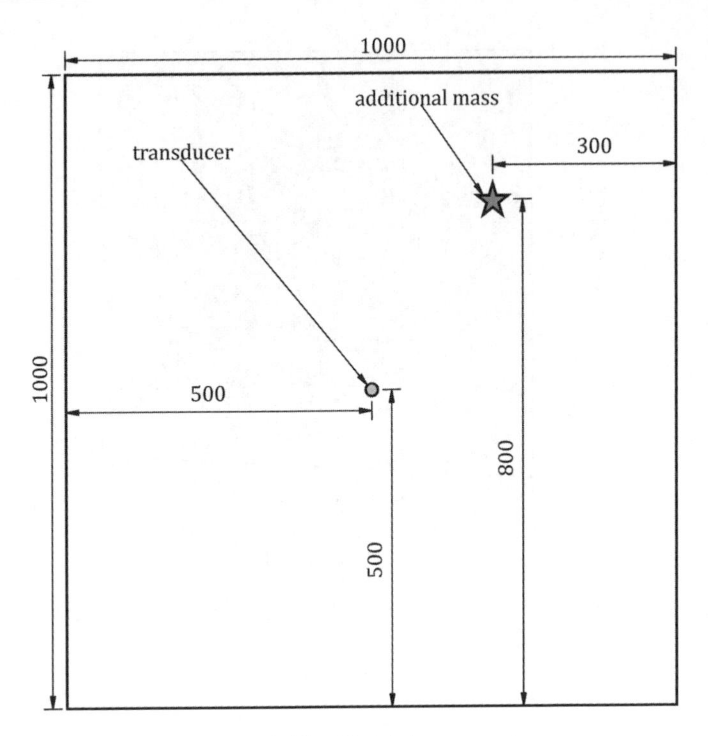

Figure 7.29 Aluminium panel

simulation. Excitation was chosen in the form of a wave packet with a length of five periods and a carrier frequency of 35 kHz (see Chapter 3).

Registered wave propagation patterns were used for determining the three mentioned coefficients, that is a map of the maximum total amplitudes, a map of the RMS coefficients and a map of the cumulative kinetic energy. The obtained results are compared in Figure 7.30.

As the measured area contained a piezoelectric element generating elastic waves, the range of detection coefficients is wide enough for subtle variations associated with the presence of discontinuities not to be noticeable. For this reason Figure 7.31 presents maps in re-scaled levels-of-grey scale, discarding the largest values of damage detection indicators. Additionally, logarithmic values of the discussed maps are presented in Figure 7.32.

The three coefficients presented in Figure 7.32 yield very similar results. Therefore further discussion will be limited to the RMS coefficient only.

In order to compare the results obtained from the numerical model with experimental ones, Figure 7.33 juxtaposes two maps of logarithmic RMS values determined for a time instant of 0.625 ms.

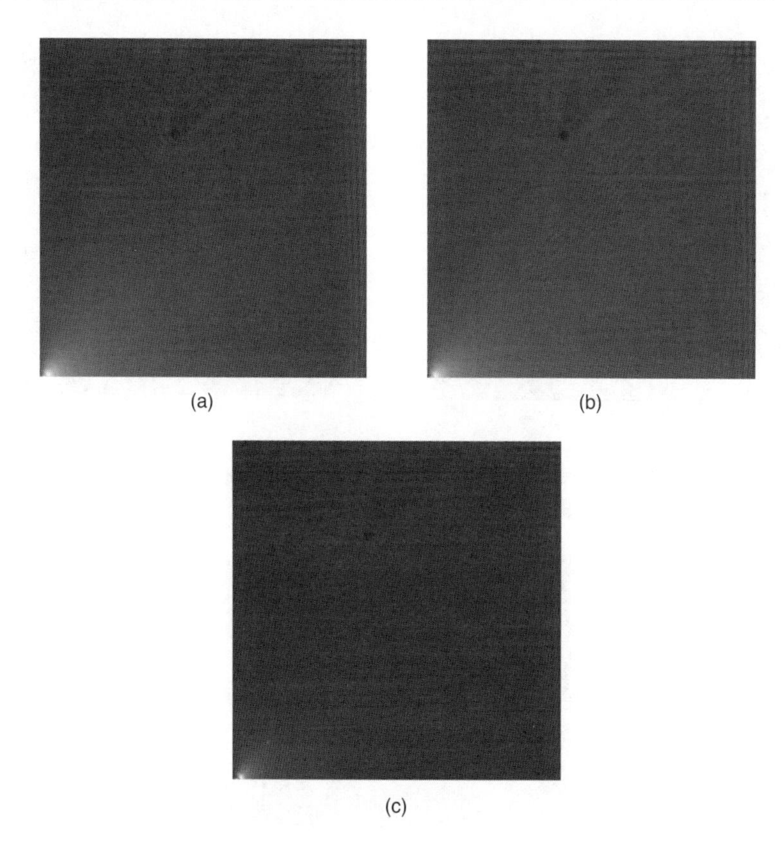

Figure 7.30 Comparison of damage detection coefficients for an aluminium plate with additional mass: (a) total amplitude, (b) RMS, (c) CKE

Both results of numerical simulation and experimental ones demonstrate good sensitivity of the method to discontinuity on the surface of the investigated element (Figure 7.33). Good agreement of numerical and experimental results should be emphasised.

The next stage of research involved attempts to employ the discussed method for analysing a fragment of an actual aircraft. The investigated real-life specimen was a helicopter tail plane made of epoxy–glass composite covered with epoxide enamel on the outside. Tail plane aerofoils were stiffened on the attack side with evenly spaced sandwich ribs. In the trailing edge part the aerofoil was reinforced with cell filler segments. Dimensions of the tail plane together with the layout of internal stiffening elements are shown in Figure 7.34. The measurement area is marked in grey.

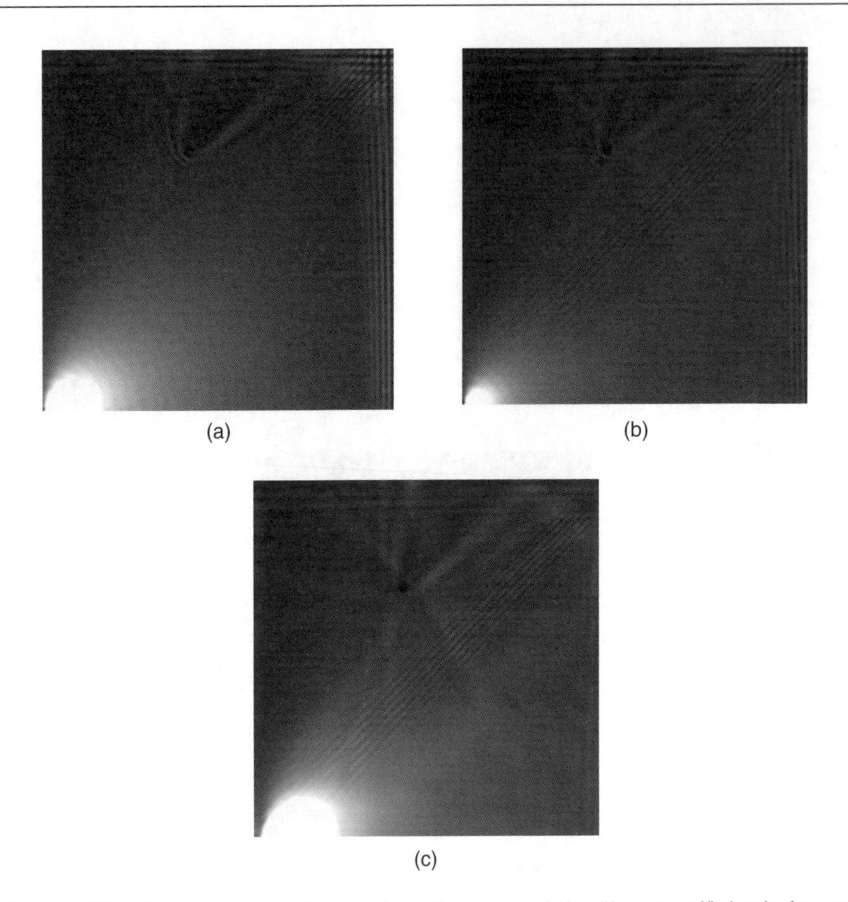

Figure 7.31 Comparison of re-scaled damage detection coefficients for an aluminium plate with additional mass: (a) total amplitude, (b) RMS, (c) CKE

Measurements were performed for three separate carrier frequencies of the excitation signal: 17.5 kHz, 35 kHz and 100 kHz. The obtained RMS maps are presented in Figures 7.35, 7.36 and 7.37, respectively. Analysis for the first two frequencies involved reference measurements (Figures 7.35(a) and 7.36(a)) and measurements with discontinuities placed on the tail plane surface (Figures 7.35(b) and 7.36(b)). Discontinuities were introduced by means of gluing seven additional masses. Discontinuities can be located thanks to visible changes of the RMS value maps corresponding to discontinuity locations. The path that a wave travels from the piezoelectric transducer shortens with a frequency increase. Therefore the method range decreases as the

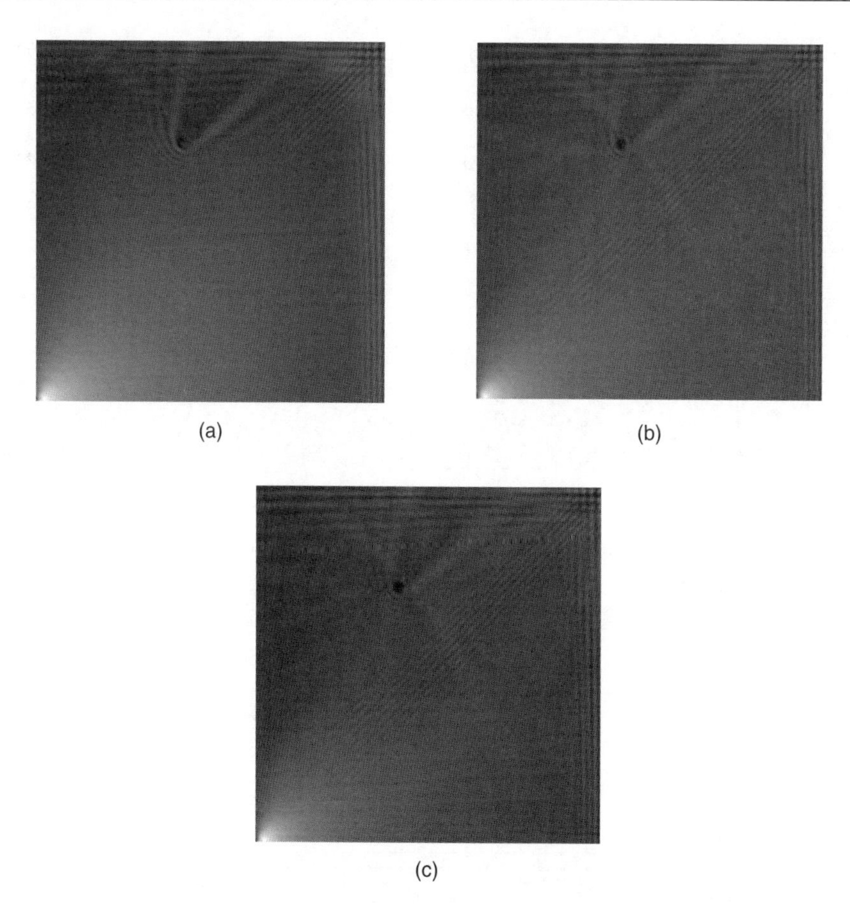

Figure 7.32 Comparison of re-scaled damage detection coefficients in logarithmic scale for an aluminium plate with additional mass: (a) total amplitude, (b) RMS, (c) CKE

frequency increases. At the same time its sensitivity to smaller damage increases. In the case of frequency of 100 kHz the method inaccurately visualises the tail plane structure (Figure 7.37). The cell filler is much less visible than in the cases of lower frequency induction, where waves propagate to longer distances.

7.3.2.1 Weighed RMS Coefficient

As the wave amplitude decreases with distance from the excitation site, one can construct a function similar to RMS by assigning increasing weights to

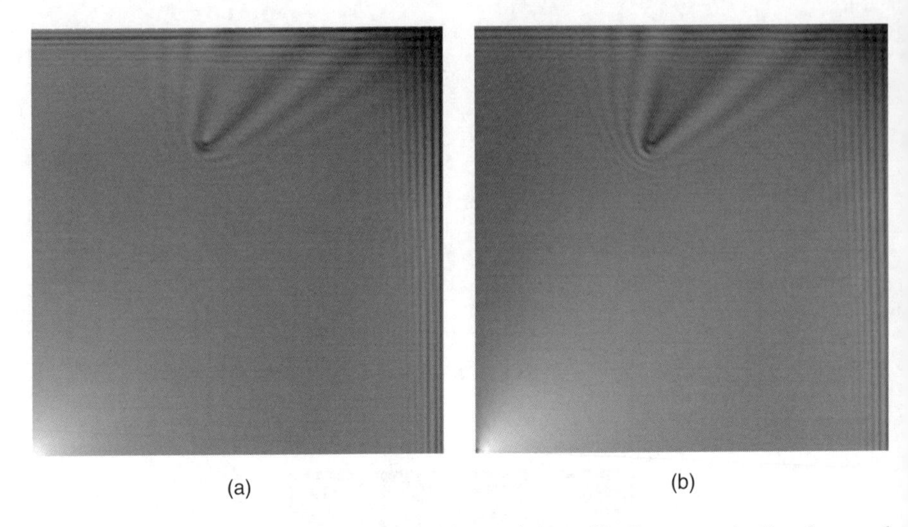

Figure 7.33 RMS maps for an aluminium plate with damage in the form of glued additional mass (1 g) determined for the time period 0–0.625 ms and log (v_{rms}): (a) numerical results, (b) experimental results

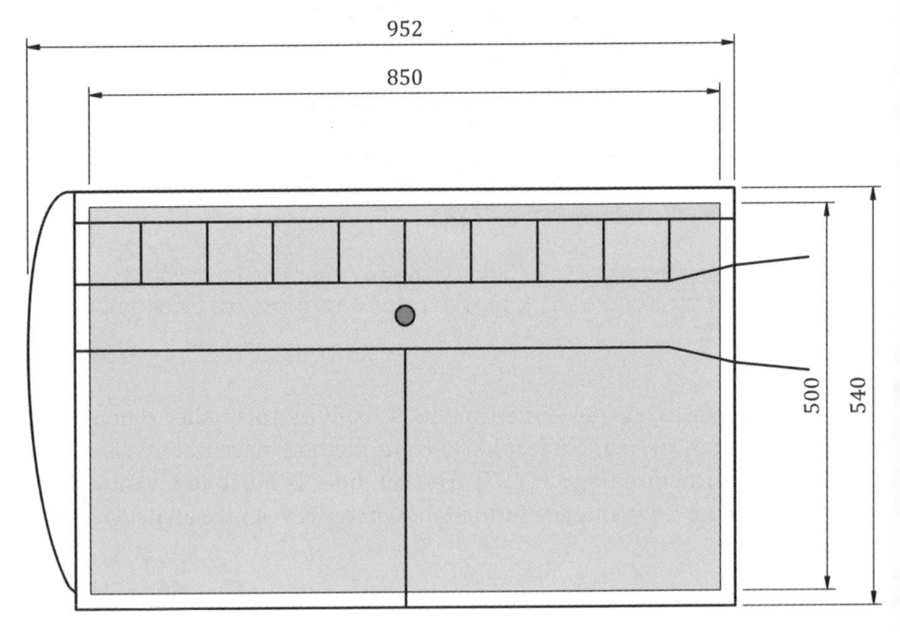

Figure 7.34 Dimensions of a PZL W-3A helicopter stabiliser together with the measurement area

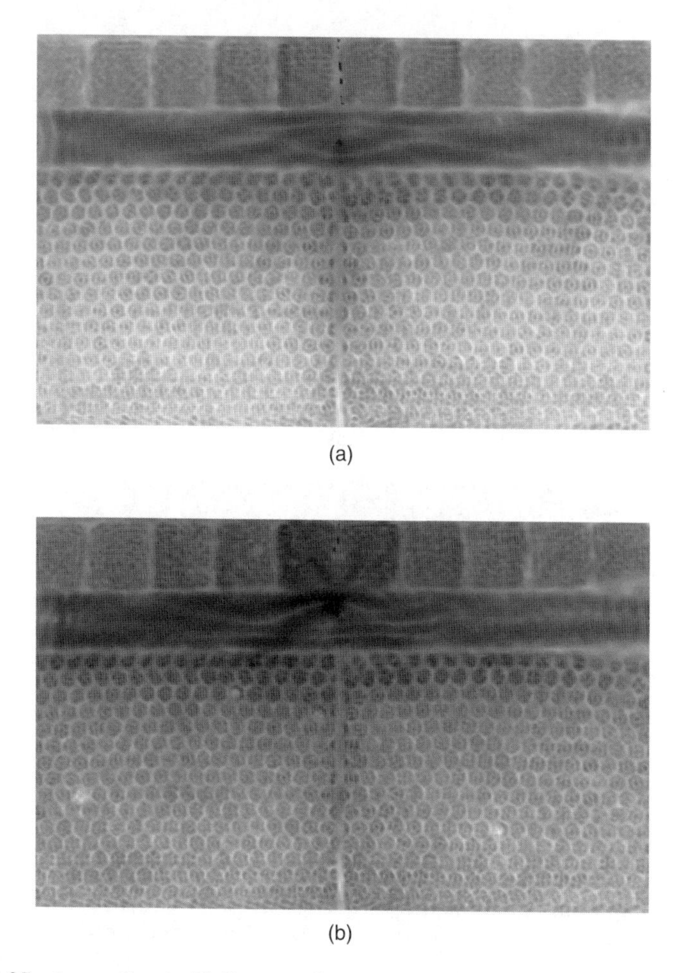

Figure 7.35 Logarithmic RMS maps for a PZL W-3A helicopter stabiliser for the excitation frequency of 17.5 kHz and $k = 0$: (a) reference measurement, (b) measurement for the tail plane with seven discontinuities on its surface

subsequent samples. The amplitude decrease is associated with dissipation of energy of the propagating wave on a growing area with the boundary defined by the circumference of a circle of radius R. This radius is equal to the distance travelled by a wave from the forcing site. However, this statement is only true as long as one considers a time period in which no wave reflections from the element edges take place. Additionally, the phenomenon of elastic wave attenuation occurs in such a case; this phenomenon is dependent on the

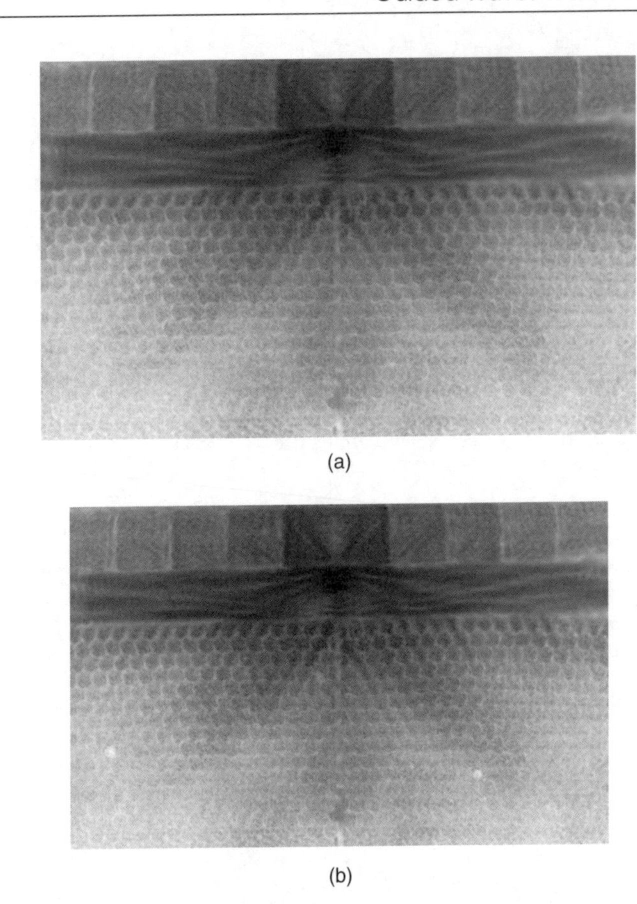

Figure 7.36 Logarithmic RMS maps for a PZL W-3A helicopter stabiliser for the excitation frequency of 35 kHz and $k = 0$: (a) reference measurement, (b) measurement for the tail plane with seven discontinuities on its surface

frequency and medium in which the wave propagates. In order to compensate for these two factors, the weight coefficient is introduced into the RMS function. In order to compare the obtained results, two different weights were assumed using the following formula:

$$v_{rms\_w^k} = \left[\frac{1}{N} \sum_{i=1}^{N} \left(v_i^2 i^k \right) \right]^{1/2} \tag{7.42}$$

where k is the weight coefficient exponent.

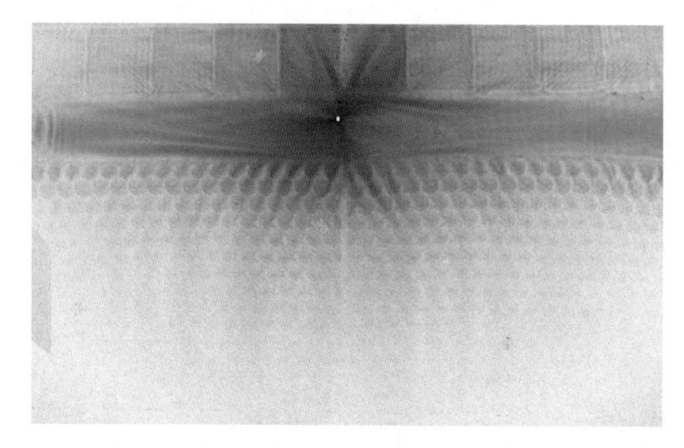

Figure 7.37 Logarithmic RMS map for a PZL W-3A helicopter stabiliser for the excitation frequency of 100 kHz and $k = 0$; measurement for the tail plane with seven discontinuities on its surface

Figures 7.38 and 7.39 juxtapose the obtained results for different coefficient values of $k = 0$ and $k = 2$ (for maps in the linear scale and in the logarithmic scale). Figure 7.40 juxtaposes values of the $v_{rms\_w^2}$ function obtained when time periods of different lengths – different numbers of time samples – were chosen for obtaining the function value.

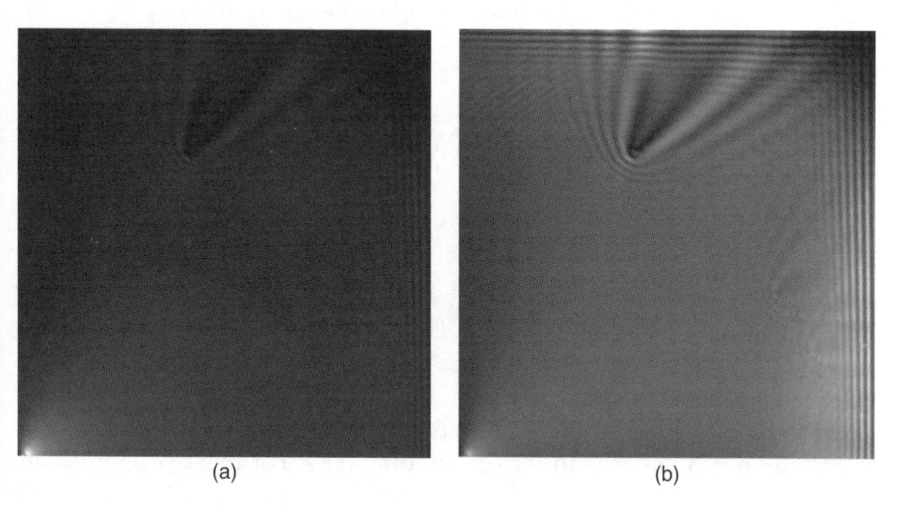

(a) (b)

Figure 7.38 Values of the $v_{rms\_w^k}$ function for 320 initial samples and a weight coefficient k of: (a) 0, (b) 2

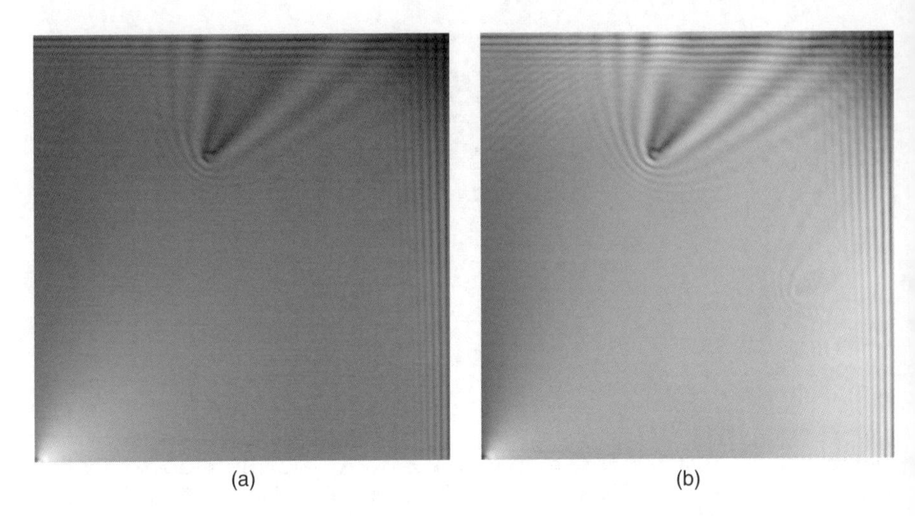

Figure 7.39 Values of the $\log(v_{rms\_w^k})$ function for 320 initial samples and a weight coefficient k of: (a) 0, (b) 2

In order to enable discontinuity detection across the whole investigated area, for the needs of deriving RMS maps one should consider such a time period in which the propagating wave travels through the whole investigated area. Increasing the number of initial samples taken into account for the RMS value determination results in waves reflected from the edges being included as well. This reduces the characteristic patterns of interfering waves appearing around damage sites. For this reason, in examples from Figures 7.38 and 7.39 RMS maps were prepared using 320 initial samples, which corresponds to the time interval for 0 to 625 μs, counting from the moment of wave excitation.

7.3.2.2 Filtering RMS Maps

Measuring elastic wave propagation by means of laser vibrometry is characterised by single points exhibiting noise levels significantly higher than for most other points. This effect results from the laser beam being unluckily reflected in the discussed points in such a fashion that a small part of it returns to the measuring head. Values of v_{rms} and $v_{rms\_w^k}$ functions for these points are much larger than for surrounding ones. For this reason, median filtering in the space domain with a 2×2 filter was employed for the plots presented above, with the aim of minimising the level of such interference. This operation allowed for removing (replacing with values from adjacent

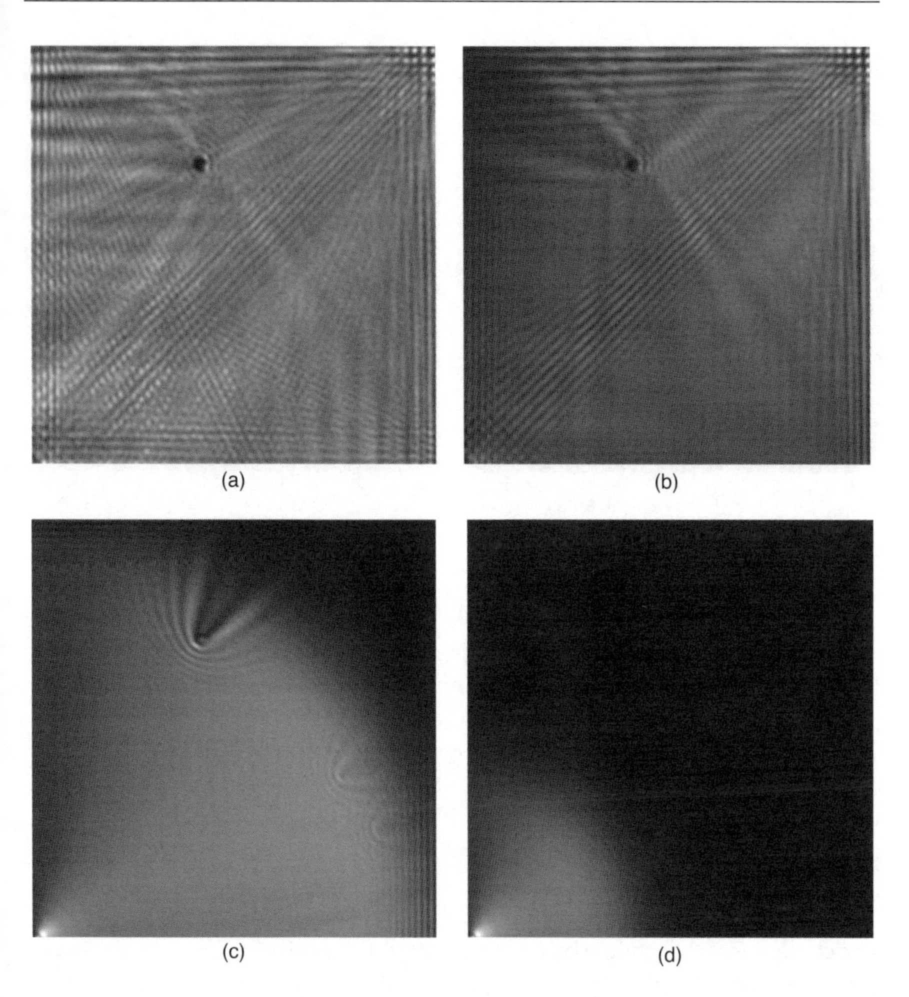

Figure 7.40 The $v_{rms\_w^2}$ function for an aluminium plate with additional mass: (a) 1024 samples, (b) 512 samples, (c) 256 samples, (d) 128 samples

points) the points in which the values were significantly different from those in their neighbourhood. At the same time, values in the remaining points did not change. Results obtained for $v_{rms\_w^2}$ without filtering and with a median filter are presented in Figure 7.41.

Additionally, filtering of RMS maps can be performed using various approaches to median filtering. One can perform filtering in the space domain for each time sample before constructing an RMS map. Filtering can also be performed in the space domain on a constructed RMS map. It is also

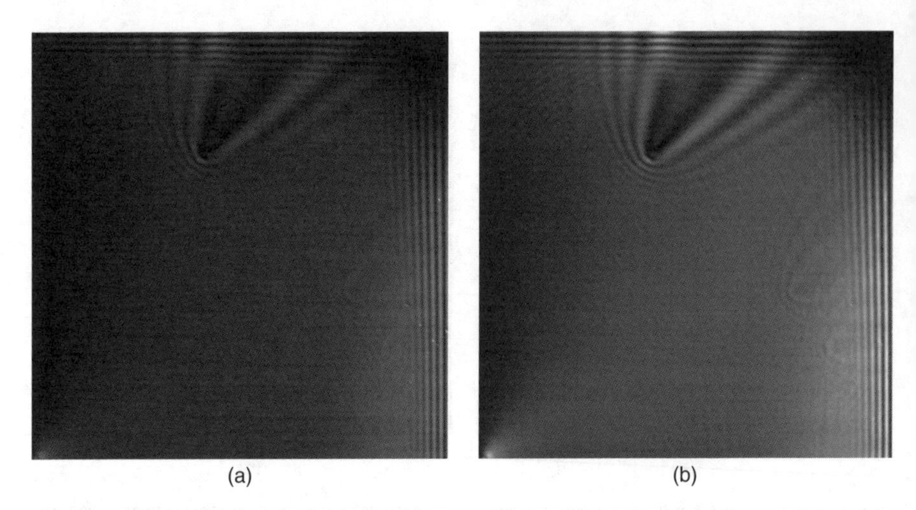

Figure 7.41 Comparison of RMS maps: (a) without filtering, (b) with 2 × 2 median filtering

possible to perform filtering in the time domain for every measurement point. Example results obtained for an aluminium plate with additional mass are presented in Figure 7.42.

Filtering in the space domain, when every temporal sample is filtered, and filtering of a ready RMS map yield very similar results. For this reason the second approach is preferred, as it requires much fewer computational operations. Signal filtering in the time domain produces no significant improvement of results.

Questions of damage detection, localisation and identification pose scientific challenges undertaken by research teams all over the world. Elastic waves have been employed in this area since the 1960s [2]. This topic still remains in progress thanks to continually improving availability of many advanced methods of generating and registering these waves in structural elements. A literature review shows many possible solutions in areas of sensor spacing and signal processing algorithms. Methods of damage detection, localisation and identification used in research papers by multiple researchers have been discussed in this chapter. Mathematical formalism, which has formed the basis for the discussed algorithms, has been presented. Performed investigations, both numerical and experimental, prove the effectiveness of the proposed methods of signal processing. Research involved simple elements, that is aluminium panels and fragments of real-life structures. Attention was focused on testing aircraft parts. The obtained results are promising and

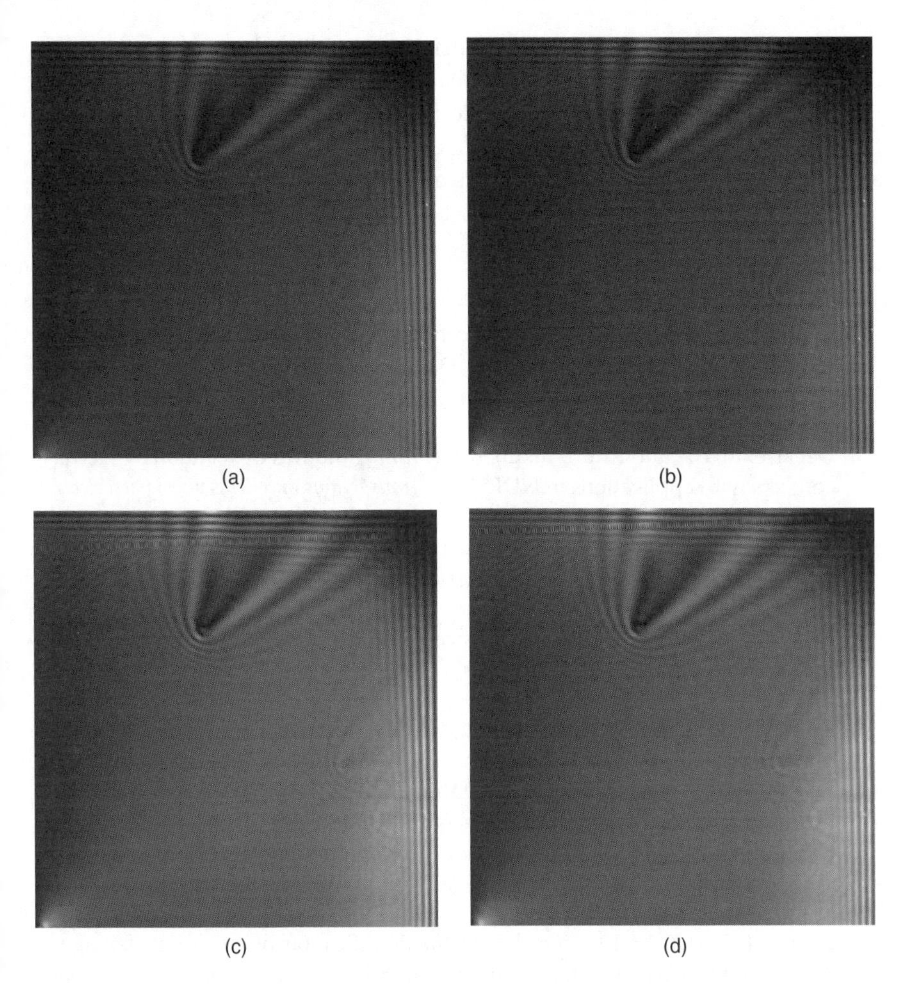

Figure 7.42 Comparison of RMS maps: (a) without filtering, (b) with temporal filtering ($n = 5$), (c) with 2×2 median filtering of each temporal sample, (d) with 2×2 median filtering of RMS only

justify attempts to employ the discussed methods in the diagnostics of real-life structures. Employing laser scanning vibrometry allows for relatively easy measurement of the whole vibration velocity field associated with elastic wave propagation, and therefore for using both sensor-system-based methods and ones involving an analysis of the whole propagation field. Employing laser vibrometry poses no problem in laboratory conditions. Operating a vibrometric setup for field measurements seems troublesome, however.

Measurement heads need to be appropriately positioned relative to the tested structure. One of the recommended head arrangements involves positioning them in points of an equilateral triangle. Systems of piezoelectric sensors seem to be more suitable for monitoring operational structures. Small dimensions and masses of these sensors constitute advantages of such an approach.

References

1. Worlton, D.C. (1961) Experimental confirmation of Lamb waves at megacycle frequencies. *Journal of Applied Physics*, **32**, 967–971.
2. Berger, H. (1967) An ultrasonic imaging Lamb-wave system for reactor-fuel-plate inspection. *Ultrasonics*, **5**, 3941.
3. Degertekin, F.L. and Khuri-Yakub, B.T. (1997) Lamb wave excitation by Hertzian contacts with applications in NDE. *IEEE Transactions on Ultrasonics, Ferroelectrics and Frequency Control*, **44**, 769–779.
4. Deutsch, W.A.K., Cheng, A. and Achenbach J.D. (1997) Self-focusing of Rayleigh waves and Lamb waves with a linear phased array. *Research on Nondestructive Evaluation*, **9**, 81–95.
5. Rose, J.L. (2000) Guided wave nuances for ultrasonic nondestructive evaluation. *IEEE Transactions on Ultrasonics, Ferroelectrics, and Frequency Control*, **47**(3), 575–583.
6. Kessler, S.S., Spearing, S.M. and Soutis, C. (2002) Damage detection in composite materials using Lamb wave methods. *Smart Materials and Structures*, **11**, 269–278.
7. Li, Z.N., Tang, J. and Li, Q.S. (2004) Optimal sensor locations for structural vibration measurements. *Applied Acoustics*, **65**, 807–818.
8. Qiu, L. and Yuan, S. (2009) On development of a multi-channel PZT array scanning system and its evaluating application on UAV wing box. *Sensors and Actuators A*, **151**, 220–230.
9. Cheng, J., Su, Z. and Cheng, L. (2010) Identification of corrosion damage in submerged structures using fundamental anti-symmetric Lamb waves. *Smart Materials and Structures*, **19**, 1–12.
10. Clarke, T., Simonetti, F. and Cawley, P. (2010) Guided wave health monitoring of complex structures by sparse array systems: influence of temperature changes on performance. *Journal of Sound and Vibration*, **329**, 2306–2322.
11. Rytter, A. (1993) Vibration based inspection of civil engineering structures. PhD Dissertation, Department of Building Technology and Structural Engineering, Aalborg University, Denmark.
12. Worden, K. and Dulieu-Barton, J.M. (2004) An overview of intelligent fault detection in systems and structures. *Structural Health Monitoring*, **3**(1), 85–98.
13. Todd, M. and Farrar, C. (2004) Embedded PZT network for non-destructive evaluation of United Airlines aircraft fuselage skins. *Course: Non-destructive evaluation* (SE163), University of California, San Diego.

14. New Aircraft Repair Technique Receives FAA Certification, Delta L-1011 Returned to Trans-Atlantic Service, 1997. Available at: http://www.sandia.gov/media/faa.htm (accessed 5 July 2011).
15. Qing, X.P., Chan, H.-L., Beard, S.-J. *et al.* (2006) Effect of adhesive on the performance of piezoelectric elements used to monitor structural health. *International Journal of Adhesion and Adhesives*, **26**, 622–628.
16. Lanzara, G., Yoon, Y., Kim, Y. and Chang, F.-K. (2009) Influence of interface degradation on the performance of piezoelectric actuators. *Journal of Intelligent Material Systems and Structures*, **20**(14), 1699–1710.
17. Cawley, P. and Alleyne, D. (2004) Practical long range guided wave inspection – managing complexity. 2nd Middle East Nondestructive Testing Conference and Exhibition Proceedings, vol. 9 (2). [online] Available at: http://www.ndt.net/article/mendt03/4/4.htm.
18. Wilcox, P.D. (2003) Omni-directional guided wave transducer arrays for the rapid inspection of large areas of plate structures. *IEEE Transactions on Ultrasonics, Ferroelectrics and Frequency Control*, **50**(6), 699–709.
19. Giurgiutiu, V. (2008) *Structural Health Monitoring with Piezoelectric Wafer Active Sensors*, Elsevier.
20. Malinowski, P., Wandowski, T., Trendafilova, I. and Ostachowicz, W. (2009) A phased array-based method for damage detection and localization in thin plates. *An International Journal of Structural Health Monitoring*, **8**(1), 5–15.
21. Rajagopalan, J., Balasubramaniam, K. and Krishnamurthy, C.V. (2006) A single transmitter multi-receiver (STMR) PZT array for guided ultrasonic wave based structural health monitoring of large isotropic plate structures. *Smart Material and Structures*, **15**, 1190–1196.
22. Wandowski, T., Malinowski, P. and Ostachowicz, W. (2011) Damage detection with concentrated configurations of piezoelectric transducers. *Smart Materials and Structures*, **20**, 025002.
23. Matt, H.M. and di Scalea, F.L. (2007) Macro-fiber composite piezoelectric rosettes for acoustic source location in complex structures. *Smart Materials and Structures*, **16**, 1489–1499.
24. Salas, K.I. and Cesnik, C.E.S. (2009) Guided wave excitation by a CLoVER transducer for structural health monitoring: theory and experiments. *Smart Materials and Structures*, **18**, 1–27.
25. Degertekin, F.L. and Khuri-Yakub, B.T. (1997) Lamb wave excitation by Hertzian contacts with applications in NDE. *IEEE Transactions on Ultrasonics, Ferroelectrics and Frequency Control*, **44**, 769–779.
26. Kessler, S.S., Spearing, S.M. and Soutis, C. (2002) Damage detection in composite materials using Lamb wave methods. *Smart Materials and Structures*, **11**, 269–278.
27. Ihn, J.-B. and Chang, F.-K. (2008) Pitch catch active sensing methods in structural health monitoring for aircraft structures. *Structural Health Monitoring, An International Journal*, **7**(1), 5–19.

28. Ihn, J.-B. and Chang, F.-K. (2004) Detection and monitoring of hidden fatigue crack growth using a built-in piezoelectric sensor/actuator network: I. Diagnostics. *Smart Materials and Structures*, **13**, 609–620.
29. Su, Z., Ye, L. and Lu Y. (2006) Guided Lamb waves for identification of damage in composite structures: a review. *Journal of Sound and Vibration*, **295**, 753–780.
30. Hu, N., Shimomukai, T., Fukunaga, H. and Su, Z. (2008) Damage identification of metallic structures using A_0 mode of Lamb waves. *Structural Health Monitoring*, **7**, 271–285.
31. Malinowski, P., Wandowski, T. and Ostachowicz, W. (2008) Multi-damage localization with piezoelectric transducers. Proceedings of the 4th European Workshop on Structural Health Monitoring, pp. 716–723.
32. Ostachowicz, W. and Kudela, P. (2007) Experimental verification of the Lamb-wave based damage detection algorithm. Proceedings of the 6th International Workshop on Structural Health Monitoring, pp. 2066–2073.
33. Ostachowicz, W., Wandowski, T. and Malinowski, P. (2010) Combined distributed and concentrated transducer network for failure indication. *Proceedings of SPIE*, **7650**. DOI: 10.1117/12.846911
34. Jin, J., Quek, S.T. and Wang, Q. (2005) Design of interdigital transducers for crack detection in plates. *Ultrasonics*, **43**, 481–493.
35. Na, J.K., Blackshire, J.L. and Kuhr, S. (2008) Design, fabrication, and characterization of single-element interdigital transducers for NDT applications. *Sensors and Actuators A*, **148**, 359–365.
36. Carazo, A.V. (2000) Novel piezoelectric transducers for high voltage measurements. Doctoral thesis, Universitat Politecnica de Catalunya, Departament of Electrical Engineering, Barcelona.
37. Diamanti, K., Soutis, C. and Hodgkinson, J.M. (2007) Piezoelectric transducer arrangement for the inspection of large composite structures. *Composites: Part A*, **38**, 1121–1130.
38. Masserey, B. and Fromme, P. (2009) Surface defect detection in stiffened plate structures using Rayleigh-like waves, *NDT&E International*, **42**, 564–572.
39. Qing, X.P., Beard, S.J., Kumar, A. *et al.* (2006) Advances in the development of built-in diagnostic system for filament wound composite structures. *Composites Science and Technology*, **66**, 1694–702.
40. Andrews, J.P., Palazotto, A.N., DeSimio, M.P. and Olson S.E. (2008) Lamb wave propagation in varying isothermal environments. *Structural Health Monitoring*, **7**, 265–270.
41. Drinkwater, B.W., Konstantinidis, G. and Wilcox, P.D. (2007) An investigation into the temperature stability of a guided wave structural health monitoring system using permanently attached sensors. *IEEE Sensors Journal*, **7**(5), 905–912.
42. Clarke, T., Simonetti, F. and Cawley, P. (2010) Guided wave health monitoring of complex structures by sparse array systems: influence of temperature changes on performance. *Journal of Sound and Vibration*, **329**, 2306–2322.

43. Park, S., Lee, C. and Sohn, H. (2010) Reference-free crack detection using transfer impedances. *Journal of Sound and Vibration*, **329**, 2337–2348.
44. Garcia, C.E. and Guemes, A. (2010) Compensation for temperature and static strain in Lamb wave propagation, in *Proceedings of the Fifth European Workshop on Structural Health Monitoring*, DEStech Publications, Inc., pp. 736–741.
45. Chen, V.C. and Ling, H. (2002) *Time-frequency Transforms for Radar Imaging and Signal Analysis*, Artech House.
46. Brandwood, D. (2003) *Fourier Transforms in Radar and Signal Processing*, Artech House.
47. Stergiopoulos, S. (2001) *Advanced Signal Processing Handbook – Theory and Implementation for Radar, Sonar and Medical Imaging Real Time Systems*, CRC Press.
48. Kahrs, M. and Brandenburg, K. (2002) *Applications of Digital Signal Processing to Audio and Acoustics*, Kluwer Academic Publishers.
49. Shin, K. and Hammond, J.K. (2008) *Fundamentals of Signal Processing for Sound and Vibration Engineers*, John Wiley & Sons, Ltd.
50. Swanson, D.C. (2000) *Signal Processing for Intelligent Sensor Systems*, Marcel Dekker.
51. Naidu, P.S. (2001) *Sensor Array Signal Processing*, CRC Press.
52. Vaseghi, S.V. (2000) *Advanced Digital Signal Processing and Noise Reduction*, John Wiley & Sons, Ltd.
53. Deng, M. (2006) Characterization of surface properties of a solid plate using nonlinear Lamb wave approach, *Ultrasonics*, **44**, 1157–1162.
54. Murayama, R. and Ayaka K. (2007) Evaluation of fatigue specimens using EMATs for nonlinear ultrasonic wave detection. *Journal of Nondestructive Evaluation*, **26**, 115–122.
55. Bermes, C, Kim, J.Y., Qu, J. and Jacobs L.J. (2008) Nonlinear Lamb waves for the detection of material nonlinearity. *Mechanical Systems and Signal Processing*, **22**, 638–646.
56. Smith, S.W. (1997) *The Scientist and Engineer's Guide to Digital Signal Processing*, California Technical Publishing, San Diego, CA, USA.
57. Su, Z. and Ye, L. (2009) Identification of damage using Lamb waves. From fundamentals to applications. *Lecture Notes in Applied and Computational Mechanics*, **48**, Springer.
58. Jeong, H. and Jang, Y.-S. (2000) Wavelet analysis of plate wave propagation in composite laminates, *Composite Structures*, **49**, 443–450.
59. Ding, Y., Reuben, R.L. and Steel, J.A. (2004) A new method for waveform analysis for estimating AE wave arrival times using wavelet decomposition. *NDT&E International*, **37**, 279–290.
60. Zak, A., Krawczuk, M. and Ostachowicz W. (2006) Propagation of in-plane waves in an isotropic panel with a crack. *Finite Elements in Analysis and Design*, **42**, 929–941.

61. Moll, J., Schulte, R.T., Hartmann, B. *et al.* (2010) Multi-site damage localization in anisotropic plate-like structures using an active guided wave structural health monitoring system. *Smart Materials and Structures*, **19**, 045022.
62. Wandowski, T., Malinowski, P., Kudela, P. and Ostachowicz, W. (2009) Experimental verification of damage localization algorithm based on triangular PZT transducers network. Proceedings of the 7th International Workshop on Structural Health Monitoring, pp. 2315–2322.
63. Schubert, F. (2008) A conceptual study on guided wave-based imaging techniques for SHM with distributed transducer array. Proceedings of the 4th European Workshop on Structural Health Monitoring, pp. 748–757.
64. Qiang, W. and Shenfang, Y. (2009) Baseline-free imaging method based on new PZT sensor arrangements. *Journal of Intelligent Materials Systems and Structures*, **20**, 1663–1673.
65. Quek, S.T., Tua, P.S. and Jin, J. (2007) Comparison of plain piezoceramics and inter-digital transducer for crack detection in plates. *Journal of Intelligent Material Systems and Structures*, **18**(9), 949–961.
66. Wang, D., Ye, L., Lu, Y. and Li, F. (2010) A damage diagnostic imaging algorithm based on the quantitative comparison of Lamb wave signals. *Smart Materials and Structures*, **19**(6), 965008.
67. Fasel, T.R. and Todd, M.D. (2010) An adhesive bond state classification method for a composite skin-to-spar joint using chaotic insonification. *Journal of Sound and Vibration*, **329**, 3218–3232.
68. Kudela, P., Ostachowicz, W. and Zak, A. (2008) Damage detection in composite plates with embedded PZT transducers. *Mechanical Systems and Signal Processing*, **22**(6), 1327–1335.
69. Stepinski, T. and Engholm, M. (2009) Piezoelectric circular array for structural health monitoring using plate waves. Proceedings of the 7th International Workshop on Structural Health Monitoring, pp. 1050–1056.
70. Yu, L., Santoni-Bottai, G., Xu, B. , Liu, W. and Giurgiutiu, V. (2008) Piezoelectric wafer active sensors for in situ ultrasonic-guided wave SHM. *Fatigue and Fracture of Engineering Materials and Structures*, **31**(8), 611–628.
71. Pena, J., Melguizo, C.P., Martinez-Ona, R. *et al.* (2004) Advanced phased array system for structural damage detection. Proceedings of the 3rd European Workshop on Structural Health Monitoring.
72. Sundararaman, A., Adams, D.E. and Rigas, E.J. (2005) Biologically inspired structural diagnostics through beamforming with phased transducers arrays. *International Journal of Engineering Science*, **43**, 756–778.
73. Criado, A., Melguizo, C.P., Macias, J.P. *et al.* (2007) Low frequency, built-in phased array system for stiffened composite structures monitoring. Proceedings of the III ECCOMAS Thematic Conference on Smart Structures and Materials.
74. Giurgiutiu, V. and Yu, L. (2007) In situ imaging of crack growth with piezoelectric-wafer active sensors. *AIAA Journal*, **45**(11), 2758–2769.

75. Ostachowicz, W., Kudela, P., Malinowski, P. and Wandowski T. (2009) Damage localisation in plate-like structures based on PZT sensors. *Mechanical Systems and Signal Processing*, **23**(6), 1805–1829.
76. Yu, L. and Giurgiutiu, V. (2008) In situ 2D piezoelectric wafer active sensors arrays for guided wave damage detection. *Ultrasonics*, **48**, 117–134.
77. Malinowski, P., Wandowski, T., Trendafilova, I. and Ostachowicz, W. (2007) Multi-phased array for damage localisation. *Key Engineering Materials*, **347**, 77–82.
78. Engholm, M. and Stepinski, T. (2010) Using 2-D arrays for sensing multimodal Lamb waves. *Proceedings of SPIE*, **7649**, 764913.
79. Wang, X. and Tang, J. (2010) An enhanced piezoelectric impedance approach for damage detection with circuitry integration. *Smart Materials and Structures*, **19**, 045001.
80. Wang, X. and Tang, J. (2010) Damage detection using piezoelectric admittance approach with inductive circuitry. *Journal of Intelligent Materials Systems and Structures*, **21**, 667–676.
81. Kuang, Y.D., Shi, S.Q., Chan, P.K.L. *et al.* (2010) Theoretical and experimental studies on the electric impedance of active piezoelectric sensors bonded on cracked beams. *Smart Materials and Structures*, **19**, 045021.
82. Park, S., Shin, H.-H. and Yun, C.-B. (2009) Wireless impedance sensor nodes for functions of structural damage identification and sensor self-diagnosis. *Smart Materials and Structures*, **18**(5), 055001.
83. Peairs, D.M., Park, G. and Inman, D.J. (2004) Improving accessibility of the impedance-based structural health monitoring method. *Journal of Intelligent Material Systems and Structures*, **15**, 129–139.
84. Mallet, L., Lee, B.C., Staszewski, W.J. and Scarpa, F. (2004) Structural health monitoring using scanning laser vibrometry: II. Lamb waves for damage detection. *Smart Materials and Structures*, **13**, 261–269.
85. Leong, W.H., Staszewski, W.J., Lee B.C. and Scarpa, F. (2005) Structural health monitoring using scanning laser vibrometry: III. Lamb waves for fatigue crack detection. *Smart Materials and Structures*, **14**, 1387–1395.
86. Staszewski, W.J., Mahzan, S. and Traynor, R. (2009) Health monitoring of aerospace composite structures – active and passive approach. *Composites Science and Technology*, **69**, 1678–1685.
87. Ruzzene, M., Jeong, S.M., Michaels, T.E. *et al.* (2005) Simulation and measurement of ultrasonic waves in elastic plates using laser vibrometry. *Review of Quantitative Nondestructive Evaluation*, **24**, 172–179.
88. Ruzzene, M. (2007) Frequency–wavenumber domain filtering for improved damage visualization. *Smart Materials and Structures*, **16**, 2116–2129.
89. Sohn, H., Dutta, D., Yang, J.Y. *et al.* (2010) A wave filed imaging technique for delamination detection in composite structures. Proceedings of the Fifth European Workshop on Structural Health Monitoring, pp. 1335–1340.
90. Grabowska, J., Palacz, M. and Krawczuk, M. (2008) Damage identification by wavelet analysis. *Mechanical Systems and Signal Processing*, **22**, 1623–1635.

91. Staszewski, W.J. (2000) Wavelets for mechanical and structural damage identification. *Zeszyty Naukowe IMP PAN w Gdańsku*, **510**, 1469.
92. Michaels, J.E., Croxford, A.J. and Wilcox, P.D. (2008) Imaging algorithms for locating damage via in situ ultrasonic sensors. Proceedings of the 2008 IEEE Sensors Applications Symposium, pp. 63–67.
93. Ostachowicz, W. and Kudela, P. (2010) Elastic waves for damage detection in structures, in *New Trends in Vibration Based Structural Health Monitoring* (eds A. Deraemaker and K. Worden), Springer Wien, New York, pp. 247–300.
94. Grabowska, J. and Krawczuk, M. (2005) Identification of discontinuities in composite rods and beams based on Lamb wave propagation. *Key Engineering Materials*, **293–294**, 517–524.
95. Chipperfield, A.J., Fleming, P.J., Pohlheim, H. and Fonseca, C.M. (1994) Genetic Algorithm Toolbox User's Guide. *ACSE Research Report 512*, University of Sheffield.

Appendix: *EWavePro* Software

A.1 Introduction

EWavePro (*elastic wave propagation*) software is used for analysing the phenomenon of propagation of longitudinal, shear and flexural waves in two- and three-dimensional thin-walled structures composed of isotropic materials or composite laminates. The abbreviation ***EWavePro*** is used here to denote the software developed by the authors. ***EWavePro*** software was developed in order to facilitate better understanding of elastic wave propagation phenomena and to be used for designing structural health monitoring systems based on changes in patterns of propagating elastic waves. In particular cases, *EWavePro* software can be employed for analysing the dynamics of isotropic and composite (laminated) structures, including analysing short-period dynamic loads.

EWavePro software includes three modules: pre-processor, computation module and post-processor. The start window and the about window are presented in Figure A.1. The pre-processor module allows structure geometry and its material properties to be modelled, generating the finite spectral element grid, and in addition guides the user through the process of imposing boundary conditions, defining excitation signals and input data for the chosen nodes and degrees of freedom, and so on. The computation module launches an external computation program (actually a Fortran module) that

Guided Waves in Structures for SHM: The Time-Domain Spectral Element Method, First Edition.
Wieslaw Ostachowicz, Pawel Kudela, Marek Krawczuk and Arkadiusz Zak.

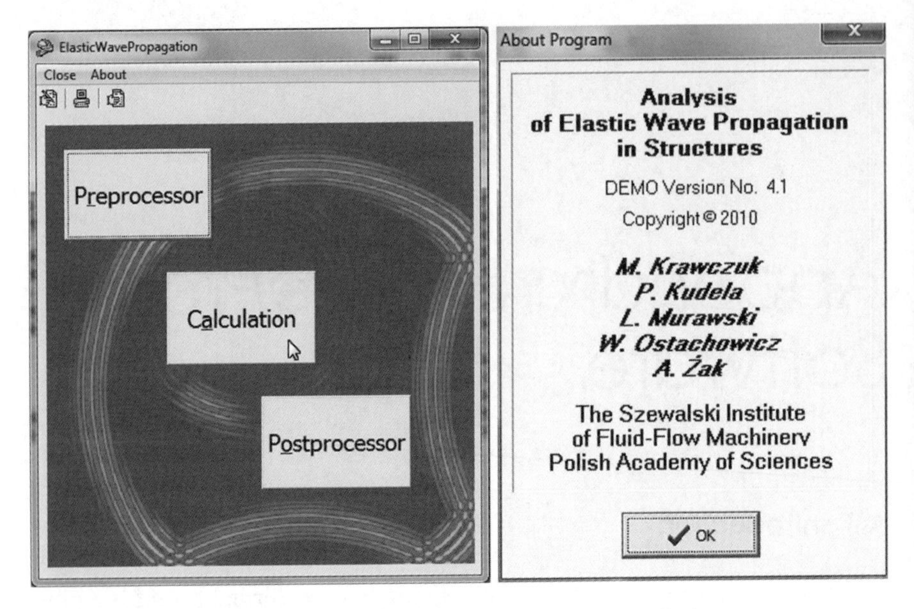

Figure A.1 Main window and the about window

solves the phenomena of elastic wave propagation. The computation program remains transparent for the user, but an *EWavePro* user should have some knowledge about the general assumptions taken when the software was designed and developed. The post-processor module was developed for result visualisation, including OpenGL-accelerated animations. Subsequent sections describe in detail each individual module from a user's point of view.

A.2 Theoretical Background and Scope of Applicability (Computation Module)

EWavePro software uses the *spectral finite element method* described in Chapter 2. This computational method takes advantage of the finite element method [1, 2] with regard to both accuracy and computational cost. Thin-walled structures can be modelled using membrane, plate or shell spectral finite elements developed specially with the aim of simulating the phenomenon of elastic wave propagation in thin-walled structures (cf. Chapters 2 and 5). Spectral finite elements used by the software are formulated as three-dimensional subparametric elements. The geometry is mapped by quadratic shape functions and the displacement field is mapped by sixth-order polynomials. In the present version of the software these are 36-node elements having 5 degrees of freedom in each node. The degrees of freedom describe three translations and two rotations. It is assumed that the translations are global in nature while the rotations are local. These elements were developed on the grounds of the Reissner–Mindlin theory of plates and shells [3]. The reason for choosing this theory is that it offers a good compromise between accuracy and computational cost.

EWavePro software is capable of modelling composite laminates comprising an arbitrary number of layers. Engineering constants of the composite laminate are calculated automatically on the basis of properties of the composite material components in a single layer (i.e. the matrix and the reinforcement fibres).

The geometry of structural elements is defined in a global Cartesian system of coordinates. As a result, coordinates of element nodes, node displacements and loads are also defined in the same coordinate system. Spectral finite elements are defined in a local coordinate system. Nodes of spectral finite elements coincide with Gauss–Lobatto–Legendre points. The node numbering scheme used in ***EWavePro*** is shown in Figure A.2. At these points shape functions based on Lagrange interpolation are defined.

As opposed to Chebyshev spectral finite elements, the shape functions based on Lagrange polynomials leads naturally to the diagonal form of the

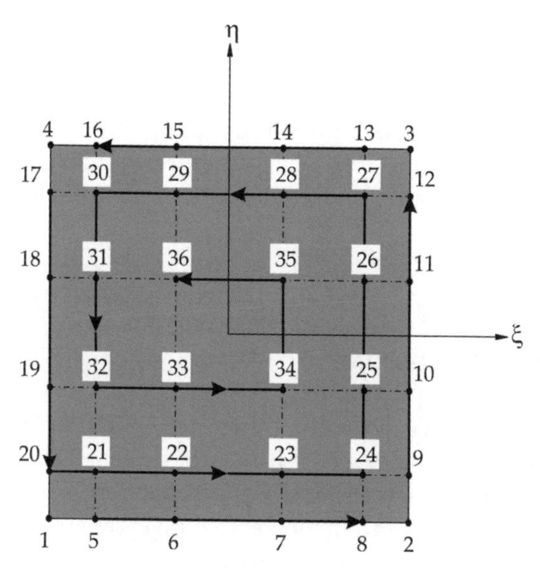

Figure A.2 Spectral finite element node numbering convention

element inertia matrix. This property is a key factor in reducing the computation time and is directly associated with the process of solving the equations of motion (details are described in Chapter 2).

The phenomenon of elastic wave propagation can be simulated using two different methods of solving the equations of motion in the time domain: the central differences method and the Newmark method. Both methods of integrating the equations of motion are available in ***EWavePro***. They were optimised for using the diagonal form of the global inertia matrix most effectively. Another important note applies to the solving algorithm. The present version of the software utilises a modification of the so-called frontal method. For this aim a special algorithm for solving the problem element-by-element was developed. This modification allowed memory resources to be saved and thus the user can solve phenomena of elastic wave propagation in structures of many degrees of freedom. In the algorithm of integrating the equations of motion the matrix of inertia is never assembled.

The software assumes displacement fields in the spectral element to follow the Reissner–Mindlin theory. As shown in Chapter 5, this theory describes the propagation of low-frequency waves, but the error increases with frequency. As a result, one can see that wave propagation velocities calculated by ***EWavePro*** are higher than those actually observed. Thus, high-frequency excitations require using higher-order theories or multi-mode theories. Such theories offer more complex displacement fields than the one derived from the Reissner–Mindlin theory (details are described in Chapter 5). Summarising, spectral finite elements developed in ***EWavePro*** are used for very good approximation of only the fundamental modes of Lamb waves. Specifically, for a 1 mm thick plate made of aluminium alloy both the S_0 and A_0 modes are well represented in the frequency range up to about 500 kHz, which constitutes the upper limit of practical application of the Reissner–Mindlin theory.

A.3 Functional Structure and Software Environment (Pre- and Post-Processors)

EWavePro consists of an execution module, a pre-processor used for preparing data for computation and a post-processor designed for result visualisation including animations of elastic wave propagation.

Installation of an ***EWavePro*** demo version involves copying all necessary files to an arbitrarily chosen directory by using the link to the demo version of the software: http://www.imp.gda.pl/en/o4/z1/Wiley-book-EWavePro-demo/. The main directory contains the basic module together with the execution procedures and database of materials. Individual subdirectories contain examples of data together with results of computations that may be read into the pre- and post-processors. A DEMO version of the software does not allow custom calculation analyses to be performed and nor does it allow modified data to be saved. The process of saving model drawings, results and animations of computed elastic waves is disabled as well.

EWavePro software is dedicated to calculation of elastic wave propagation in three-dimensional isotropic and composite (laminate) structures. However, any part of the analysed structure must be made of *isotropic material.* Parts made of composite materials are modelled by means of defining material constants for each layer of the composite (matrix and reinforcement fibres). One also needs to define the direction, thickness and volume fraction of reinforcing fibres of each composite layer. The software automatically computes the necessary material constants of the whole composite material. One should note that mechanical properties (material constants) of the composite material must always be defined in the local coordinate system of each element. For this reason, the sought solution expressed through stresses and deformations are derived in the same, local, system of coordinates, unlike the global displacements and reactions, which are derived in a global coordinate system (except for moments).

For any part of the structure made of composite material each individual layer of material is integrated separately and the global matrices of inertia and stiffness of the given part are computed as sums of global matrices of inertia and stiffness of the individual layers. Even though the procedure described above is more time-consuming than simple integration along the thickness, the obtained result describes the variation of stiffness of the analysed structural element along the normal direction much better. This results from the fact that the proposed method utilises more integration points.

EWavePro software is fitted with a management algorithm and both pre- and post-processors. The pre-processor is designed for constructing and analysing models in 3D space. It offers the standard functions of this type of software, such as set operations, previewing the data set being created and saving bitmap images.

The pre-processor offers 11 tabs – main modules for constructing models for analysis. In order to construct the model to be used for analysing elastic wave propagation one needs to open individual tabs in sequence and define the model parameters that they offer.

The procedure of creating a numerical model using ***EWavePro*** software is presented for the case of an aluminium pressure tank. The tank wall thickness equals 10 mm. A general view of the physical tank model is presented in Figure A.3. Elastic wave excitation is chosen to be activated at the height

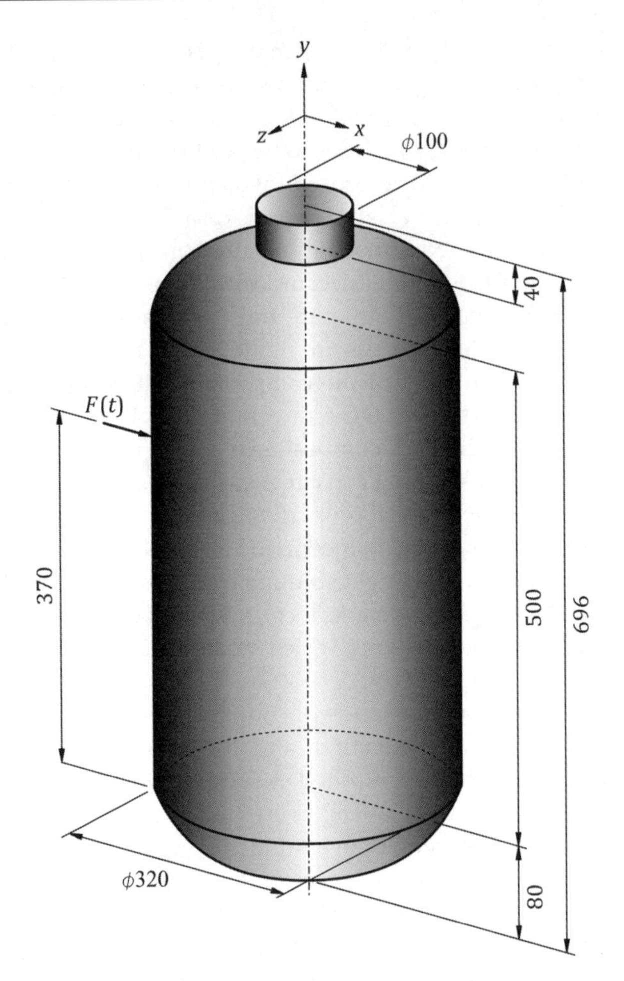

Figure A.3 General view of the physical model of a pressure tank

of 370 mm from the tank base (Figure A.3) in the direction normal to the tank surface. The excitation is realised by means of a 75 kHz periodic force modulated by a Hanning window with four cycles. The total excitation time is assumed as 0.5 ms.

The first tab, '***MATERIALS***' (Figure A.4), contains a database of material data used for defining properties of both composite structures and elements made of isotropic materials ('*MAT. TYPE*' tab). If a material necessary for analysing the given model is not present in the database, a new material

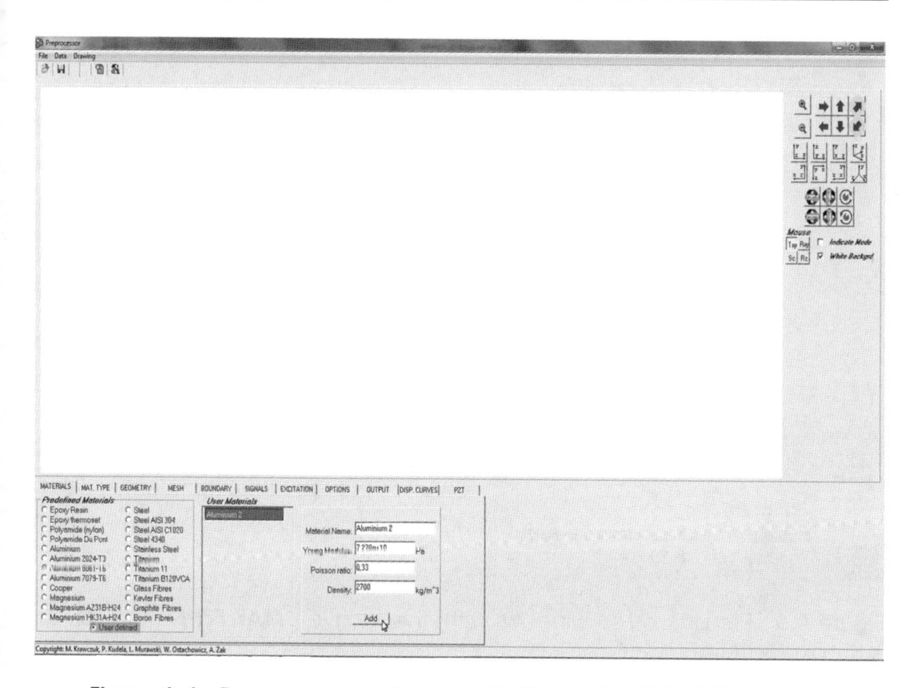

Figure A.4 Pre-processor window with the active '***MATERIALS***' tab

can be defined. The user can define any number of custom materials, which will automatically be stored in the file 'MATERIAL.base'. In the case of the analysed pressure tank the aluminium it is made of has slightly different properties from any of the aluminium materials available in the database. For this reason the new material named Aluminium 2 was defined (Figure A.4). Its Young's modulus equals 7.27×10^{10} Pa, Poisson's coefficient is 0.33 and the density is 2700 kg/m^3.

On the '***MAT. TYPE***' tab the user defines the type of material the analysed structure is made of. The choice is limited to items defined on the '*MATERIALS*' tab. The material can be isotropic, as in the case of the pressure tank being analysed (Figure A.5). The user can also define an arbitrary composite material incorporating materials defined on the '*MATERIALS*' tab.

In Section A.4 later in the Appendix (see Figures A.28 to 41), a composite material-based aircraft wing structure will be discussed.

The fundamental step of constructing a model for analysing elastic wave propagation involves defining its geometrical parameters on the '***GEOMETRY***' tab. Model construction begins with defining locations of characteristic points of the structure. Figure A.6 presents the first eight points used for

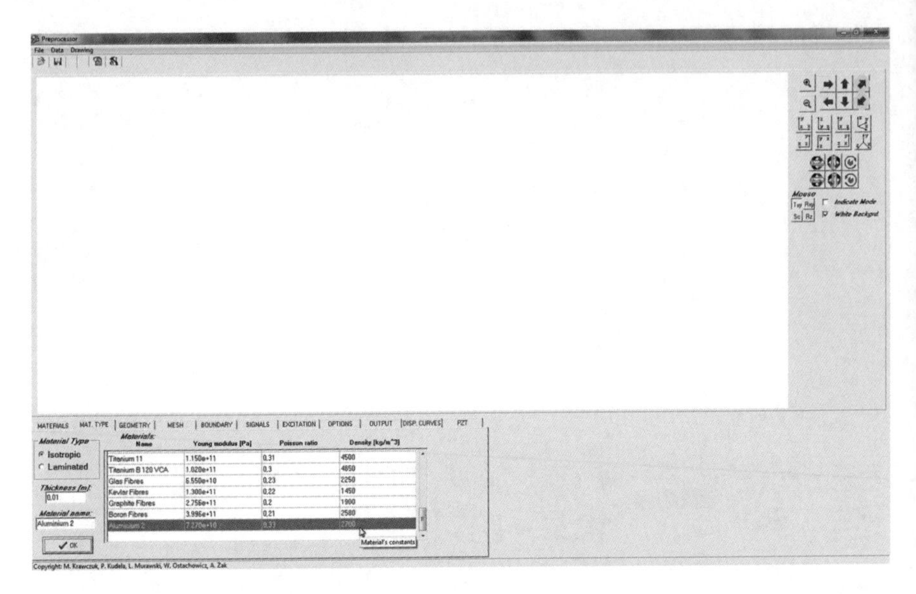

Figure A.5 Pre-processor window with the active '***MAT.TYPE***' tab for an isotropic material

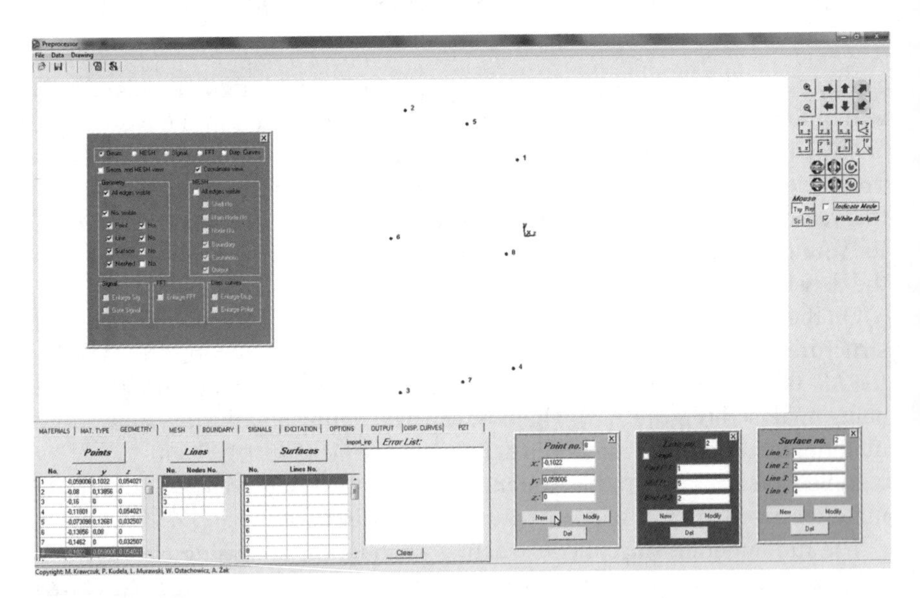

Figure A.6 Pre-processor window with the active '***GEOMETRY***' tab – creating points

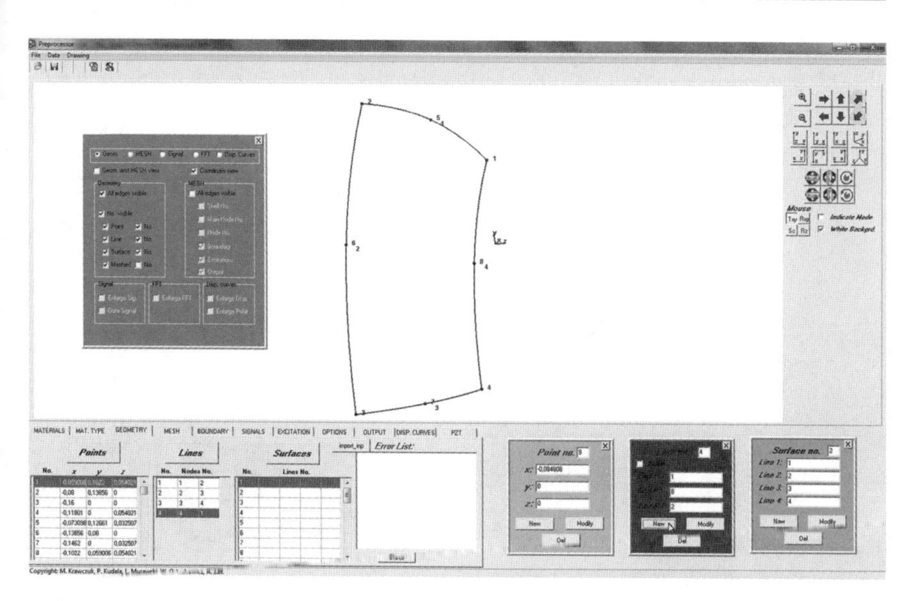

Figure A.7 Pre-processor window with the active '***GEOMETRY***' tab – creating curvilinear sections

modelling the pressure tank shell. In the next step, straight lines or second-order curves are spanned on the characteristic points (Figure A.7). All the points and sections necessary for modelling the pressure tank geometry are presented in Figure A.8. Defined sections are used for constructing surfaces (Figure A.9). The software allows one to use shell elements with curvilinear edges. The model of the complete geometry of the pressure tank is presented in Figure A.10. The pre-processor offers several methods for manipulating visualisations of the model being constructed: arbitrary rotations and translations are possible, as well as zooming. This functionality is available in the upper right area of the pre-processor window.

The '***MESH***' tab (Figure A.11) is designed for creating the mesh of spectral finite elements on the surfaces defined earlier. The software allows the density of the finite element mesh to be chosen. The current version implements 36-node spectral finite elements. The pre-processor offers extended functionality for model element visualisation; these are available in the '*Drawing Settings*' window. The '*Indicator Box*' window allows one to magnify selected fragments of the model and to search for nodes in the selected area. Found nodes can be sent to text fields on other tabs. Such fields define other node parameters (e.g. boundary conditions or excitation).

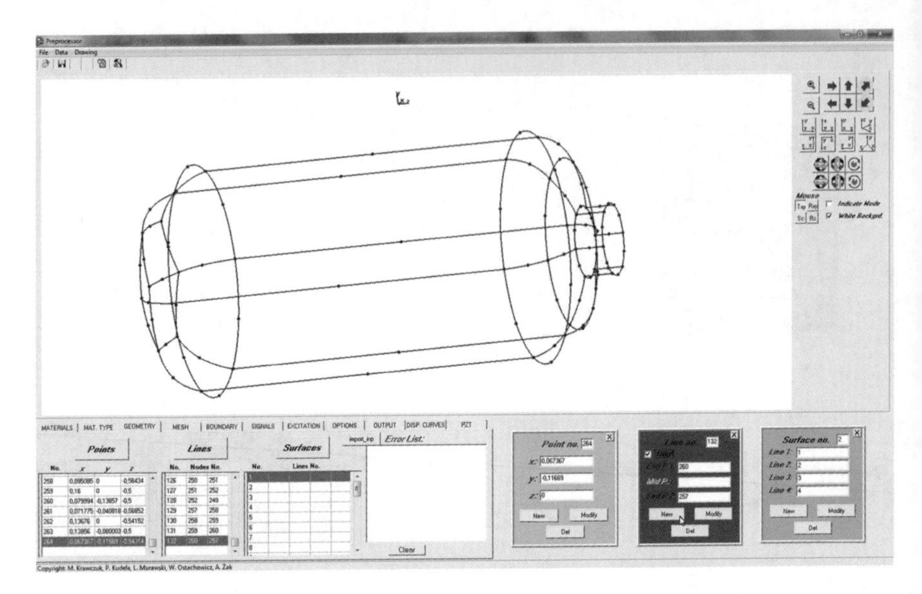

Figure A.8 Pre-processor window with the active '***GEOMETRY***' tab – viewing all points and lines of the tank model

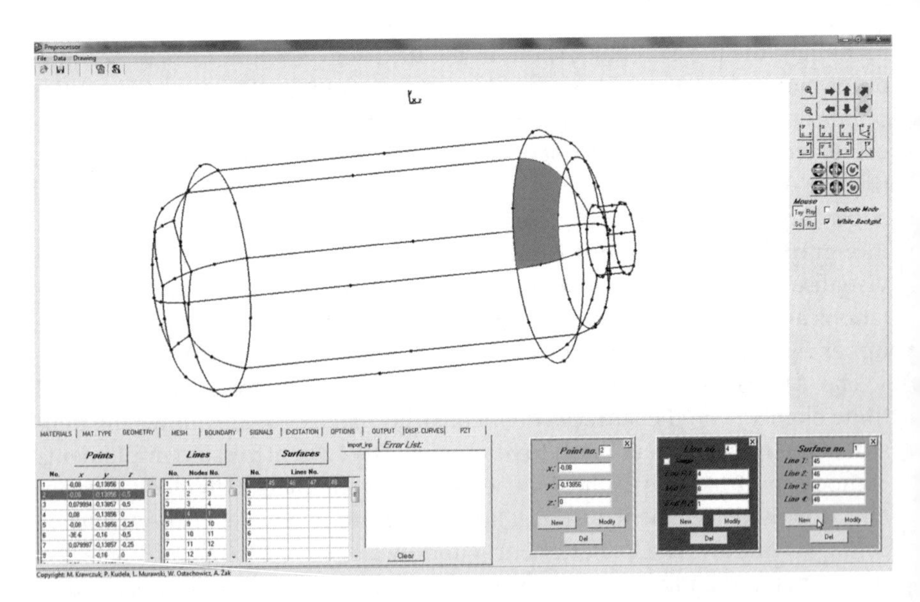

Figure A.9 Pre-processor window with the active '***GEOMETRY***' tab – creating surfaces

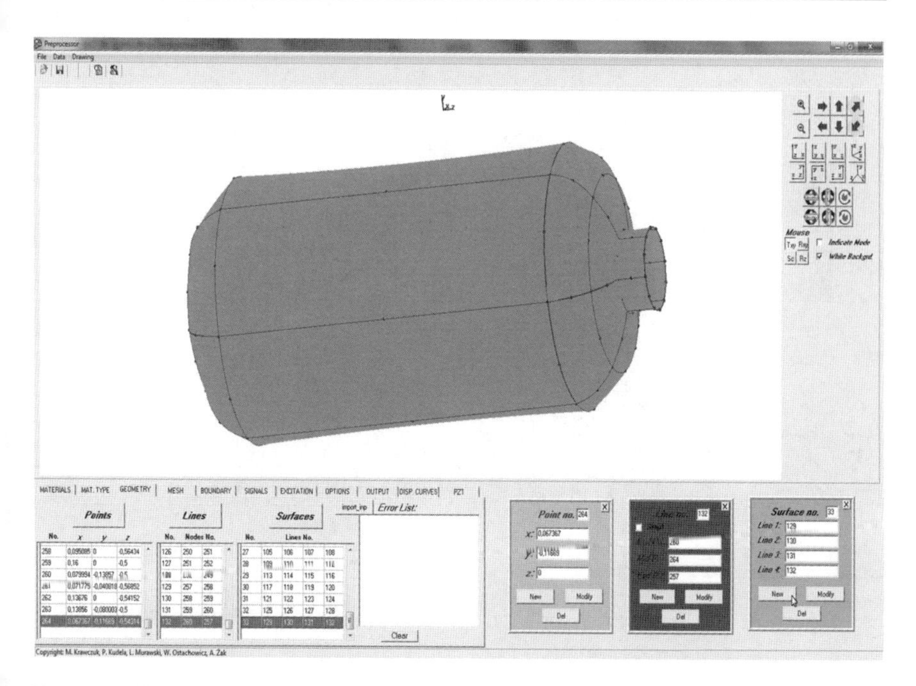

Figure A.10 Pre-processor window with the active '***GEOMETRY***' tab – viewing the whole geometry of the tank model

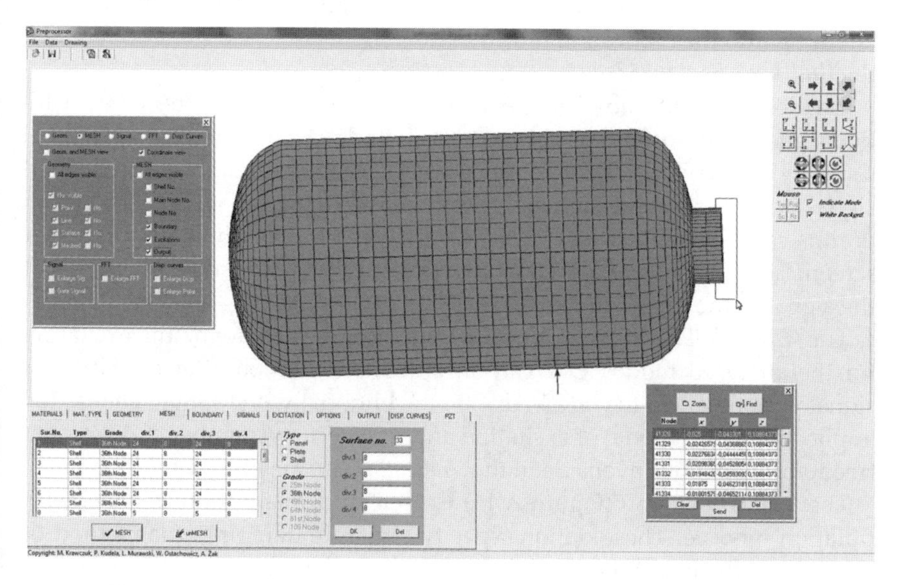

Figure A.11 Pre-processor window with the active '***MESH***' tab

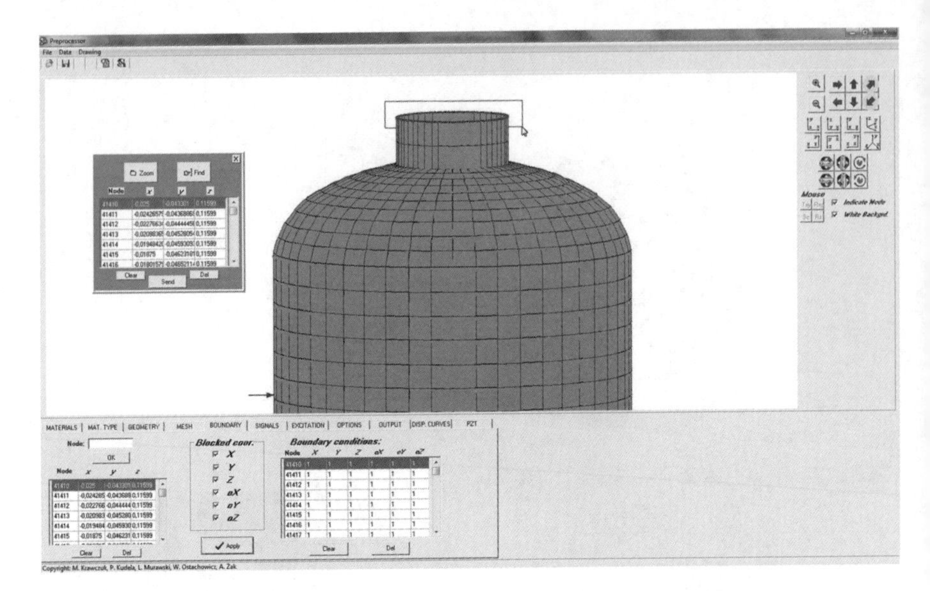

Figure A.12 Pre-processor window with the active '***BOUNDARY***' tab

The '***BOUNDARY***' tab (Figure A.12) is designed for defining boundary conditions on the analysed model. Figure A.12 presents the functionalities of zooming into the model and quickly searching for selected nodes offered by the '*Indicator Box*' window.

On the '***SIGNALS***' tab (Figure A.13) one can define parameters of the excitation signal. Three different types of signal window can be imposed: *Hanning*, *Gaussian* and *Triangle*. The pre-processor presents the graph of the defined signal both in the time domain and in the frequency domain (FFT). Both graphs can be enlarged (see '*Drawing Settings*') to the whole pre-processor screen. The '*Gate Signal*' function allows one to view the modulated part of the signal in the time domain.

The '***EXCITATION***' tab (Figure A.14) is designed to define the excitation for the analysed model. One can only define excitation after definition of signal parameters in the '*SIGNALS*' tab. Additionally, one should choose the application point (node) of excitation and its direction (number of degrees of freedom). Functions available in the '*Indicator Box*' may be useful here.

The '***OPTIONS***' tab (Figure A.15) is used to define options for the computation module. These options specify structure damping, the method of integrating the equations as well as the way of imposing excitation on the model.

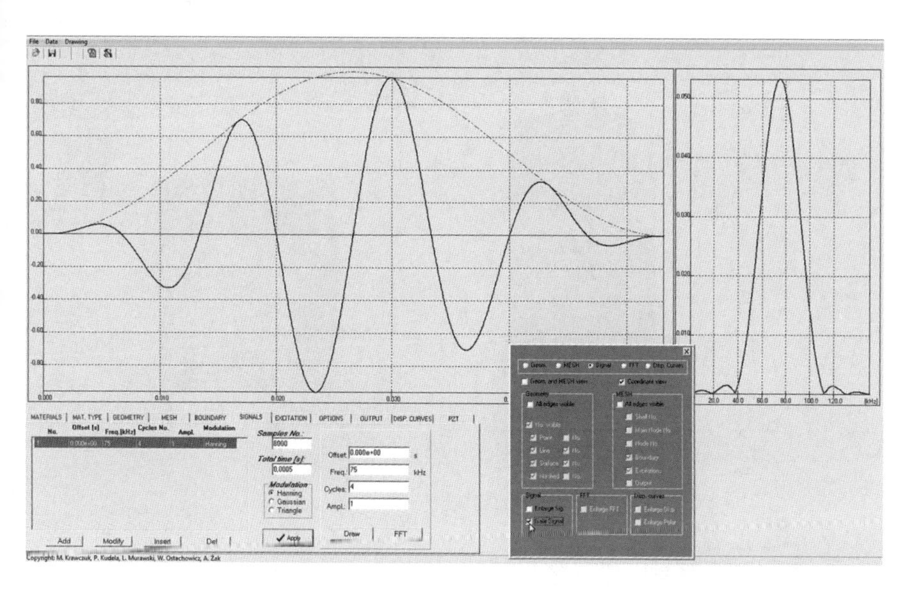

Figure A.13 Pre-processor window with the active '***SIGNALS***' tab

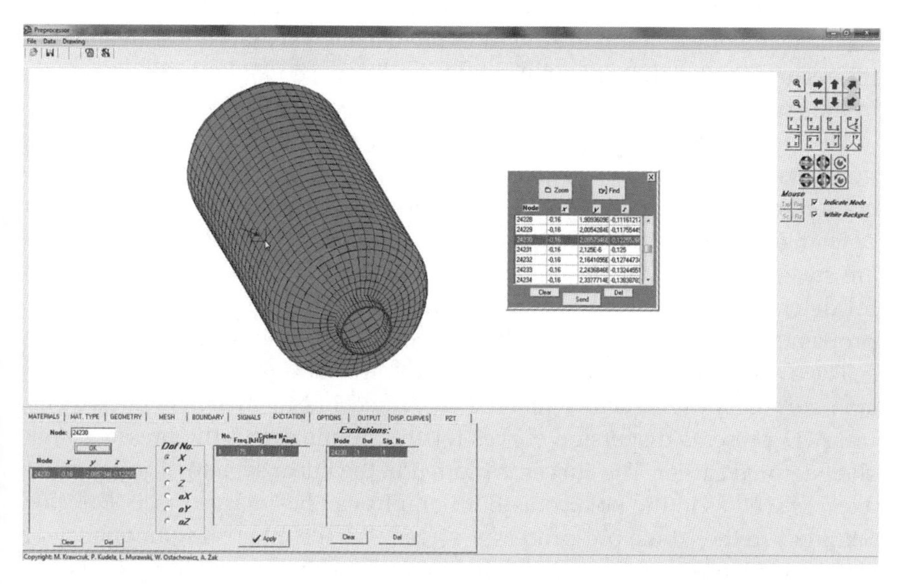

Figure A.14 Pre-processor window with the active '***EXCITATION***' tab

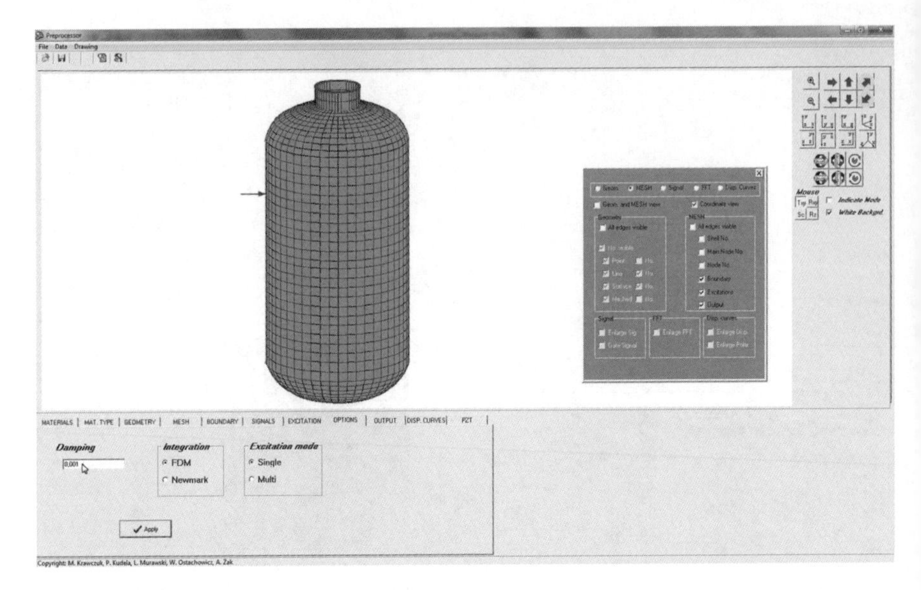

Figure A.15 Pre-processor window with the active '***OPTIONS***' tab

Data files with results of calculations obtained from numerical analyses of elastic wave propagation in structures are usually very large. For this reason it is recommended that only interesting results are chosen for further analysis in the post-processor.

The '***OUTPUT***' tab (Figure A.16) is used for specifying the interesting numbers of nodes and of degrees of freedom. Displacement (velocity, acceleration) data for the given nodes are determined in order to prepare a graph of changes of these parameters as a function of time. Changes for all nodes are saved separately with the defined time step. These data form the basis for determining the number of frames to be used for animating elastic wave propagation. The animation can be played and saved in the post-processor.

The '***DISP.CURVES***" tab (Figure A.17) allows one evaluation of the dispersion curves of the structure material specified in the '*MAT.TYPE*' tab. Figure A.17 presents dispersion curves for the aluminium of the pressure tank shell defined earlier. The curves is plotted in the linear system of rectangular axes, as well as in the polar one. Both graphs can be enlarged (see '*Drawing Settings*') to the whole pre-processor screen. For the interesting frequency of elastic wave propagation one can read the numerical values of propagation velocity of the given form of the elastic wave.

EWavePro software is enhanced with the capability to model piezoelectric elements. Piezoelectric elements can be used both as sensors and as actuators

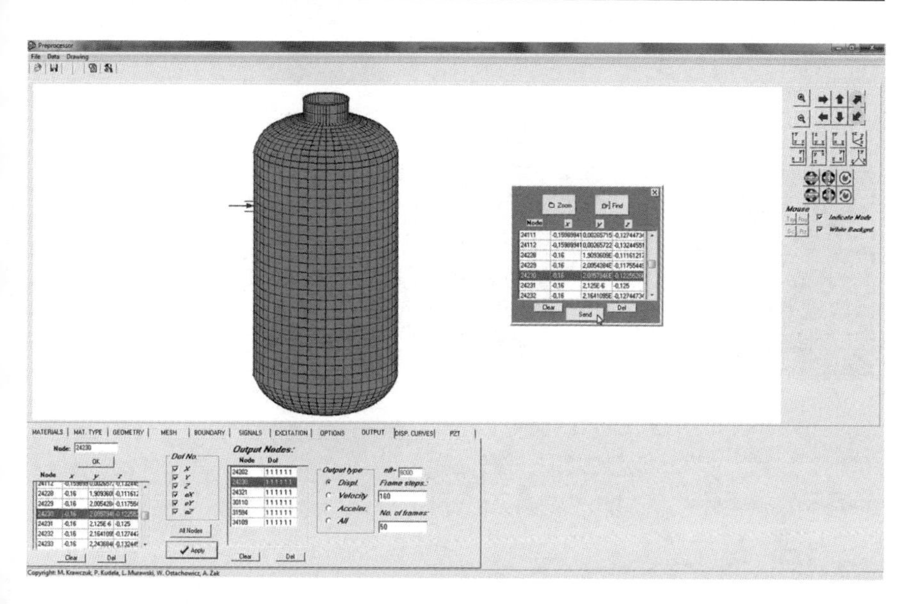

Figure A.16 Pre-processor window with the active '***OUTPUT***' tab

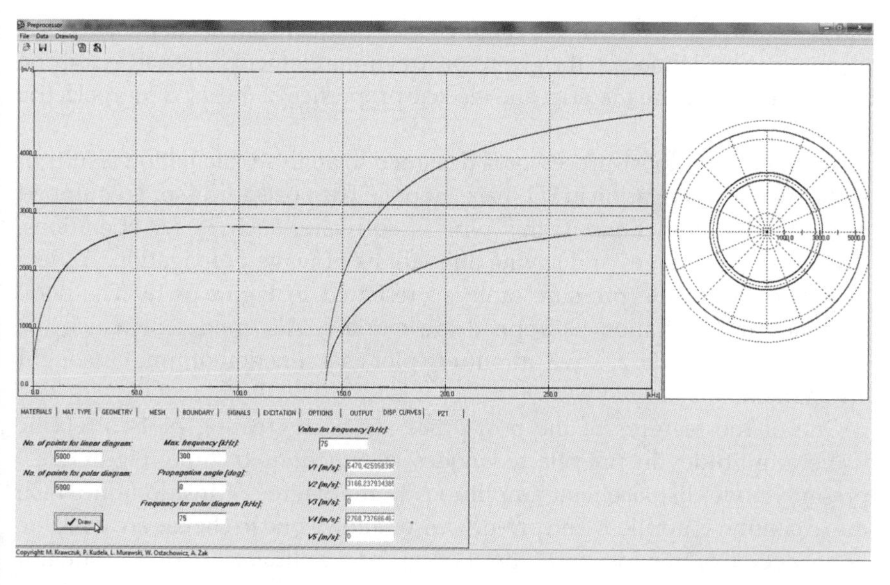

Figure A.17 Pre-processor window with the active '***DISP. CURVES***' tab

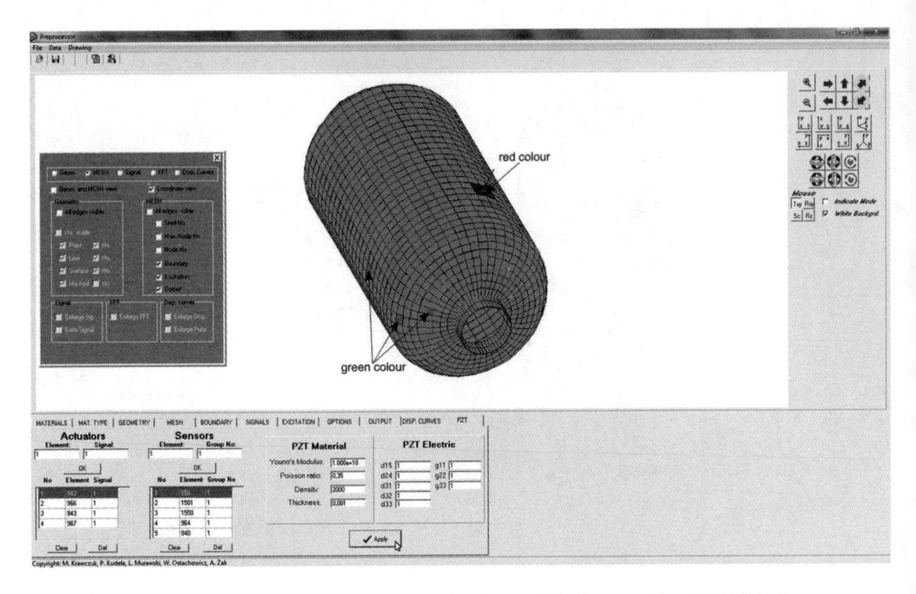

Figure A.18 Pre-processor window with the active '***PZT***' tab

of elastic waves. The '*PZT*' tab (Figure A.18) allows locations of piezoelectric elements to be defined on the given spectral finite elements. In such a case one also has to define the material and electric properties of the used piezoelectric elements.

The ***EWavePro*** post-processor is designed to analyse calculation results of elastic wave propagation in 3D space models. The post-processor is equipped with standard functions of this type of software, such as set operations, saving bitmap images and saving animations of wave propagation. A view of the model of the pressure tank is presented in Figure A.19. The node in which the excitation us applied are marked. Also nodes in which the system response (i.e. a wave amplitude plot as a function of time) is sought are marked.

One of the features of the post-processor is in creating plots of elastic wave amplitudes in pre-selected nodes as functions of time. Figure A.20 presents node displacement amplitudes in the degree of freedom in which the excitation is applied. The pre-processor allows one to choose an available node, degree of freedom and type of observed amplitudes (of displacements, velocities or accelerations).

An important feature of the post-processor is the capability to present animations of elastic wave propagation in the analysed structures. The

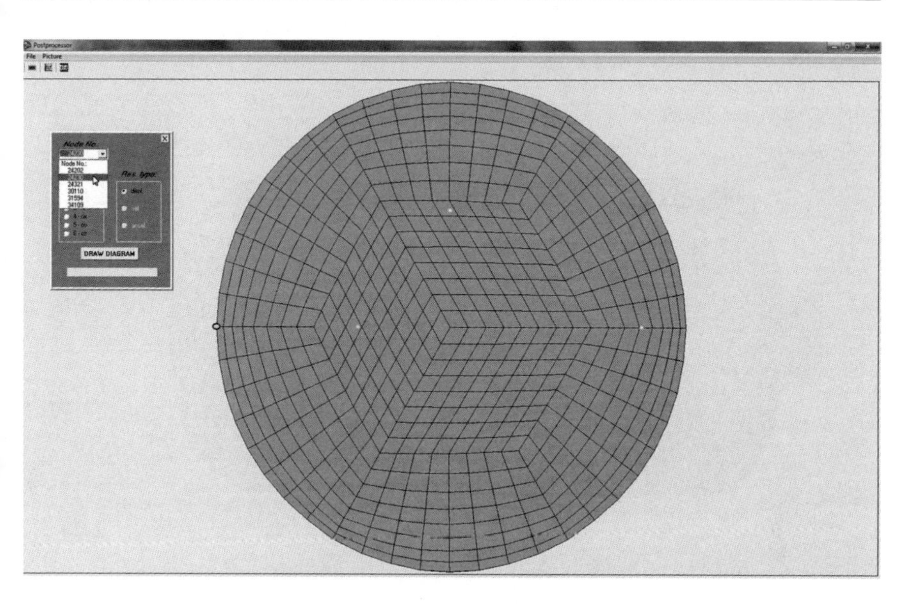

Figure A.19 Post-processor with the model of the pressure tank

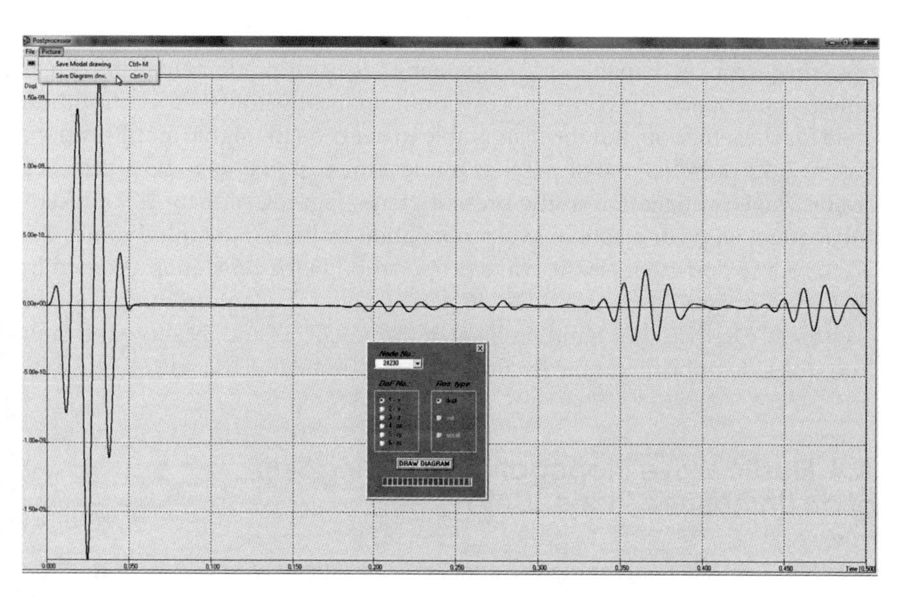

Figure A.20 Plot of elastic wave amplitudes as a function of time, for the node in which the excitation was applied, for the pressure tank model

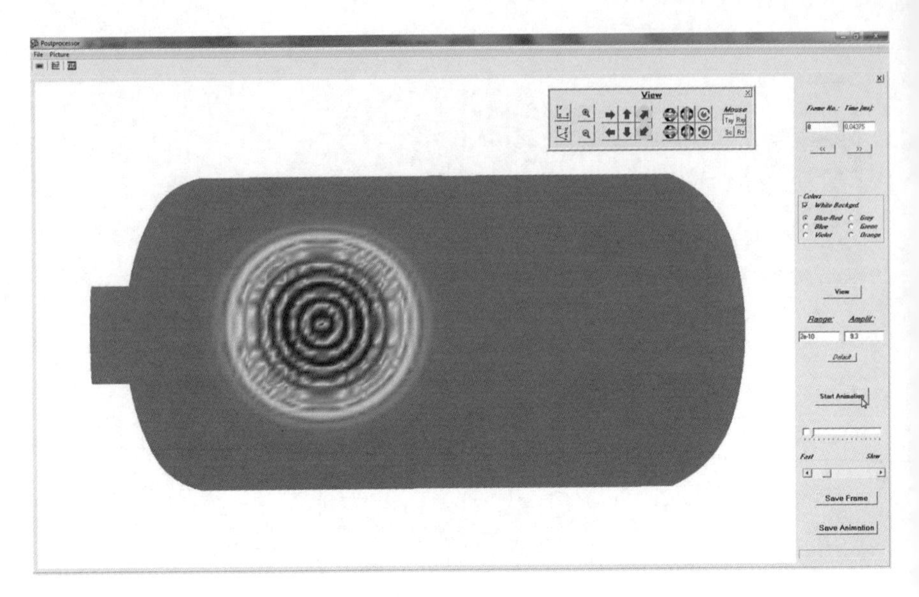

Figure A.21 The eighth frame of elastic wave propagation in the pressure tank model; view of the tank from the side of excitation

post-processor is capable of visualising structural elements in three-dimensional space. The post-processor automatically adjusts the amplitude scale for presentation, but the user is able to override the signal amplification. Figure A.21 presents one of the first frames (the eighth one, to be precise) of elastic wave propagation in the pressure tank. Figures A.22 to A.27 present subsequent steps of elastic wave propagation in the same tank. Frames 21, 27, 32, 39, 56 and 60 of elastic wave propagation in the tank are presented in sequence. Images of frames 8, 21, 56 and 60 show the tank from the side of excitation. On the other hand, images of frames 27, 32 and 39 show the tank from the side opposite to the excitation.

A.4 Elastic Wave Propagation in a Wing Skin of an Unmanned Plane (UAV)

Another example concerning the analysis of elastic wave propagation in the wing skin of an unmanned plane (UAV) is presented. Figure A.28 presents a physical model of a UAV wing skin. The skin is made of 10 mm thick

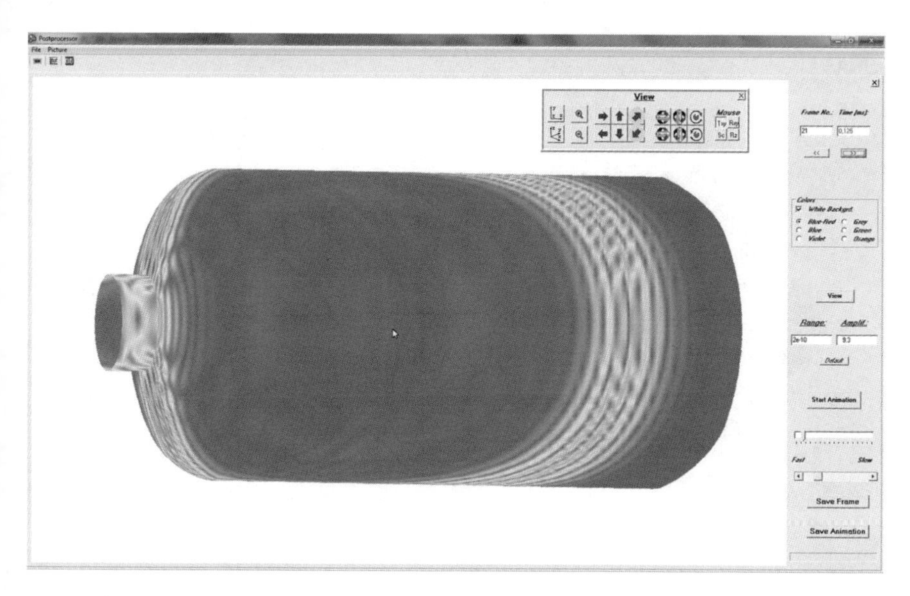

Figure A.22 The twenty-first frame of elastic wave propagation; view of the tank from the side of excitation

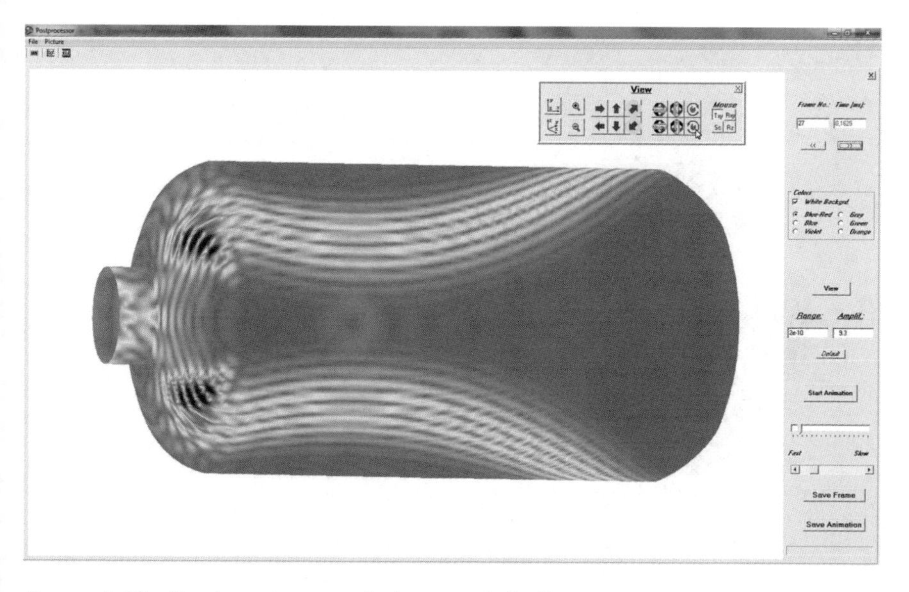

Figure A.23 The twenty-seventh frame of elastic wave propagation; view of the tank from the side of excitation

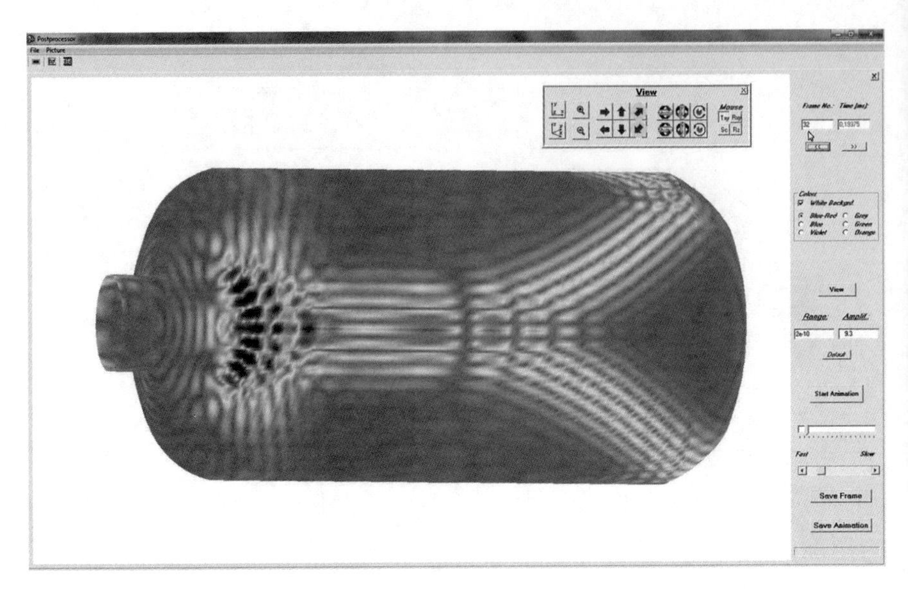

Figure A.24 The thirty-second frame of elastic wave propagation; view of the tank from the side opposite to excitation

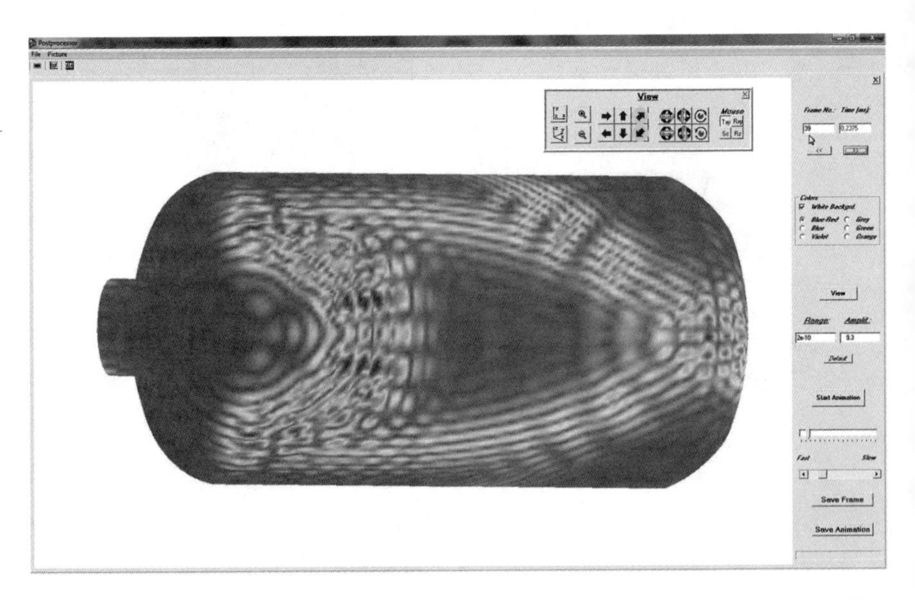

Figure A.25 The thirty-ninth frame of elastic wave propagation; view of the tank from the side opposite to excitation

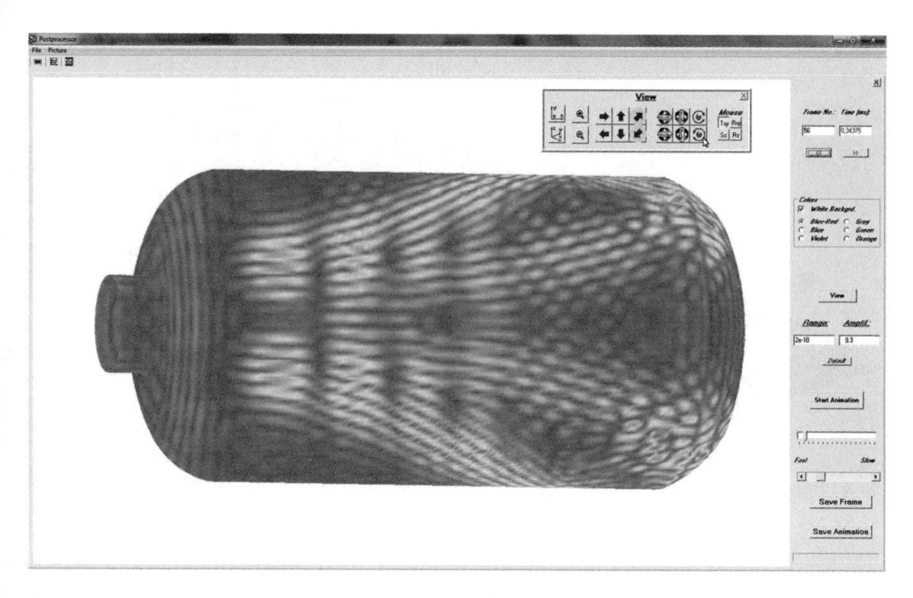

Figure A.26 The fifty-sixth frame of elastic wave propagation; view of the tank from the side of excitation

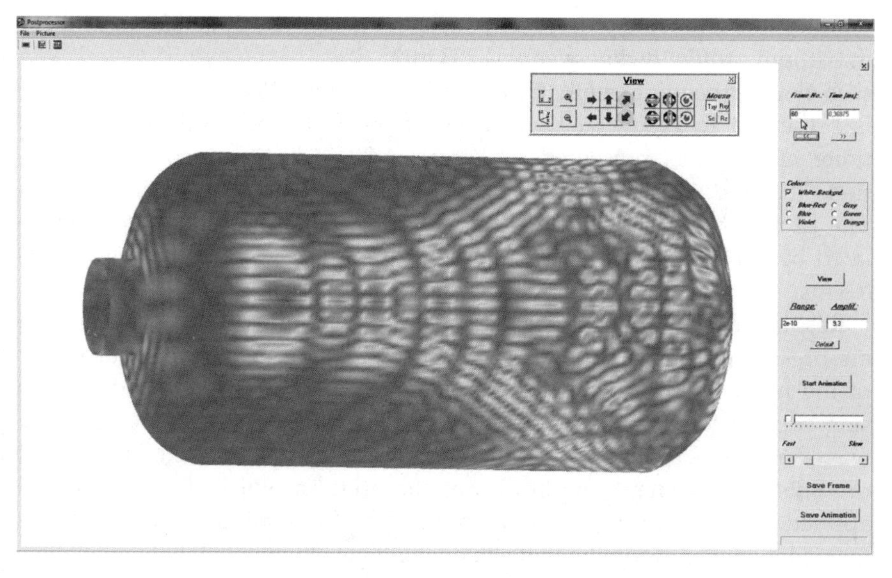

Figure A.27 The sixtieth frame of elastic wave propagation; view of the tank from the side of excitation

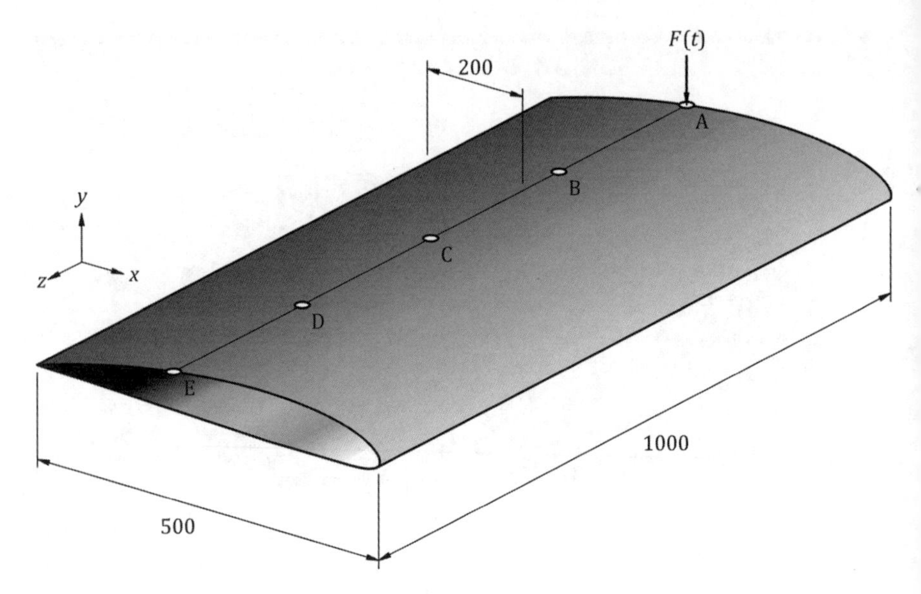

Figure A.28 General view of the physical model of the wing skin

aluminium. The wing skin span equals 0.5 m, its width is 0.25 m and the maximum profile thickness equals 233 mm.

A view of the wing geometry as modelled in the ***EWavePro*** pre-processor is presented in Figure A.29. A view of the full model utilising spectral finite elements is presented in Figure A.30. It is assumed for calculations that the elastic wave excitation is activated on the wing skin edge. The excitation is realised by means of a 75 kHz sinusoidal force modulated by a Hanning window with four cycles. The total excitation time is 0.5 ms.

Results of calculations of elastic wave propagation in the UAV wing skin are presented in the ***EWavePro*** post-processor. Figure A.31 presents a view of the model with a marked node in which excitation is applied (on the side edge of the wing). Also nodes in which the system response (i.e. the wave amplitude plot as a function of time) is sought are marked. Figure A.32 presents displacement amplitudes of the node in which the excitation is applied. Subsequent Figures A.33 to A.35 present displacement amplitudes for nodes spaced along the wing skin length (see Figure A.31).

Another feature of the post-processor is the capability to present animations of elastic wave propagation in the analysed structures. Individual frames from animation of elastic wave propagation in the skin of the analysed wing

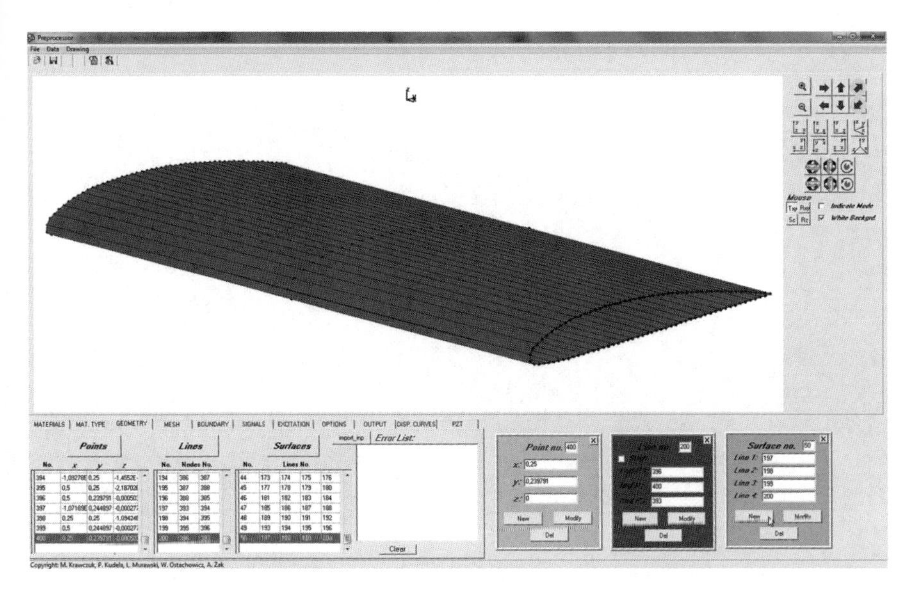

Figure A.29 View of the UAV wing skin geometry

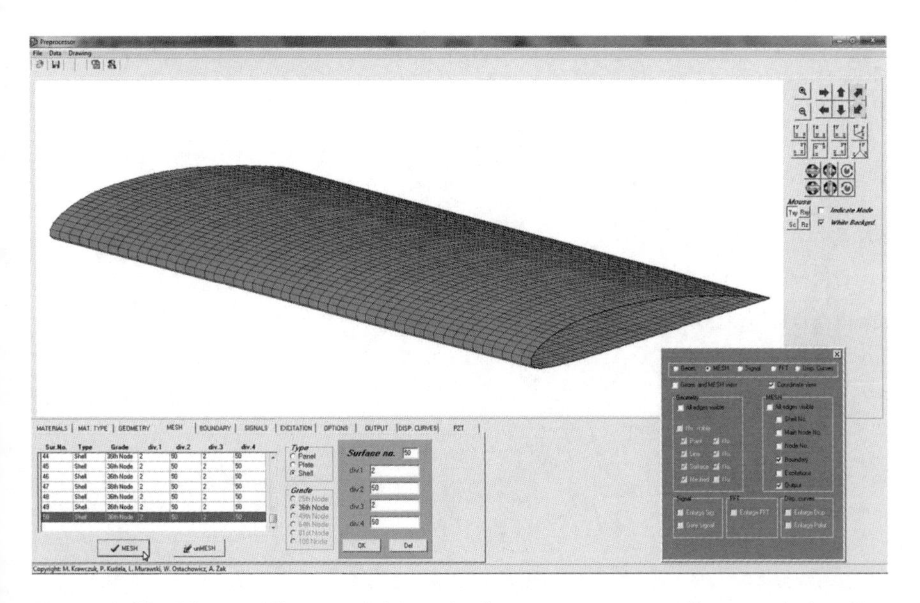

Figure A.30 View of the model for elastic wave propagation analysis – the UAV wing skin

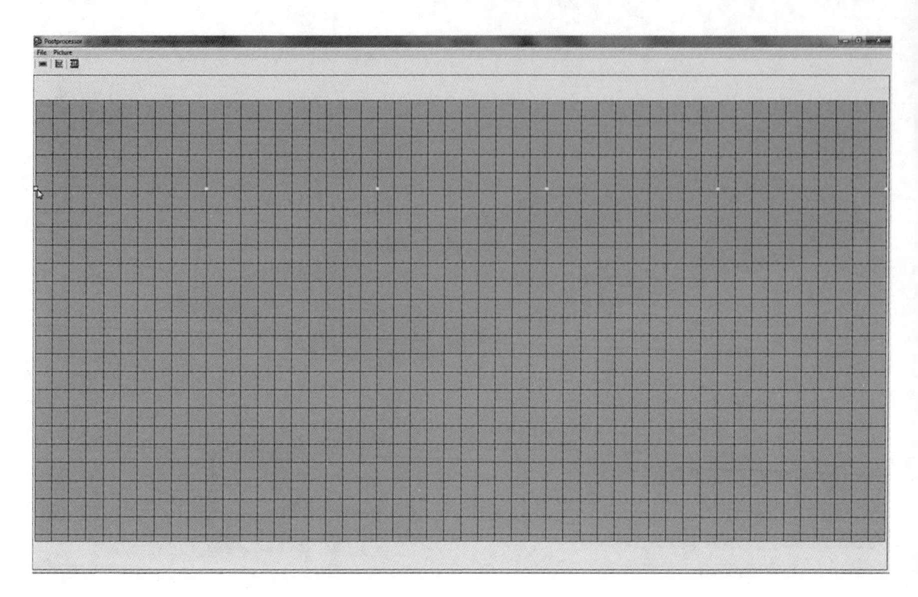

Figure A.31 View of the UAV wing skin model constructed in the post-processor

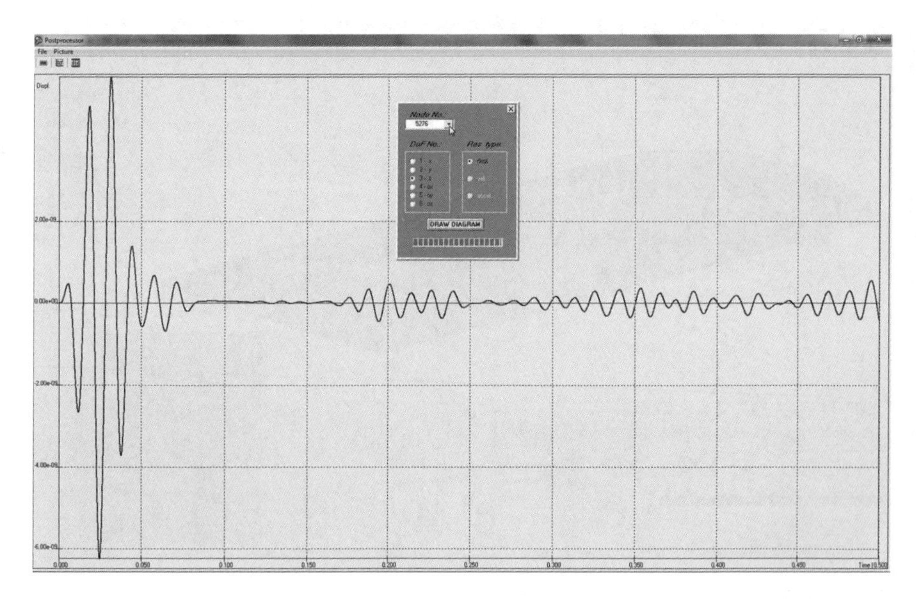

Figure A.32 Amplitudes of the elastic wave as a function of time, for the node in which excitation was applied – 0.0 L

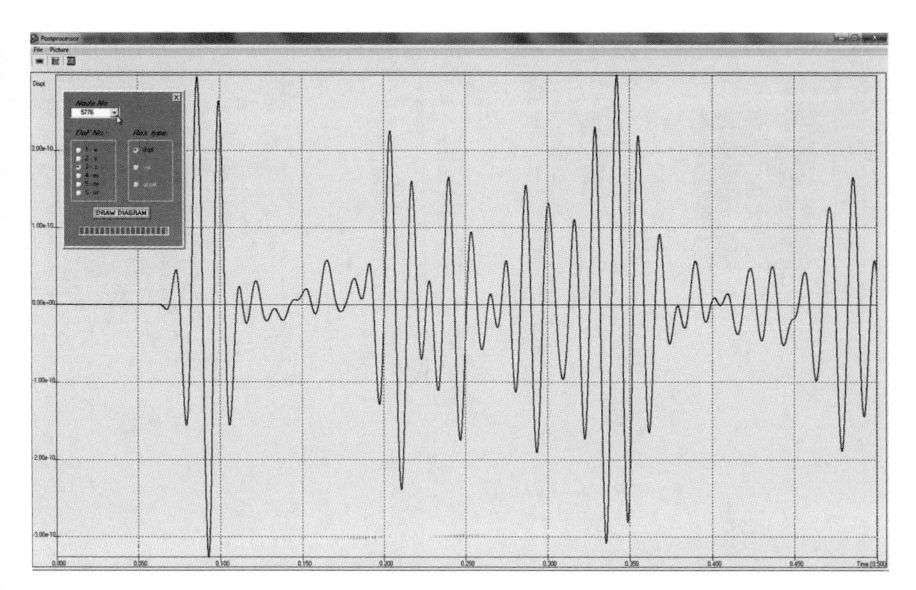

Figure A.33 Amplitudes of the elastic wave as a function of time, for the node located 0.4 L away from the point of excitation

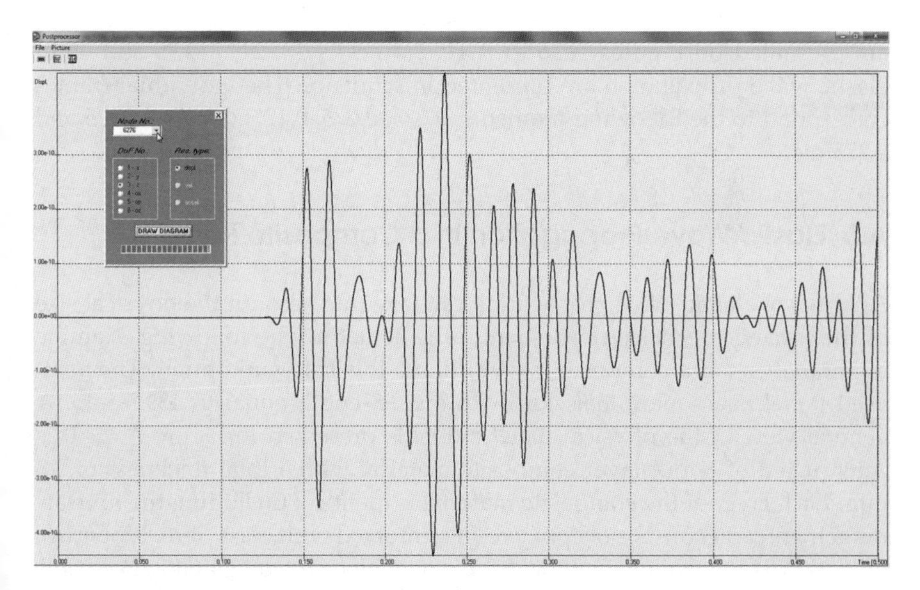

Figure A.34 Amplitudes of the elastic wave as a function of time, for the node located 0.8 L away from the point of excitation

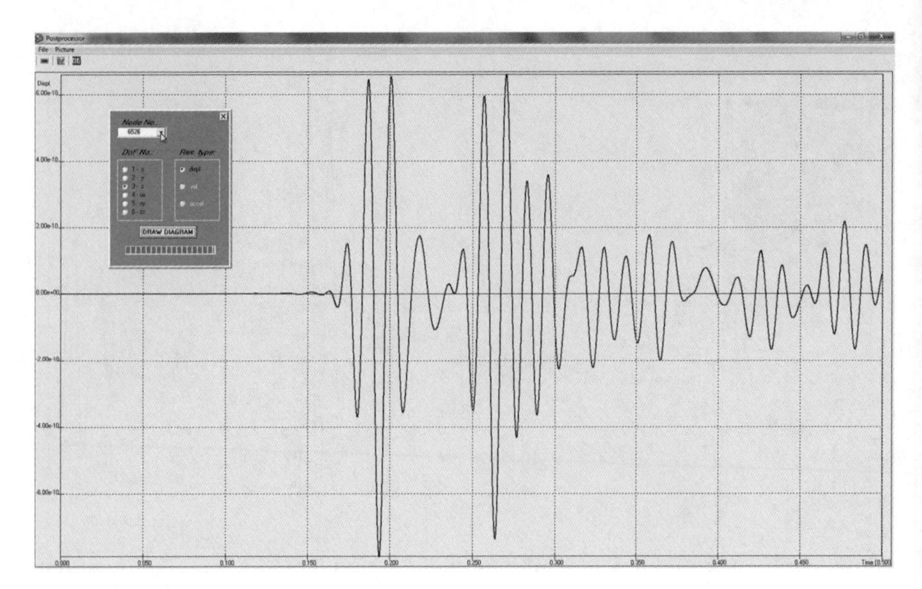

Figure A.35 Amplitudes of the elastic wave as a function of time, for the node located 1.0 L away from the point of excitation

are presented on Figures A.36 to A.41. Frames 5, 13, 21, 31, 35 and 45 of elastic wave propagation are presented in sequence. The individual frames correspond to the following moments: 0.025, 0.075, 0.125, 0.1875, 0.2125 and 0.275 ms.

A.5 Elastic Wave Propagation in a Composite Panel

Analysis of elastic wave propagation can take into account the physical size of the forcing piezoelectric element – the actuator. The following example of an analysis of a composite panel utilises this software option. The analysed panel is a square, flat skin with a side length equal to 439.5 mm. A general view of the physical panel model is presented in Figure A.42. It is constructed of a four-layer composite material with a total thickness of 2.8 mm. Each layer of the composite material is identical (including the fibre arrangement) and consists of epoxy resin reinforced with glass fibre. Properties of the epoxide resin are as follows: Young's modulus of 3.4 GPa, Poisson's coefficient of 0.35, density of 1250 kg/m^3. Properties of the glass fibre are as follows: Young's modulus of 85 GPa, Poisson's coefficient of 0.23, density of 2250 kg/m^3.

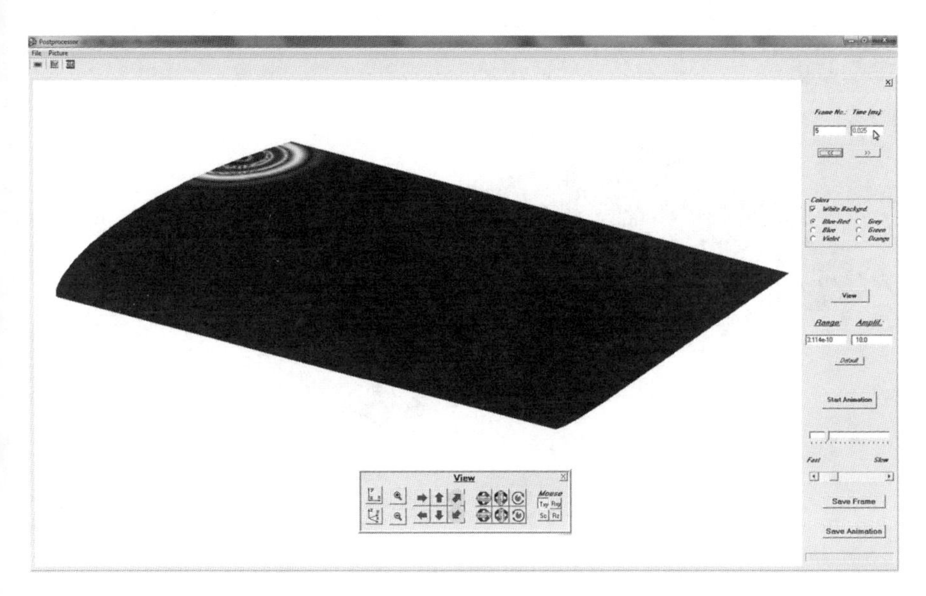

Figure A.36 Propagation of elastic waves in the wing skin for $t = 0.025$ ms

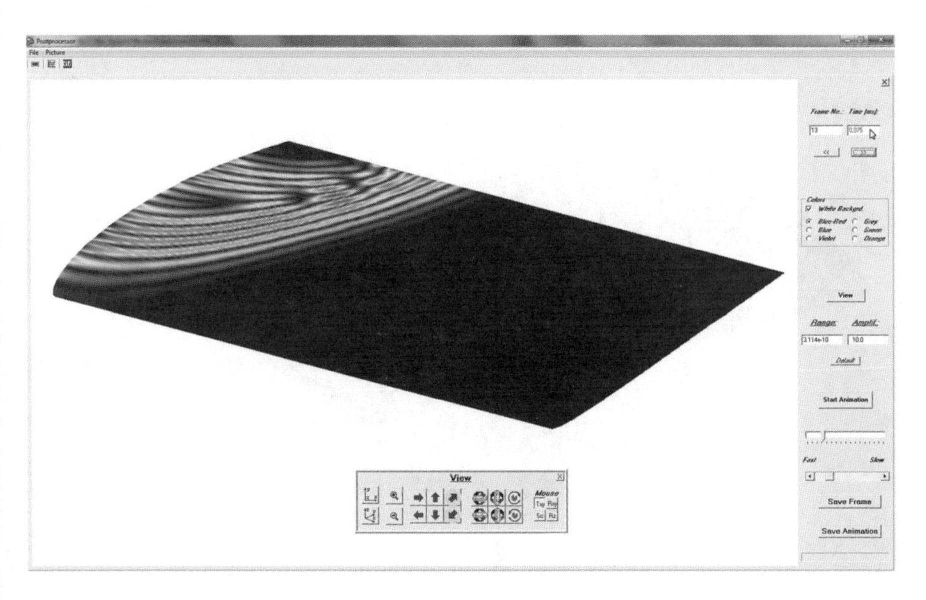

Figure A.37 Propagation of elastic waves in the wing skin for $t = 0.075$ ms

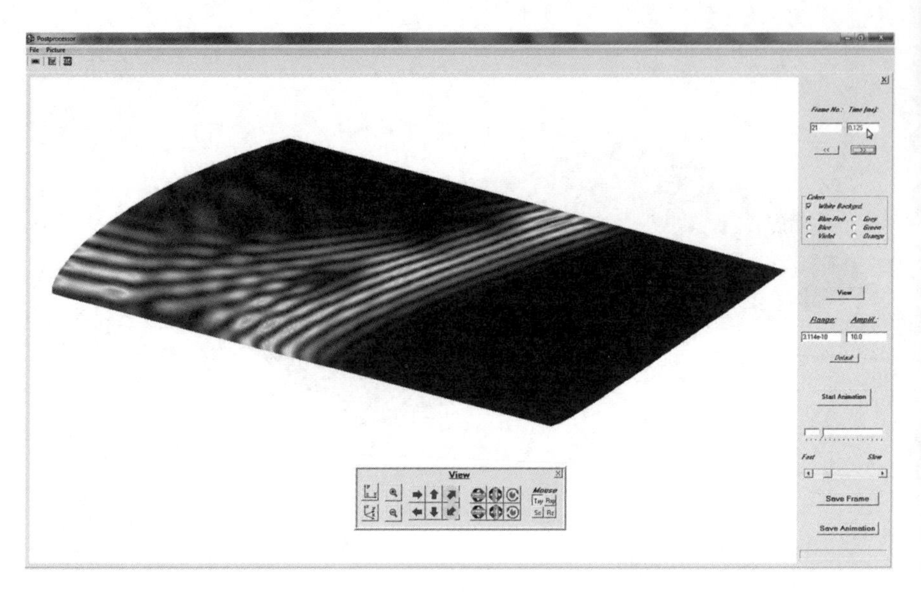

Figure A.38 Propagation of elastic waves in the wing skin for $t = 0.125$ ms

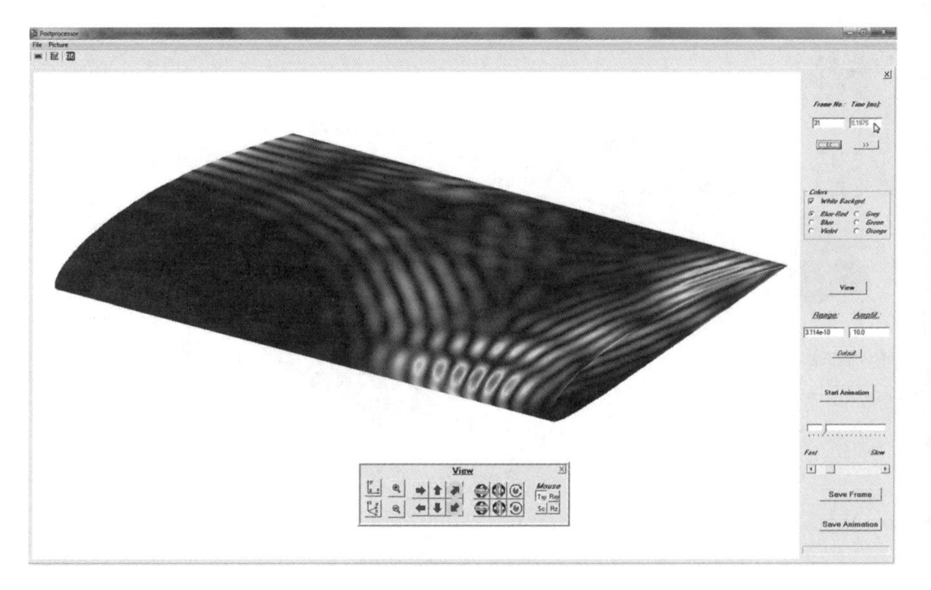

Figure A.39 Propagation of elastic waves in the wing skin for $t = 0.1875$ ms

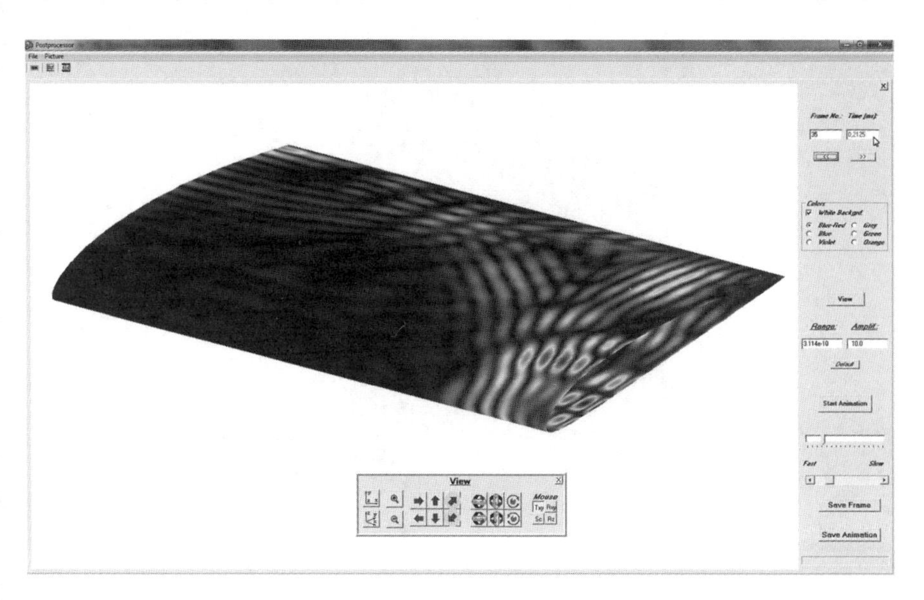

Figure A.40 Propagation of elastic waves in the wing skin for $t = 0.2125$ ms

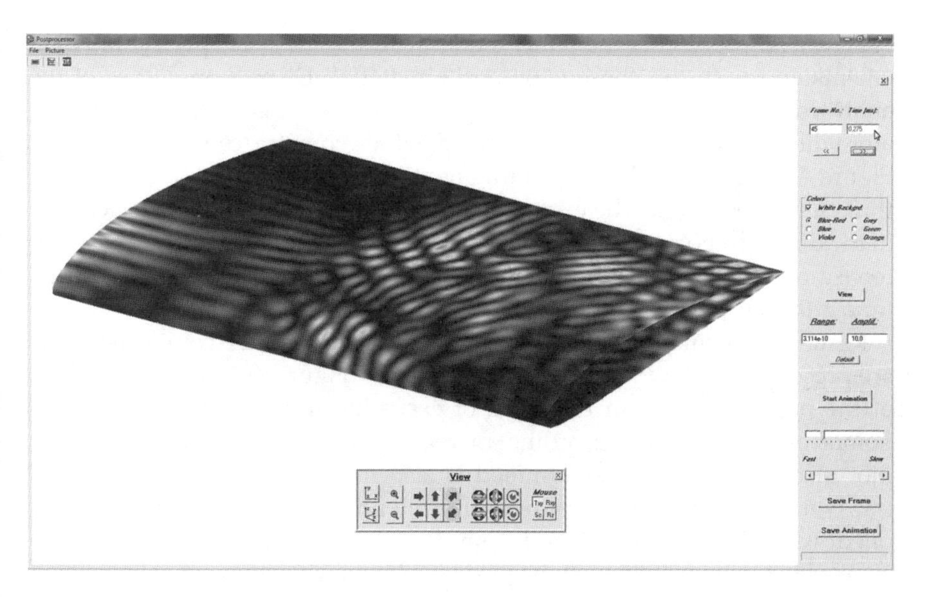

Figure A.41 Propagation of elastic waves in the wing skin for $t = 0.275$ ms

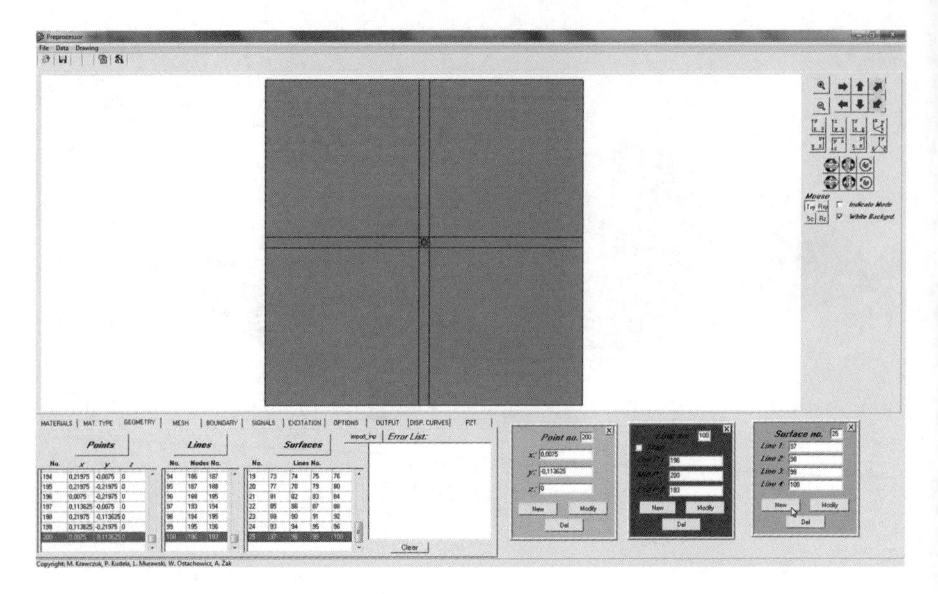

Figure A.42 View of the composite panel model geometry

Figure A.43 presents the panel model with the mesh of spectral finite elements and with the chosen composite material. Dispersion curves of the analysed material are presented in Figure A.44. A circular piezoelectric actuator with a diameter of 10 mm is bonded to the geometric centre of the panel. Figure A.45 presents an enlarged fragment of the panel with the actuator and three sensors marked. The excitation is realised by the piezoelectric actuator in the direction normal to the panel surface, by means of a 35 kHz sinusoidal force modulated by a Hanning window with three cycles. The total excitation time is 0.5 ms.

Results of calculations of elastic wave propagation in the composite panel are presented in the ***EWavePro*** post-processor. Figure A.46 presents a view of the panel model with the point of excitation (centre of the panel) and points of elastic wave recording marked. The point for which the time plot of elastic waves is determined is also marked. Figure A.47 presents transverse displacement amplitudes for the node in which excitation is applied, while Figure A.48 plots the amplitudes for the point marked in Figure A.46. Figures A.49 to A.54 present subsequent steps of elastic wave propagation – individual frames from the animation of elastic wave propagation in the composite panel. Frames 10, 17, 33, 49, 59 and 65 of elastic

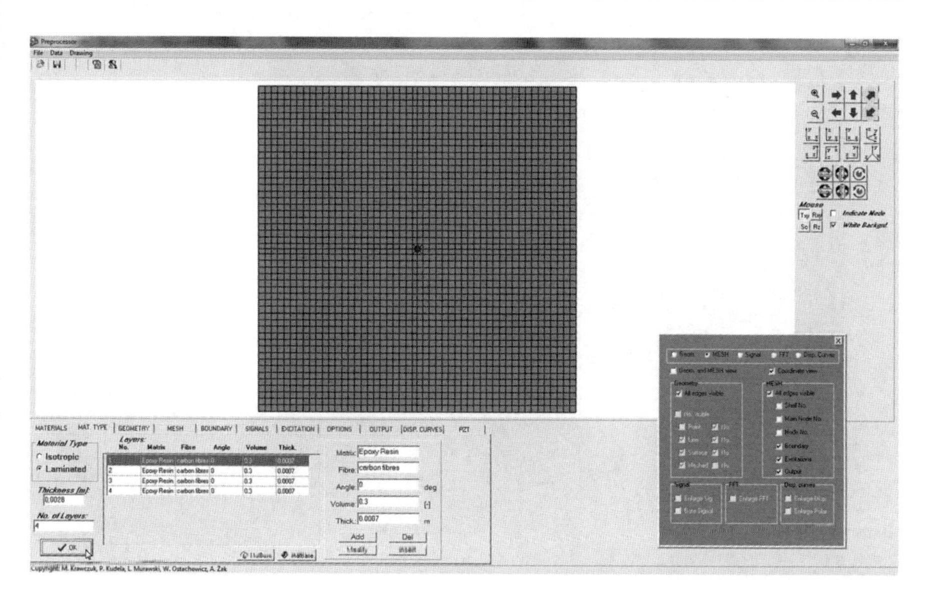

Figure A.43 Mesh of spectral finite elements of the panel with the defined composite material

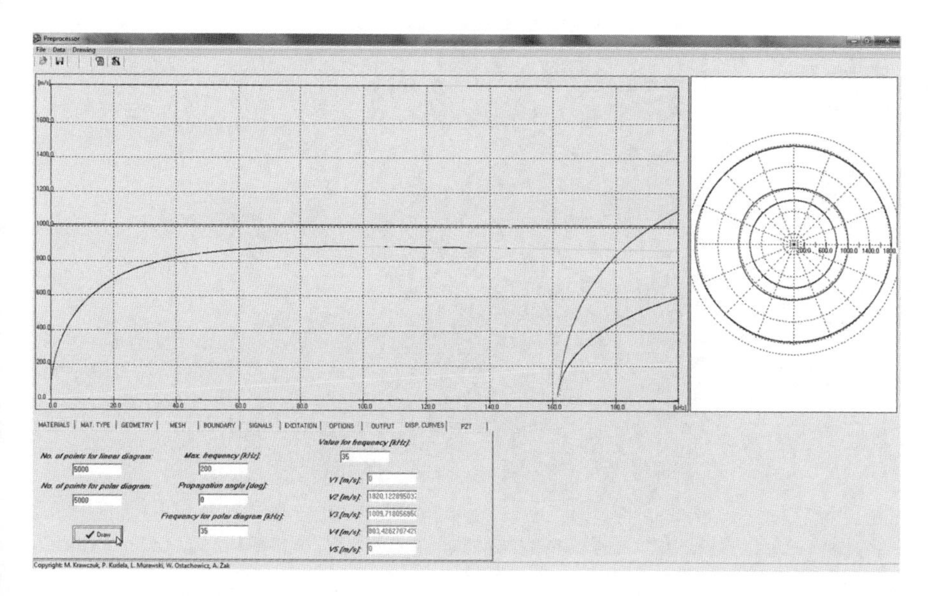

Figure A.44 Dispersion curves of the material of the composite panel

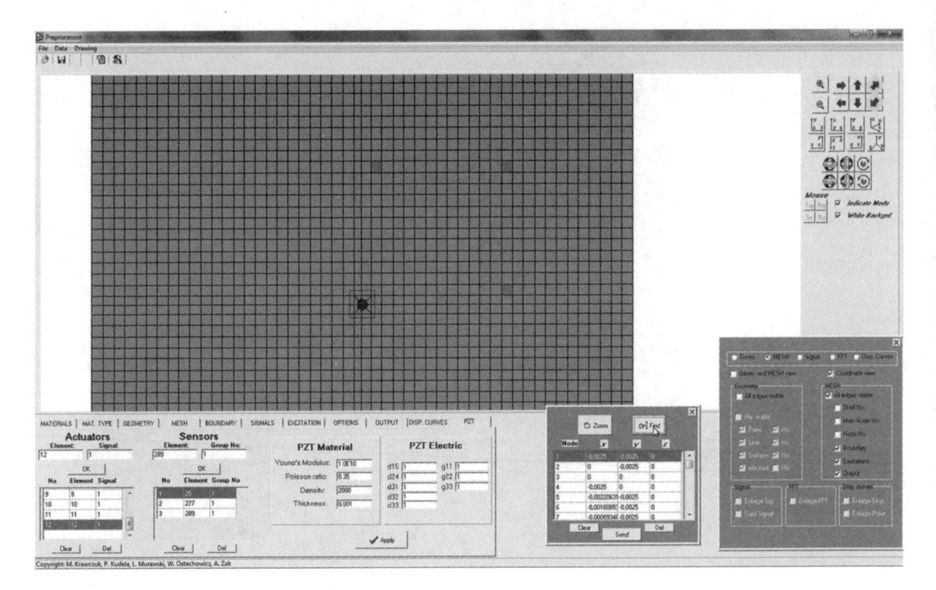

Figure A.45 View of a fragment of the panel model with modelled actuator and sensors

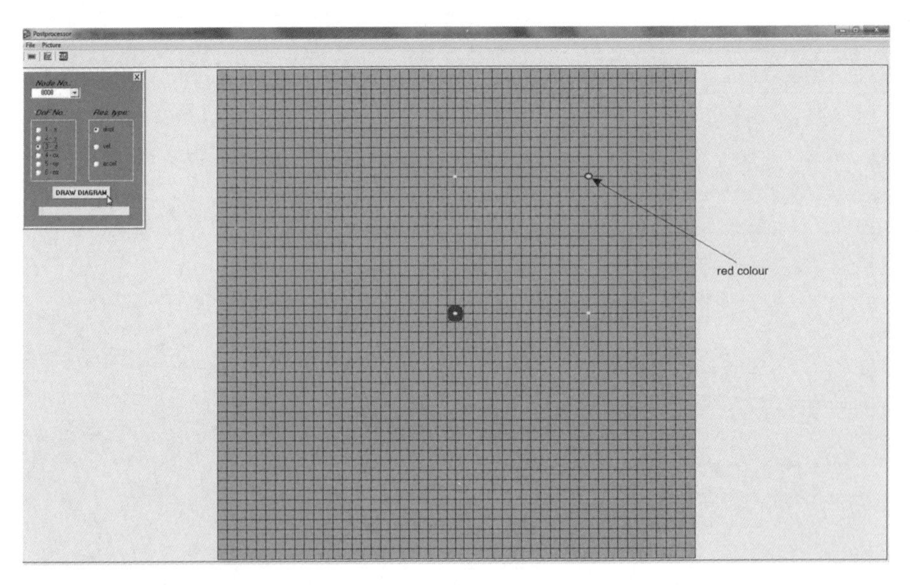

Figure A.46 Model of the composite panel with the point of excitation and points of elastic wave recording marked

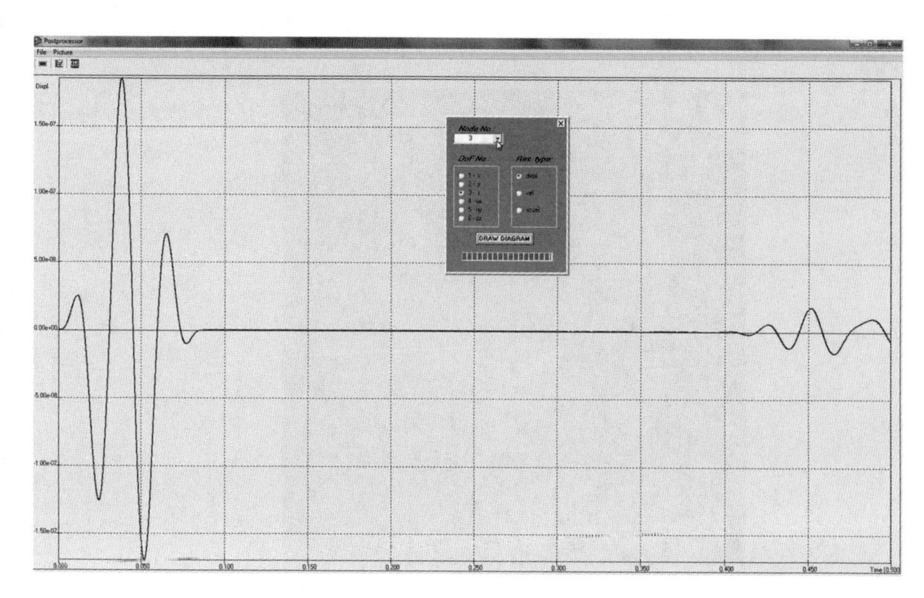

Figure A.47 Amplitudes of the elastic wave as a function of time, for the node in which excitation was applied

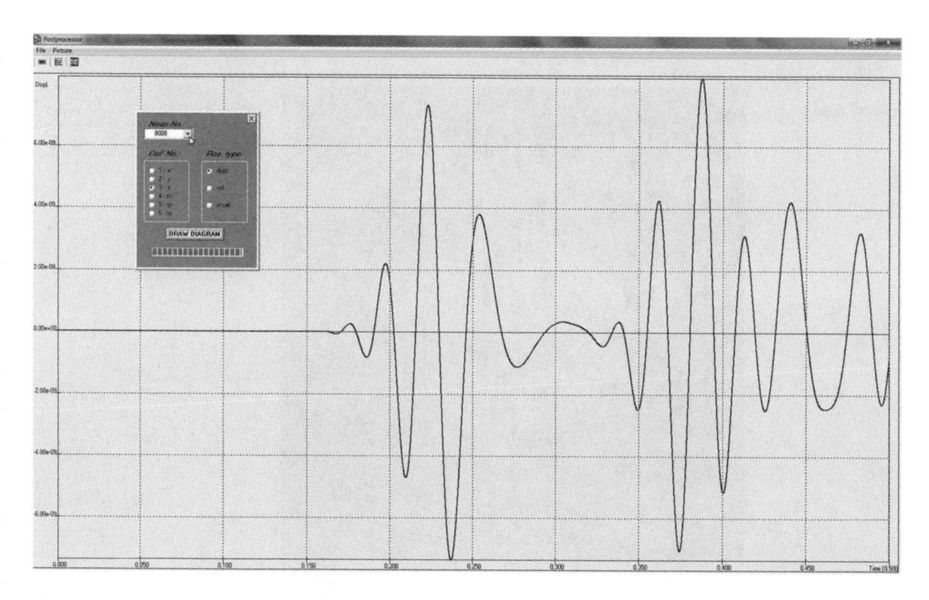

Figure A.48 Amplitudes of the elastic wave as a function of time, in the centre of the panel semi-diagonal

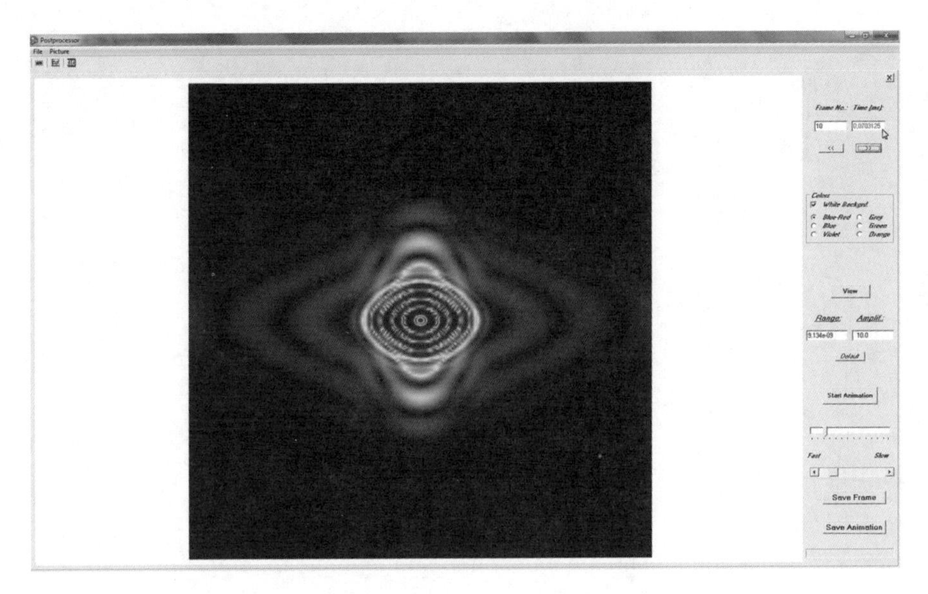

Figure A.49 Propagation of elastic waves in the panel for $t = 0.070$ ms

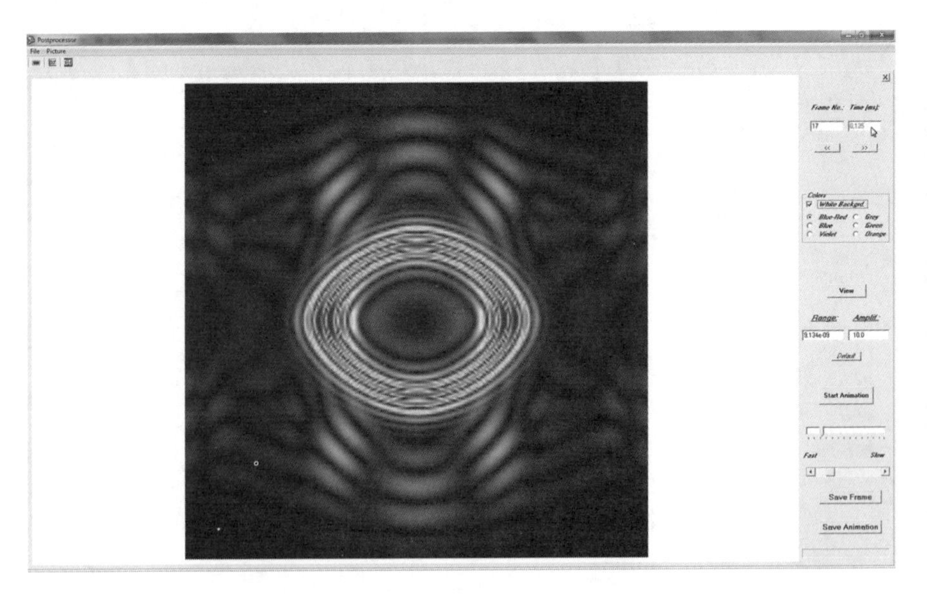

Figure A.50 Propagation of elastic waves in the panel for $t = 0.125$ ms

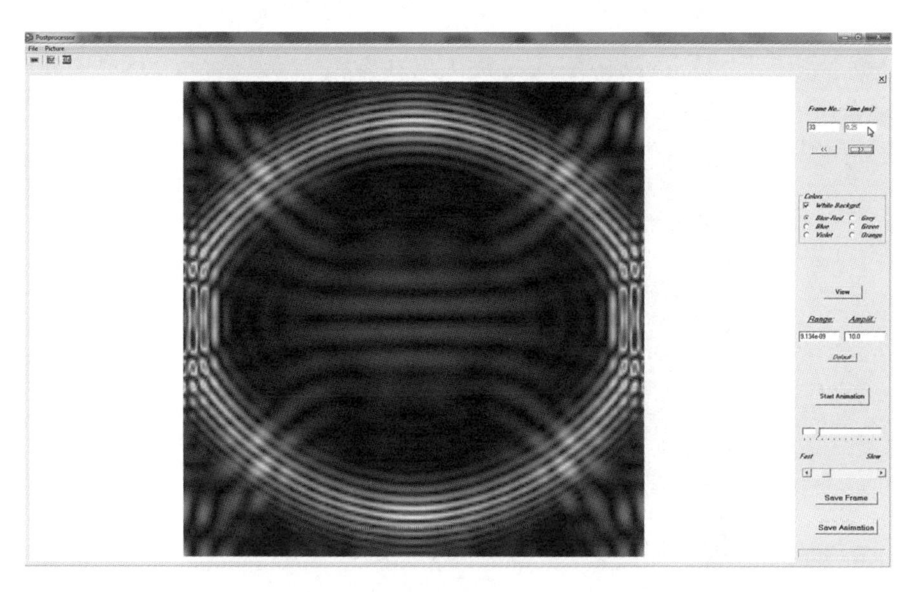

Figure A.51 Propagation of elastic waves in the panel for $t = 0.250$ ms

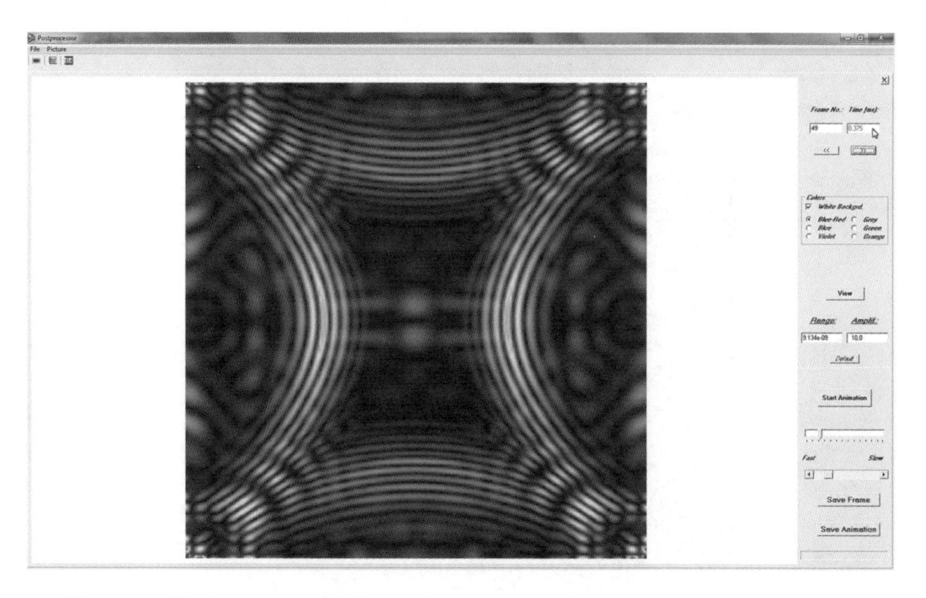

Figure A.52 Propagation of elastic waves in the panel for $t = 0.325$ ms

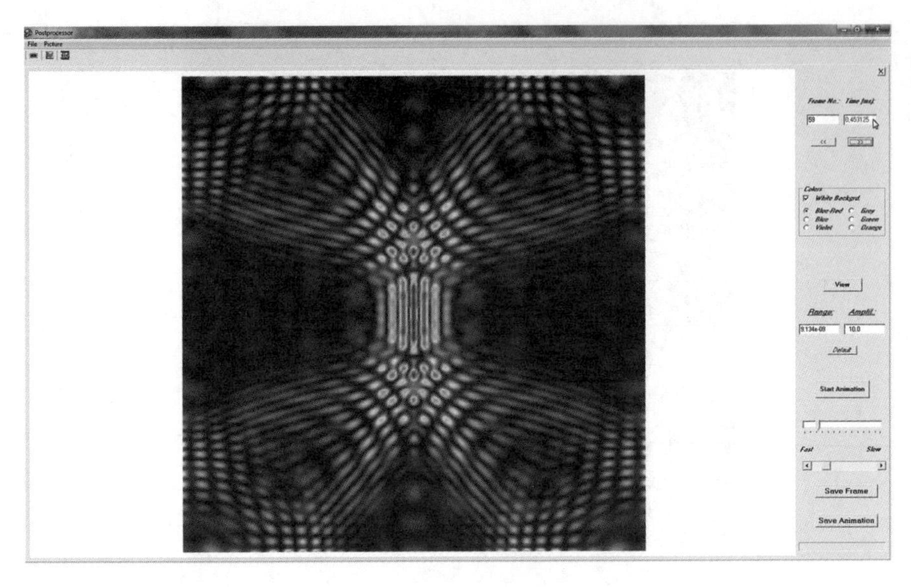

Figure A.53 Propagation of elastic waves in the panel for $t = 0.453$ ms

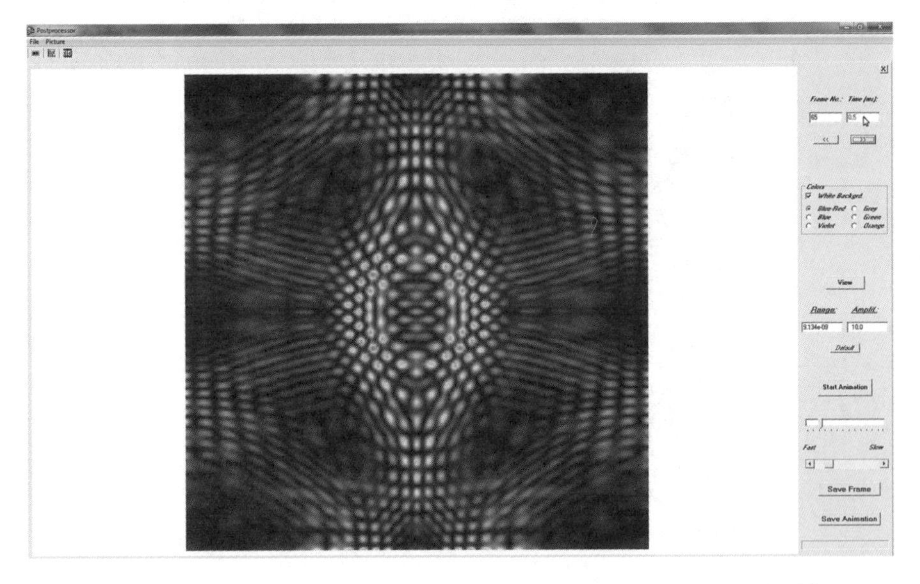

Figure A.54 Propagation of elastic waves in the panel for $t = 0.500$ ms

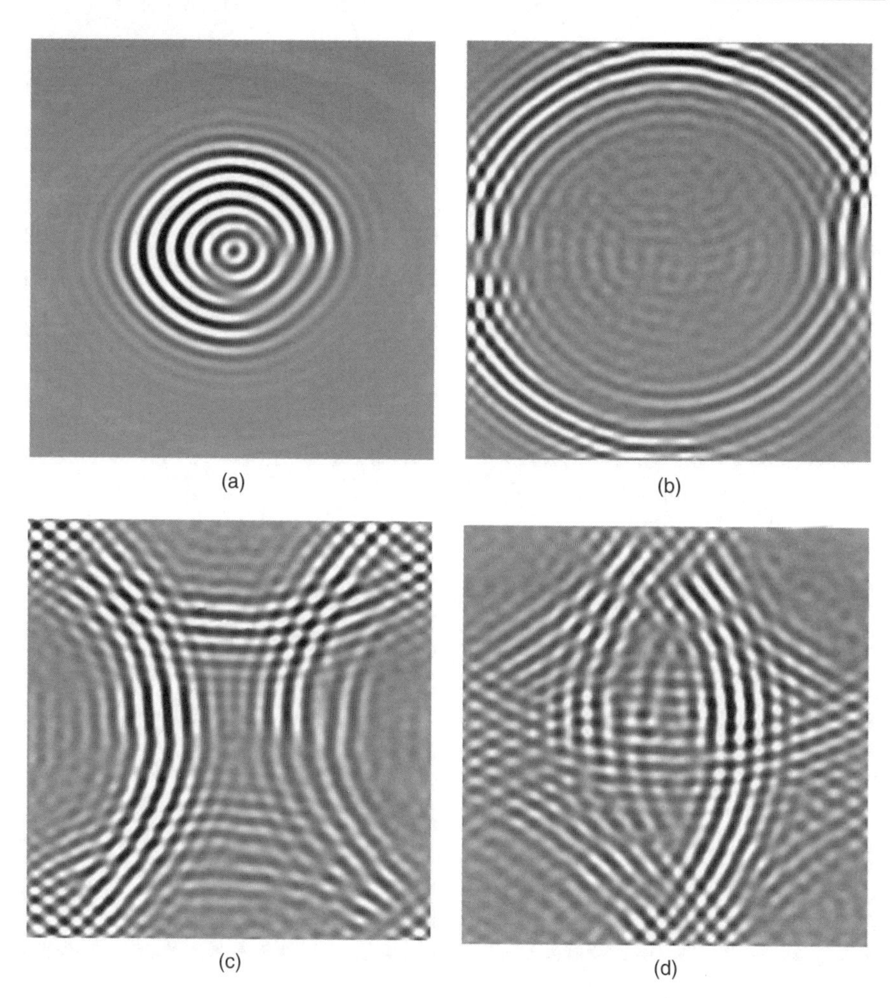

Figure A.55 Snapshots of wave propagation patterns in a composite panel for time instances a) 0.070 ms, b) 0.125 ms, c) 0.250 ms, d) 0.375 ms and e) 0.500 ms – experiment

wave propagation in the panel are presented in sequence. These correspond to moments of 0.070, 0.125, 0.250, 0.375, 0.453 and 0.500 ms, respectively.

The numerical calculations presented above are verified against experimental research. Figure A.55 presents the results of experimental measurements which comply with the numerical results presented in Figure A.56.

The Appendix presents the capabilities of the software developed by the authors. The ***EWavePro*** software is used for analysing elastic wave

Figure A.56 Snapshots of wave propagation patterns in a composite panel for time instances a) 0.070 ms, b) 0.125 ms, c) 0.250 ms, d) 0.375 ms and e) 0.500 ms - numerical simulation

propagation. It is equipped with pre- and post-processors. Basic functionality of the software is presented. The procedure of developing a three-dimensional model is presented, together with the available functionality for presenting the analysis results. Three independent examples of analysing elastic wave propagation are presented: in a pressure tank, in a UAV wing skin and in a panel constructed of composite material. Finally, the results of the analyses are compared with experimental ones, proving the usefulness of the developed software.

References

1. Zienkiewicz, O.C. and Taylor, R.L. (1991) *The Finite Element Method*. 4th edn, vol. 2, *Solid and Fluid Mechanics. Dynamics and Non-linearity*, McGraw-Hill Book Company.
2. Rao, S.S. (1989) *The Finite Element Method in Engineering*, 2nd edn, Pergamon Press.
3. Ochoa, O.O. and Reddy, J.N. (1992) *Finite Elements Analysis of Composite Laminates*, Kluwer Academic Publishers.

[illegible] developed software.

References

[illegible]

Index

Aliasing, mode, 148–9, 153–5

Beam, 133–41
Bonding Layer, 227–30
  modeling, 227–30
Boundary Conditions, 70–72

Chebyshev Polynomials, 48, 56–9
Christoffel Equations, 13
CKE Damage Indicator, 265–6
Cumulative Kinetic Energy Damage Indicator, *see* CKE Damage Indicator

Damage Detection, 237, 247–52
Damage Identification, 250, 251, 252–3, 266–9
  Inverse Methods, 268–9
Damage Localisation, 250–51, 269–88
Damping, 69
Dispersion Curves, 26
  of c_g/c_p for aluminium rod, 40–41
  for Lamb waves, 27–9
  for membranes, 171–3
  for one-dimensional elements, 141–3
  for plates, 177–9
  for Shear Horizontal waves, 33–4
  for shells, 184
  for solid elements, 204, 205,
Dispersion Relationship, 8

Elastic Wave Excitation, 94–104
Elastic Wave Registration, 104–14
Electromechanical Coupling, 208–19
Error Analysis,
  for membranes, 185–7
  for plates, 187–8
  for rod natural frequency, 145–7

Finite Elements, 48
Flexural Waves, 43–5
  in rod element, 158–61, 162–4
Full Field Measurements, 105, 107

Gauss Quadrature, 76–8
Gauss-Laguerre Quadrature, *see* Laguerre Quadrature

Guided Waves in Structures for SHM: The Time-Domain Spectral Element Method, First Edition.
Wieslaw Ostachowicz, Pawel Kudela, Marek Krawczuk and Arkadiusz Zak.

Gauss-Legendre Quadrature, *see* Gauss Quadrature
Gauss-Lobatto Quadrature, *see* Lobatto Quadrature
Gauss-Lobatto-Legendre Quadrature, *see* Lobatto Quadrature
Group Velocity, 8–10

Higher Order Models
 for Membranes and Plates, 171, 174, 175, 180, 181, 189
 for Rods and Beams, 132, 140–41

Inertia Matrix, 69, 81

Laguerre Polynomials, 60–62
Laguerre Quadrature, 78–81
Lamb Waves, 4–5, 6, 7
 anti-symmetric modes, 21, 24–5
 distribution of displacements and stresses, 29–32
 equations, 17–25
 methods for generating/registering, 115
 symmetric modes, 21, 22–4
Laser Vibrometry, 109–14, 252
Lobatto Quadrature, 75–6
Longitudinal Waves, 2, 36–9
 in cracked rod element, 156–8
 in rod element, 147–55, 162–4
Love Waves, 4

Membranes, 175–81

Natural Frequency, 144–7
Navier equations, 11
Nine-Mode Theory, for solid elements, 203–4
Non-contact Measurements, 93, 103, 107, 109

Numerical Integration, 75
 using Gauss quadrature, 76–8
 using Laguerre quadrature, 78–81
 using Lobatto quadrature, 75–6

Phase Velocity, 8, 9–10
Phased Arrays, 236–7, 251, 255–61
Piezoelectric Actuator, 219
Piezoelectric Sensor, 217–19
Piezoelectric Transducer, 98–102, 105
Pitch-catch methods, 247
Plates, 169–75
Pochhammer Frequency Equation, 39
 solution for rod elements, 39–41
Pulse-echo methods, 247

Rayleigh Waves, 3–4, 16–17
Rayleigh-Lamb equation,
 for antisymmetric modes of Lamb waves 25
 solving for Lamb waves, 26–9
 for symmetric modes of Lamb waves 23
RMS Damage Indicator, 264–5
 maps, 277–84
  filtering, 284–7
Rod Elements, 127–32
Root Mean Square Damage Indicator, *see* RMS Damage Indicator

Sensor configuration
 concentrated, 250–52
 distributed, 247–50
Sensor layout, *see* Sensor configuration
Shear Horizontal Waves, 32–4
Shear Waves, 2–3
Shells, 181–4
SHM, *see* Structural Health Monitoring

Six-Mode Theory, for solid elements, 202–3
Solid Spectral Finite Elements, 201–2
Spectral Finite Element Method, 47–92
  characteristics, 48
  steps, 49–52
Spectral Finite Elements, 48–9
Stiffness Matrix, 69, 81
Structural Health Monitoring (SHM), 235–47

Torsional Waves, 42–3

Wave Packet, 8–9
Wavelength, 6
Wavenumber, 6